Advanced Structured Materials

Volume 210

Common engineering materials are reaching their limits in many applications, and new developments are required to meet the increasing demands on engineering materials. The performance of materials can be improved by combining different materials to achieve better properties than with a single constituent, or by shaping the material or constituents into a specific structure. The interaction between material and structure can occur at different length scales, such as the micro, meso, or macro scale, and offers potential applications in very different fields.

This book series addresses the fundamental relationships between materials and their structure on overall properties (e.g., mechanical, thermal, chemical, electrical, or magnetic properties, etc.). Experimental data and procedures are presented, as well as methods for modeling structures and materials using numerical and analytical approaches. In addition, the series shows how these materials engineering and design processes are implemented and how new technologies can be used to optimize materials and processes.

Advanced Structured Materials is indexed in Google Scholar and Scopus.

Visakh P. M.
Editor

Rubber Based Bionanocomposites

Characterisation

Editor
Visakh P. M.
School of Marine Sciences
Cochin University of Science
and Technology
Kochi, Kerala, India

ISSN 1869-8433 ISSN 1869-8441 (electronic)
Advanced Structured Materials
ISBN 978-981-10-2977-6 ISBN 978-981-10-2978-3 (eBook)
https://doi.org/10.1007/978-981-10-2978-3

This Springer imprint is published by the registered company Springer Nature Singapore Pte Ltd.
The registered company address is: 152 Beach Road, #21-01/04 Gateway East, Singapore 189721, Singapore

Contents

Chapter 1
Rubber Based Bionanocomposites: Characterization: State of Art and New Challenges

Visakh P. M. and Marina Arias

Abstract This chapter provides short version of all chapters. Here I am writing about the different chapter topics such as cellulose-based rubber bionanocomposites, cellulose-based rubber blends and microcomposites, chitin based rubber nanocomposites, chitin in rubber-based blends and microcomposites, starch-based rubber nanocomposites, bacterial cellulose (BC) based rubber nanocomposites, lignin-based rubber composites and bionanocomposites, advancements in green nanocomposites: a comprehensive review on cellulose-based materials in biocomposites and bionanocomposites and taraxacum koksaghyz rodin as an alternative source of natural rubber and inulin.

Keywords Bionanocomposites · Cellulose · Rubber · Rubber blends · Microcomposites · Chitin nanocomposites · Chitin rubber blends · Starch nanocomposites · Bacterial cellulose · Lignin bionanocomposites · Green nanocomposites · Taraxacum koksaghyz rodin natural rubber · Inulin

1.1 Cellulose-Based Rubber Bionanocomposites

Natural rubber was first obtained from *Hevea brasiliensis* in South America and it has long been used for the manufacture of a wide variety of products. The first records of the natives of the island of Haiti playing with "balls made of gum from a tree" date back to the sixteenth century during Columbus's second voyage [1]. To improve the physical, chemical and mechanical properties of rubber, polymer chains are subjected to crosslinking processes [2]. Two or more phases, consisting of a matrix

Visakh P. M. (✉)
School of Marine Sciences, Cochin University of Science and Technology, Cochin, Kerala 682016, India
e-mail: visagam143@gmail.com

M. Arias
Neiker Tecnalia, Campus Agroalimentario de Arkaute. Apto 46, 01080 Vitoria-Gasteiz (Araba), Spain

Visakh P. M. *Rubber Based Bionanocomposites*, Advanced Structured Materials 210,
https://doi.org/10.1007/978-981-10-2978-3_1

(rubber in this case) and other compounds, usually called fillers, are introduced to strengthen the mechanical and thermal properties of the matrix [3]. In the rubber industry, carbon black (CB) is the most widely used reinforcing filler, with 90% of its production being used in the manufacture of tires [4]. Cellulose is traditionally obtained from wood, but there are other alternative sources. These include cotton, ramie, sisal, flax, wheat, potato tubers, beet pulp, soybeans, as well as some marine organisms such as seaweed and tunicates. In addition, there are specific bacteria that synthesize pure nanofibrillated cellulose [5]. Nanocelluloses can be classified into three types, (1) cellulose nanofibers (CNFs) (2) cellulose nanocrystals (CNCs), and (3) bacterial nanocellulose (BNC) [6].

CNCs are rod-like or whisker-shaped, 3–50 nm in diameter and 100–500 nm in length, and are obtained by acid or enzymatic treatment of cellulose of plant origin. The hydrophilic nature of nanocellulose becomes a challenge when trying to introduce it into a rubbery hydrophobic polymer matrix. In this case, surface functionalization is required to complete the interaction between filler and matrix [7]. Most materials used in medicine are still of synthetic origin [8], but natural polymers have gained popularity in recent years.

However, composites including CNCs and bioglass particles as reinforce fillers of NRL matrices for tissue engineering was recently published with promising results [9]. There are studies which indicate that the addition of antimicrobial nanofillers can confer antimicrobial attributes to the final rubber nanocomposites [10]. Furthermore, antimicrobial properties can also be imparted to the bionanocomposites by incorporating antibiotics into their composition (i.e., ampicillin, gentamycin, tetracycline, chloromycetin) [11].

Worldwide awareness of the environmental problems posed by packaging waste has led to major efforts to develop packaging incorporating environmentally friendly materials. Cellulose has received increasing interest in recent years due to its renewable origin and biodegradability. Also, cellulose is odorless and tasteless, which are highly desirable properties for food-packing. In particular, nanocellulose (NC) in nanocrystalline or fibrillated form, as mentioned in the introduction section, has excellent properties such as low toxicity and density, high tensile strength, and high aspect ratio.

Rubber concrete is a cement-based composite material that combines waste-tire rubber particles in a specific ratio to modify its internal structure. Replacing part of fine aggregates with rubber in conventional concrete. Rubber bullets are non-lethal weaponry where various types of rubber (including PB, SBR, and BR) are filled with CB, clay, calcium carbonate, and other metal fillers. Considering previous discussions about the benefits of NC as an alternative to CB, CNCs, or CNF, it is worth evaluating as fillers to create rubber bullets that increase the sustainability of the military industry. The excellent oxygen and gas barrier properties of CNFs and their excellent mechanical properties have encouraged studies aimed at using them in the development of biodegradable films for coating applications. However, there are some disadvantages that must be overcome before they can be efficiently used for this purpose, such as their high sensitivity to moisture and their poor water vapor barrier properties.

These disadvantages are mainly due to the high hydrophilicity of cellulose, which can establish strong hydrogen bond interactions with water molecules, and the preparation of composites with less hydrophilic material will make the passage of water through the mixture more difficult. Moisture sensitivity enhances its vapor barrier properties. Membranes have a relevant role in chemical technology and are currently increasingly used in a wide range of applications. Permeation control of chemical species in contact with membranes is an important property to consider when developing new materials for membrane technology applications.

Drug delivery is the control of the rate of permeation of a specific drug from a reservoir into the body. In separation processes, membranes need to allow some components to permeate, while completely restricting the entry of others [12].

Biocomposites of rubber and nanocellulose improve the barrier properties to non-polar organic solvents due to the hydrophilic nature of the latter. Due to the known distribution and high hydrophilicity of CNFs in the rubber matrix, small amounts of CNFs have been shown to significantly reduce the permeant diffusion of non-polar organic solvents by creating a tortuous path. The self-healing properties were achieved due to the dynamic hydrogen bond supramolecular networks between the oxygen groups of ENR and the hydroxyl groups on the surface of CNCs. Considering that chain interdiffusion is necessary to obtain self-healing properties, this application requires slightly crosslinked rubber [13].

1.2 Cellulose-Based Rubber Blends and Microcomposites

Blends are molecular mixtures of two or more different polymers that combine the advantages and disadvantages of the original polymers to form a new material with hybrid properties. It has attractive properties such as processability, cost-effectiveness, and customized features [14]. These blends are made of polymers that do not mix easily due to differences in chemical compatibility. They form multiple distinct phases within the material, each with unique properties. Vegetable nanofibers, cellulose nanowhiskers (CNWs), and bacterial cellulose fibers have been used as reinforcements for synthetic and natural polymers. All of these nanocelluloses are more widely available, inexpensive, biodegradable, renewable, and low-density.

Cellulose, the world's most common polymer, is non-toxic, renewable and biodegradable. Cellulose or its derivatives, especially cellulose nanocrystals (CNC) [15] or cellulose nanofibrils (CNF) [16], cellulose nanowhiskers (CNs), and microfibrillated cellulose [17] have been widely used as model fillers in various polymeric matrices since early times reported the use of nanocellulose as a reinforcing agent in a polymeric matrix [18]. Chemical modification of cellulose that reduces the hydroxyl groups present in biofillers is expected to result in improved strength and higher crystallinity of the composites [19]. Cellulose is used in a variety of products, including flexible displays, textiles, sensors, composites, membranes, light-emitting diodes, and batteries. Factors that determine whether an application is suitable for them are their source, synthesis method, size, and processing conditions.

The properties of the natural rubber-cellulose composite are halfway between the two. By supporting the natural rubber matrix, the addition of cellulose to natural rubber can improve the composite's tensile characteristics [20].

Natural rubber-cellulose microcomposite typically has a lower glass transition temperature (T_g) than pure natural rubber. Cellulose microfibrils in the NR matrix act as physical crosslinks, preventing the NR chains from moving freely. NR content, cellulose content, and composite shape all affect the T_g of NR-cellulose microcomposite. DSC analysis was performed by incorporating cellulose nanoparticles (CNPs) in a blend of synthetic rubber (SR)/polylactic acid (PLA) to examine the composition of PLA with the addition of SR and CNPs [21]. It is observed that the T_g of all PR and PNR blends are found to be almost at the same temperature (70 °C). NR/cellulose whisker nanocomposites with 10 wt% cellulose whiskers showed a slightly lower onset degradation temperature (~265 °C) compared to pure natural rubber [22].

The lower onset degradation temperature of bagasse whiskers compared to rubber may be the reason for the lower degradation observed in rubber/cellulose whiskers nanocomposites.

The surfaces of pure NR and ENR exhibit a smooth appearance, while the nanocomposite of NR and ENR exhibits typical broken appearances as a result of the presence of CNCs [23]. Even at 1% CNC concentration in NR nanocomposites, there is some incorporation of CNCs into the NR matrix, and this incorporation becomes stronger as the CNC content increases. AFM is a useful tool for understanding the fundamental properties of NR, the surface morphology of cellulose, and novel NR-based materials with enhanced capabilities for various applications. It offers quantitative information and high-resolution images on the adhesion, hardness, roughness, and topography of NR surfaces [24, 25].

The diameter of cellulose fibers and the distance between individual molecules can be measured using AFM. Cellulose crystals captured by AFM typically display a lattice-like structure with repeating units of 0.6 nm. In addition, the size and orientation of cellulose crystals can be determined using AFM [26]. AFM can distinguish areas of strong and weak adhesion, revealing information about the efficiency of interfacial bonding and potential areas for development. The presence of crystalline cellulose peaks in the XRD pattern of a natural rubber-cellulose composite indicates that cellulose is dispersed throughout the NR matrix and that large, discrete phases are not formed. Ensuring proper interfacial interaction between the two components is critical to improving the properties of the composite. SAXS is used to examine the structure of NR at the nanoscale, typically 1–100 nm. It is useful to determine the distribution and size of filler particles in NR composites and the size and distribution of crosslinks in NR networks. SAXS can also be used to show that the crosslink density of NR increases with increasing radiation dose. This is because irradiation creates new crosslinks between NR chains [27]. Rheological data can be used to predict how natural rubber will flow during injection molding, optimize natural rubber processing conditions, and in-use performance of rubber products. Consequently, it can be applied to describe the structure and properties of natural rubber [28, 29]. Cellulose has the ability to reduce viscous dissipation, leading to more elastic behavior. This is because the cellulose fibers restrict the movement of the rubber chain. Increased viscosity and lower flowability of cellulose inclusions may cause processing problems that require adjustments in processing parameters.

1.3 Chitin Based Rubber Nanocomposites

Chitin is the second most important natural polymer in the world. It is poly(β-(1 → 4)-N-acetyl-d-glucosamine), an important natural polysaccharide first identified in 1884. It is the most significant polymer synthesized annually in the world after cellulose [30]. The main commercial sources of chitin are the shells of crabs and shrimp. Bleaching is often included to get rid of remaining pigments and achieve a colorless result. Due to differences in the ultrastructure of the starting materials, these procedures must be adapted for each chitin source. The resulting chitin must be assessed for purity and color, as any remaining protein or pigment may be problematic for further use, especially for biomedical devices [31]. Demineralization and deproteinization are usually the first two processes in the chemical extraction of chitin. Demineralization is often performed before deproteinization to improve the deproteinization surface area [32]. The chitin obtained by this method had a lower molecular weight than the chitin obtained by the standard acid–base method, and its morphology and purity were comparable. This method involves the lyophilization of well-distributed aqueous mixtures of rubber latex and chitin to produce nanocomposite powders, which are then converted into samples by hot pressing. Gopalan et al. [33] extracted chitinous whiskers from crab shells and reinforced them with natural rubber. The crab (red crab) limb used in this study is the main source of chitin production as it is relatively rich in chitin. Nanocomposites have also been prepared using NR latex and regenerated chitin (R-chitin) [34]. The stress–strain curves of NR/ regenerated chitin composites exhibited improved tensile strength and modulus with the addition of regenerated chitin. The rigidity of the composites tends to improve with the addition of R-chitin. Kawano et al. prepared porous and self-assembled chitin nanofibers (CNFs) and further reinforced them with natural rubber (CNF-NR) to form composite sheets. The nanofiller formed a network structure on natural rubber (NR) particles through self-assembly above them. The addition of chitin increased tensile strength, elastic modulus, and tensile strength. The reinforcing effect of CNF in sheets was confirmed by tensile tests [35].

Yin et al. prepared nanocomposites with high ChNF/NR content for polymers with reversible plasticity and shape memory [36]. The mechanical properties of NR can be significantly affected by the addition of nanofillers. Steam explosion and mild hydrolysis of oxalic acid were used to create CHNF chitin nanofibers (12–30 nm) from chitin powder [37]. To create the NR/CB/CHNF composites, CHNF was uniformly dissolved in natural rubber (NR) latex, dried, and mixed with carbon black (CB) in a two-roll mill. In the case of XSBR-CNC, the peak value of the tan delta shows a diminishing characteristic on increasing the chitin nanocrystals (CNCs) due to the restriction imparted by the crystals on the flexibility of the polymer chain leading to lower mobility. In the case of XSBR-CNCs, the peak value of delta tan shows a decreasing characteristic with increasing number of chitin nanocrystals (CNCs) due to the limitation imposed by the crystals on the flexibility of the polymer chain, resulting in decreased mobility. Incorporation of regenerated chitin (R-chitin) into the latex medium reduces the thermal stability of NR [34].

As R-chitin content increases, there is a tendency for thermal stability to decrease. There is only one major mass loss stage in the thermal degradation behavior of all NR/R-chitin composites, and no clear degradation stage is visible in any DTG curve. Another study found that TGA and DTG analyses indicate that the incorporation of chitin does not significantly change the thermal stability of the styrene butadiene rubber (XSBR) matrix [38]; in other words, the annealing process does not significantly affect the maximum rate of weight loss.

Midhun et al. observed that the curing time of NBR-CHNW (chitin nanowhiskers) is higher than that of NBR gum due to the surface absorption of accelerator molecules by the reactive functional groups of CHNW, retarding the curing effect [39].

1.4 Chitin in Rubber-Based Blends and Microcomposites

Chitin, a versatile natural biopolymer, has emerged as a focus of scientific interest due to its remarkable properties. Abundant in the exoskeletons of arthropods and the cell walls of fungi, this polysaccharide is composed of linear chains of N-acetylglucosamine. Chitin-enhanced properties include, but are not limited to, its high tensile strength, biodegradability, and low toxicity [40], Chitin's biodegradability has emerged as a beacon of hope, reducing the burden of plastic waste and contributing to a circular economy, while chitin's thermal properties further expand its horizons of applications. The structural composition of chitin gives it remarkable mechanical properties [41].

The linear chains of N-acetylglucosamine create a strong and resilient framework that gives chitin a high degree of tensile strength. Chitin can undergo deacetylation to form chitosan, which contains amino groups and can be positively charged under certain conditions. However, chitin in its natural state is not inherently positively or negatively charged. The biocompatibility of chitin makes it suitable for applications where contact with living tissues cannot be avoided. This feature has inspired research to incorporate chitin into medical devices [42], wound dressings, and other healthcare materials, increasing its compatibility with biological systems. Incorporation of chitin into rubber-based mixtures carefully achieves uniform dispersion and ensures optimal interaction between the two materials. Various processing techniques such as melt mixing, solution blending, and in situ polymerization have been explored to create well-defined microstructures within the composite material. Although rubber is known for its elasticity at room temperature, the addition of chitin can result in improved thermal stability. This improvement is particularly relevant in applications that consider exposure to high temperatures, such as automotive components that may experience heat during operation. Chitin is a naturally abundant aminopolysaccharide composing the exoskeletons of crustaceans and fungal cell walls with intriguing structural varieties based on differences in molecular chain packing and hydrogen bonding patterns, α-chitin leading to three polymorphic crystal forms known as β-chitin and γ-chitin.

DMA results demonstrated a steady shift of the storage modulus peak to higher temperatures with increasing chitin loading for natural rubber composites this corresponds to the restricted molecular mobility of the rubber chains induced by the chitin networks thermal degradation onset temperature also increased by over 15 °C with a chitin content above 5 phr the incorporated chitin [43]. Several processing techniques have been explored for fabricating chitin-rubber microcomposites [44], each presenting distinctive advantages and challenges. Melt mixing involves blending chitin powder with rubber and crosslinking agents at elevated temperatures, allowing thermal phase mixing but risking uneven dispersion and thermal degradation. Electrospinning uses electric fields to create non-woven mats with microscale fiber diameters to enhance mechanical reinforcement through fiber orientation. Each method contributes to the diverse landscape of chitin-rubber microcomposites, offering tailored solutions for specific applications [45]. The chitin chains intertwine with the rubber matrix, creating a material that inherits the tensile strength of chitin while retaining the elasticity and resilience intrinsic to rubber. Elevated temperatures, ranging between 120 and 140 °C, catalyze the thermal phase mixing of chitin and rubber [46].

Advanced techniques such as rheological studies, thermal analysis, and microscopy provide insights into the behavior of chitin and rubber during the melt mixing process.

1.5 Starch-Based Rubber Nanocomposites

Starch belongs to the family of polysaccharides that can be extracted from natural resources like potatoes, corn, rice, cassava, barley, wheat, and so on. The structural composition of starch includes amylose, which is a water-soluble linear unbranched chain constituting nearly 30% of the starch, and amylopectin having a branched structure made of α-D-glucose units comprising nearly 70% of starch, the details of which will be mentioned in the following sections. Starch is a naturally occurring carbohydrate formed in plants by photosynthesis and is present in abundance in plant parts like roots, seeds, and stalks. Depending on the biological function, starch is categorized into two—storage starch and transitory starch. The starch-based rubber composites have gained much attention during the past few years since it is abundantly available in nature as well as biodegradable. The structural details of starch were confirmed by analyzing the residue left from its total acid hydrolysis. Two types of α-glucans are present in starch—amylose and amylopectin. Starch is synthesized in a granular form and its biosynthesis starts in the hilum. The starch granules are found to exhibit a variety of shapes like ellipsoids, platelets, polygons, and tubules. Many methodologies are available to evaluate the properties of starch, among which microscopy has gained importance [47].

Atomic force microscopy (AFM) and scanning electron microscopy (SEM) are some of the well-known tools to study the size and morphological details of starch granules. Modification can also be carried out chemically, in which new functional groups are introduced into starch molecules. Acid modification is carried out by treating starch below its gelatinization temperature in aqueous acid solutions, while hydrolysis reduces the size of starch granules and the viscosity of the paste. Nanostarch particles were precipitated by Najafi et al. by dropwise adding the aqueous phase of starch to acetone. They performed the synthesis using acetylated corn starch and the final particle size was in the range of 221–324 nm [48].

Starch nanoparticles (SNPs), produced by recrystallizing debranched waxy maize starch, were modified using oxygen and ammonia vacuum cold plasma (CP). Fourier transform infrared spectroscopy measurements of the modified SNPs revealed a new carbonyl or carboxyl signal at 1720 cm^{-1}. Modification of SNPs by grafting using the NMP method demonstrated in this work provides the ability to graft a huge range of monomers for a wide range of applications. Starch-rubber nanocomposites can also be prepared by incorporating starch and a nano-sized filler such as clay into a rubber matrix. Melt blending involves mixing the starch and rubber components in a molten state, followed by the addition of nanoparticles and subsequent cooling to form a solid material. Chemical and physical treatments and grafting are some of the commonly implemented modifications. In particular, modification methods that reduce the hydrophilic properties of SNPs have attracted much attention. Starch-rubber nanocomposites can also be prepared by incorporating starch and a nano-sized filler such as clay into the rubber matrix. In the molten mixture, starch and rubber components are mixed in a molten state, and then nanoparticles are added to form a solid material. Deborah et al. synthesized starch nanocrystals from various botanical sources and then prepared natural rubber nanocomposites with these synthesized starch nanoparticles. The method they adopted for the synthesis was casting or evaporation method [49].

Starch-based hydrogels of natural rubber are getting greater interest, but phase separations are usually seen due to poor compatibility. Vudjung et al. synthesized crosslinked natural rubber (XNR) and cassava starch (CSt) by IPN method [50]. The carbon black and vulcanization ingredients were mixed in an internal mixer and incorporated into the coagulated rubber filler compound, and then the different compounds were vulcanized at 145° to prepare the composite [51]. Starch-based rubber nanocomposites have potential applications in various fields, including packaging, automotive, manufacturing, and biomedical industries. In the packaging industry, these materials can be used to make environmentally friendly biodegradable and compostable materials.

1.6 Bacterial Cellulose (BC) Based Rubber Nanocomposites

Bacterial cellulose, which is chemically identical to plant cellulose is produced by the fermentation of bacteria, the most noteworthy is the non-pathogenic species of *Komagateibacter* such as *K. Xylinus*, formerly known as *Acetobacter* or *Gluconacetobacter*. Alike Plant cellulose is free from BC lignin, hemicellulose, pectin, and other components of lignocellulosics, providing high purity. It has a unique 3D nanofiber network structure with fibers that are 100 times thinner in diameter than plant-based fibers [52]. Treatment with acids, e.g., sulphuric acid leads to the elimination of regions of low lateral order resulting in the formation of rod-like cellulose nanocrystals, also referred as whiskers [53–57].

The BC is synthesized in the form of a ribbon composed of 10–100 microfibrils that protrude from the pores of bacterial rod. The rate at which the ribbon elongates is 2 μ/min. There are nearly 50 sites arranged in rows along the longitudinal axis of the outer envelope of the bacterial cell that can produce BC microfibrils [58]. Bacterial cellulose shows high thermal stability and can withstand temperatures up to 220 °C without degradation. TGA studies have revealed that BC shows a one-step degradation process and ~65% weight occurs at 280 °C with a maximum degradation rate at 300 °C. The purity of BC, high crystallinity, and orientation of cellulose chains within fibrils induce high thermal stability. It exhibits a T_g of nearly 40 °C and T_m of 85 °C [59].

Trovatti et al. have modified the surface of bacterial cellulose nanofibers (BC) via in situ admicellar polymerization of styrene using hexadecylpyridinium chloride as the surfactant referred as BCPS [60]. The process involves the adsorption of a double layer of cationic surfactant onto the surface of BC nanofibers followed by adsolubilization of the styrene monomer into the surfactant layer. The use of a water-soluble initiator promotes the polymerization of styrene in the aqueous medium. Akin to NR/BCW composite films, SBR/BCW films exhibit the water-stimuli responsive dynamic mechanical behavior where the films are produced by mixing of SBR latex and BCW suspension followed by evaporation. The storage modulus increases rapidly with BCW content which is attributed to the three-dimensional network of rigid whiskers connected via hydrogen bonding and immobilisation of numerous rubber chains around the whisker surface. In the rubbery region, there is an augmentation of the storage modulus (E′) from 2.8 to 1161 MPa at 20 phr of BC [61].

The cellulase complex contains endoglucanase which promotes the hydrolysis of amorphous domains that act as structural defects causing the random scission of the chains to smaller fragments. Nonetheless, the crystalline regions are less vulnerable to hydrolysis as a result of the strong hydrogen bonding that exists between them. BC with higher porosity, lower film thickness, and lower density than *G. xylinus*. Moreover, ultrafine-dried BC from *G. europaeus* exhibits better water absorption capacity, higher Young's modulus, and higher hardness compared to room temperature freeze-dried BC [62]. Industrial and agricultural wastes have been identified as the cost-effective feedstock for the growth of bacteria [63–66].

For instance, tea infusions [67], corn steep liquor [68], fruit and vegetable juices [69], molasses [70, 71], wheat straw acid hydrolysate [72], waste cotton fabrics [73], rice bark [74], konjac powder hydrolysate [75], etc., have been extensively investigated for optimizing BC production. Yin et al. have developed SBR/ BCW nanocomposites by mixing SBR latex with BCW suspensions and subsequently coagulated using aqueous sodium chloride (NaCl) solution. The dried coagulum is melt mixed with compounding ingredients and compression molded [76].

XNBR/BCW nanocomposites can be prepared by combining suspensions of BCW and XNBR latex followed by casting into films or by coagulating using aqueous NaCl solution (6.5%). The coagulated rubber is dried, further compounded, and vulcanized.

1.7 Lignin-Based Rubber Composites and Bionanocomposites

Lignin is considered an excellent substitute for raw materials in the preparation of chemicals and polymers [77]. Lignin valorization is a research area that seeks to convert lignin into useful chemicals and materials, including biofuels, plastics, and chemicals for use in industries such as pharmaceuticals, cosmetics, and agriculture [78]. Annamalai et al. [79] carried out a study of the preparation and addition of nanoscale organosolv lignin (OSL) to NR latex to improve its mechanical properties. SEM studies of dried OSL indicated that the structure occurs layer by layer with particle sizes larger than 2 μm. Lignin content has been limited to about 30 wt%, often requiring chemical modification, solvent fractionation of lignin, or prohibitively expensive additives. Each of these factors is a deterrent to industrial adoption of lignin-based polymers, limiting the potential of this renewable resource.

Wang et al. [80] conducted a comparative study on the synergistic effect of carbon black/lignin and silica/lignin hybrid filled natural rubber composites. The authors used a co-precipitation method to prepare the mixtures to guarantee good dispersion of lignin in NR matrices and the potential for industrial scale-up. It is important to note that the addition of lignin plays an important role in the reinforcing effect of silica-filled NR composites. The enhanced interaction between silica/lignin and NR and the improved dispersion of the filler contributed significantly to the improved reinforcement. Boonying et al. developed a coating composite of lignin and natural rubber modified with natural rubber-graft-polyacrylamide. The composite material released a total of N substances over a period of 112 days. The nitrogen release in Li/NR-g10 coated urea is limited due to the restricted hydrophilic channels within the coating. Nguyen et al. [81] reported earlier that combining lignin with the nitrile group containing rubbers produced a new class of melt-processable thermoplastic elastomers. These exhibited unique yield stress and high toughness primarily due to morphology control of the dispersed lignin phase [82]. In 2017, Feng et al. [83] used lignin-coated CNC (L-CNC) to reinforce a methacrylate-based resin (MA) photopolymerized via SLA and post-cured using UV and heat treatment at 120 °C for 40 min.

Tran et al. [84] incorporated nanoscale dispersed lignin into nitrile rubber. Temperature-controlled miscibility between the lignin and rubber during high-shear melt phase synthesis enables tuning of material morphology and performance. The sustainable material has an unprecedented yield strength (15–45 MPa), is strain-hardened under large deformation, and has excellent recyclability.

1.8 Advancements in Green Nanocomposites: A Comprehensive Review on Cellulose-Based Materials in Biocomposites and Bionanocomposites

Future trends in nanocomposite are related to smart fibers, multifunctionality, biomedical use, energy sector, and sustainable nanocomposite development. Detailed progress and future trends are mentioned below [85]. Some of the research work in the field of nanobiocomposites, green composites, and biohybrids have gained attention for the convergence of nanotechnology and biology [86]. The future trends suggest multipurpose focus in the functionality, smart innovative technologies, and multipurpose applications such as in health care, energy, and in the environmental sector. The detailed analysis is as follows [87]. Cellulose, which is the main component of plant cell walls, possesses a molecular structure that highlights its unique qualities [88]. Cellulose is composed of linear chains of glucose units, which exhibit both crystalline and amorphous regions. Cellulose-based biocomposites exhibit exceptional resistance to heat degradation, making them highly suitable for applications that require strong temperature resistance. This provides opportunities in various industries, such as automotive manufacture, construction, and aerospace, where materials need to endure different heat conditions while maintaining their structural integrity [89]. When combined with other polymers, cellulose creates packaging materials that redefine industry norms and promote environmental consciousness [90]. Cellulose, when introduced into the automotive industry, revolutionizes it by increasing the ratio of strength to weight in components. The successful exploration of cellulose-based biocomposites requires a joint endeavor between material scientists and engineers to fully exploit their capabilities [91]. Explore the domain of nanocellulose, a minuscule entity that possesses the potential to improve the properties of cellulose-based biocomposites.

Cellulose nanocrystals have exceptional mechanical strength, establishing unprecedented benchmarks in the field of nanocomposites. Envision materials enhanced with the power of Computer Numerical Control (CNC), offering extraordinary strength and durability, and enabling applications that require exceptional resilience. The manufacturing of cellulose-based bionanocomposites involves the complex task of incorporating nanocellulose into polymer matrices, a process that extends beyond simple blending. Envision a realm where the amalgamation of nanocellulose and polymer transcends into an artistic endeavor, molding the attributes of the bionanocomposite in a manner that redefines the realm of potential [92]. Nanocellulose plays a prominent role in the field of electronics, serving as the foundation for pliable and lightweight conductive materials. Envision wearable gadgets that effortlessly adapt to the body, utilizing the mechanical properties of nanocellulose for power. This is more than just a progression in technology; it signifies a complete transformation in our interaction with electronic devices, opening up possibilities for a new era characterized by adaptability and conductivity [93].

Researchers and environmental experts collaborate to understand the impact caused by these materials, guaranteeing that the progress made in innovation is in line with sustainability objectives. It is important to prioritize responsible progress that aligns with the well-being of the planet, rather than focusing solely on progress itself [94]. The combination of PVC and cellulose in construction is a significant advantage, offering environmentally friendly options for pipes, fittings, and structural elements [95]. The long-lasting nature of PVC combines effectively with the renewable properties of cellulose, offering structures that not only endure over time but also contribute to a more environmentally friendly construction industry. PVC/cellulose biocomposites are becoming more and more popular among automakers as component suppliers because of their exceptional strength and sustainability [96]. Similarly, these biocomposites are used in the building industry to create materials like panels and profiles, which helps the sector in its search for environmentally friendly substitutes.

1.9 Taraxacum Koksaghyz Rodin as an Alternative Source of Natural Rubber and Inulin

Natural rubber is one of the most important polymers, for human life, produced by plants. It is used in thousands of products and hundreds of medical devices converting it into an essential raw material. NR combines high strength with incredible resistance to fatigue. It has moderate resistance to environmental damage by heat, light, and ozone which are one of its drawbacks. Natural rubber compounds are exceptional for their flexibility, good electrical insulation, low internal friction, and resistance to most inorganic acids, salts, and alkali, even though it has poor resistance to petroleum products such as oil, gasoline, and Naphtha [97].

Rubber was also strategic for the industrial revolution. Rubber demand increased constantly from 1850 through 1912 in Industrial Europe and the United States. The search for high-quality rubber trees to be exploited was very important, the reason why businessmen pushed entrepreneurs and traders to penetrate Amazonian rainforests [98]. Humans have tried to replace it with synthetic rubber, but NR consumption cannot be replaced [99]. It is essential in the use of heavy goods vehicles, agricultural vehicles, and aircrafts and in civil engineering, due to qualities like low heating and the ability to regain its original shape. 65–70% of the global consumption of NR is done by the World's tire manufacturing industry [100]. *Hevea brasiliensis* has got very allergenic proteins and the continued contact with them can cause this allergy. The first appearances of rubber allergies were reported in the late 1980s and it's now widespread [101]. Mountains of Kazakstan in an expedition that was part of a Russian Government program. The aim of the expedition was to discover new or better plants that could supply all critical materials originating from plants [102]. Few years later it became the most promising alternative rubber source in Russia and was planted in a quite considerable extension, yielding 28.35 kg/Ha, with some of the plantings producing several times that much. Inulin, a non-digestible carbohydrate; it is a fructan that is not only found in many plants, but has also been part of human's daily diet for several centuries [103].

It is a mixture of poly- and oligosaccharides which almost have the same chemical structure GFn (G: glucose, F: fructose, and n: number of fructose units linked to one another). The degree of polymerization (DP) goes between 2 and 60 or higher. The links between the molecules are of a very special type: the β (2–1) form, which makes this molecule indigestible for all higher animals [104]. Many factors such as plant source, climate and growing conditions, harvesting time, and storage conditions, determine inulin's DP [105]. Inulin with low DP can be added to food as a low-calorie sweetener, whereas inulin with higher DP can be used as a fiber-type prebiotic with several health promoting effects [106].

In plants, dicot species belonging to Asteraceae are the main sources of inulin-type fructans [107]. Important species are chicory (*Chicorium intybus* L.), artichoke (*Cynara scolymus*), Jerusalem artichoke (*Helianthus tuberosus* L.), dandelion (*Taraxacum officinale*), dahlia (*Dahlia variabilis*), yacon (*Polymnia sonchifolia*) [108].

References

1. Browne EA (2014) Rubber, second. A & C Black Ltd, London
2. Thorn AD, Robinson RA (1994) Compound design. In: Bhowmick AK, Hall MM, Benarey HA (eds) Rubber products manufacturing technology. Marcel Dekker, Inc., New York, pp 1–79
3. Yasin S, Hussain M, Zheng Q, Song Y (2021) Effects of ionic liquid on cellulosic nanofiller filled natural rubber bionanocomposites. J Colloid Interface Sci 591:409–417
4. Roy K, Pongwisuthiruchte A, Chandra Debnath S, Potiyaraj P (2021) Application of cellulose as green filler for the development of sustainable rubber technology. Curr Res Green Sustain Chem 4:100140
5. Ávila Ramírez JA, Cerrutti P, Bernal C, Errea MI, Foresti ML (2019) Nanocomposites based on poly(lactic acid) and bacterial cellulose acetylated by an α-hydroxyacid catalyzed route. J Polym Environ 27:510–520
6. Shojaeiarani J, Bajwa DS, Chanda S (2021) Cellulose nanocrystal based composites: a review. Compos Part C Open Access 5:100164
7. Sinclair A, Zhou X, Tangpong S, Bajwa DS, Quadir M, Jiang L (2019) High-performance styrene-butadiene rubber nanocomposites reinforced by surface-modified cellulose nanofibers. ACS Omega 4:13189–13199

8. Andrade KL, Ramlow H, Floriano JF, Acosta ED, Faita FL, Machado RAF (2023) Latex and natural rubber: processing techniques for biomedical applications. Braz J Chem Eng. https://doi.org/10.1007/s43153-023-00317-y
9. Silva MJ, Dias YJ, Zaszczyńska A, Kołbuk D, Kowalczyk T, Sajkiewicz P, Yarin AL (2023) Three-phase bio-nanocomposite natural-rubber-based microfibers reinforced with cellulose nanowhiskers and 45S5 bioglass obtained by solution blow spinning. J Appl Polym Sci. https://doi.org/10.1002/app.54661
10. Sethulekshmi AS, Jayan JS, Saritha A, Joseph K, Aprem AS, Sisupal SB (2022) Antimicrobial studies in rubber nanocomposites—a mini review. Ind Crops Prod. https://doi.org/10.1016/j.indcrop.2022.115374
11. Li J, Cha R, Mou K, Zhao X, Long K, Luo H, Zhou F, Jiang X (2018) Nanocellulose-based antibacterial materials. Adv Healthc Mater. https://doi.org/10.1002/adhm.201800334
12. Baker RW (2005) Membrane technology. Kirk-Othmer Encycl Chem Technol. https://doi.org/10.1002/0471238961.1305130202011105.a01.pub2
13. Cao L, Yuan D, Xu C, Chen Y (2017) Biobased, self-healable, high strength rubber with tunicate cellulose nanocrystals. Nanoscale 9:15696–15706
14. Ghovvati M, Guo L, Bolouri K, Kaneko N (2023) Advances in electroconductive polymers for biomedical sector: structure and properties. Mater Chem Horizons 2:125–137
15. Dufresne A (2013) Nanocellulose: a new ageless bionanomaterial. Mater Today 16:220–227
16. Thomas SK, Parameswaranpillai J, Krishnasamy S, Begum PMS, Nandi D, Siengchin S et al (2021) A comprehensive review on cellulose, chitin, and starch as fillers in natural rubber biocomposites. Carbohyd Polym Technol Appl 2:100095
17. Parambath Kanoth B, Claudino M, Johansson M, Berglund LA, Zhou Q (2015) Biocomposites from natural rubber: synergistic effects of functionalized cellulose nanocrystals as both reinforcing and cross-linking agents via free-radical thiol–ene chemistry. ACS Appl Mater Interfaces 7:16303–16310
18. Gopi S, Balakrishnan P, Chandradhara D, Poovathankandy D, Thomas S (2019) General scenarios of cellulose and its use in the biomedical field. Mater Today Chem 13:59–78
19. Shaghaleh H, Xu X, Wang S (2018) Current progress in production of biopolymeric materials based on cellulose, cellulose nanofibers, and cellulose derivatives. RSC Adv 8:825–842
20. Visakh PM, Thomas S, Oksman K, Mathew AP (2012) Effect of cellulose nanofibers isolated from bamboo pulp residue on vulcanized natural rubber. BioResources 7:2156–2168
21. Thomas MS, Koshy RR, Mary SK, Thomas S, Pothan LA (2019) Starch, chitin and chitosan based composites and nanocomposites. Springer
22. Bras J, Hassan ML, Bruzesse C, Hassan EA, El-Wakil NA, Dufresne A (2010) Mechanical, barrier, and biodegradability properties of bagasse cellulose whiskers reinforced natural rubber nanocomposites. Ind Crops Prod 32:627–633
23. Yihun FA (2022) Nanochitin preparation and its application in polymer nanocomposites: a review. Emerg Mater 5:2031–2060
24. Murthy NS (2011) Techniques for analyzing biomaterial surface structure, morphology and topography. Surface modification of biomaterials. Elsevier, pp 232–255
25. Gaboriaud F, de Gaudemaris B, Rousseau T, Derclaye S, Dufrêne YF (2012) Unravelling the nanometre-scale stimuli-responsive properties of natural rubber latex particles using atomic force microscopy. Soft Matter 8:2724–2729
26. Abraham E, Pothan LA, Thomas S (2012) Preparation and characterization of green nanocomposites based on cellulose nanofibre and natural rubber latex. In: 15th European conference on composite materials, Venice, Italy
27. Shanks RA (2014) Characterization of nanostructured materials. Nanostructured polymer blends. Elsevier, pp 15–31
28. Lima P, da Silva SPM, Oliveira J, Costa V (2015) Rheological properties of ground tyre rubber based thermoplastic elastomeric blends. Polym Test 45:58–67
29. Polychronopoulos ND, Vlachopoulos J (2018) Polymer processing and rheology. In: Jafar Mazumder M, Sheardown H, Al-Ahmed A (eds) Functional polymers polymers and polymeric composites: a reference series, pp 133–180

30. Rudall KM, Kenchington W (1973) The chitin system. Biol Rev Camb Philos Soc 48:597–633. https://doi.org/10.1111/j.1469-185x.1973.tb01570.x
31. Soetemans L, Uyttebroek M, Bastiaens L (2020) Characteristics of chitin extracted from black soldier fly in different life stages. Int J Biol Macromol 165:3206–3214. https://doi.org/10.1016/j.ijbiomac.2020.11.041
32. Moussian B (2019) Chitin: structure , chemistry and biology a cuticle a body shape a evolution á barrier. Springer, Singapore. https://doi.org/10.1007/978-981-13-7318-3
33. Gopalan Nair K, Dufresne A (2003) Crab shell chitin whisker reinforced natural rubber nanocomposites. 1. Processing and swelling behaviour. Biomacromolecules 4:657–665. https://doi.org/10.1021/bm020127b
34. Yu P, He H, Luo Y, Jia D, Dufresne A (2017) Elastomer reinforced with regenerated chitin from alkaline/urea aqueous system. ACS Appl Mater Interfaces 9:26460–26467. https://doi.org/10.1021/acsami.7b08294
35. Kawano A, Yamamoto K, Kadokawa JI (2017) Preparation of self-assembled chitin nanofiber-natural rubber composite sheets and porous materials. Biomolecules 7:18–21. https://doi.org/10.3390/biom7030047
36. Yin J, Hu J, Han Y, Chen Y, Hu J, Zhang Z, Huang S, Duan Y, Wu H, Zhang J (2022) Facile fabrication of high nanofiller-content natural rubber nanocomposites for reversible plasticity shape memory polymers. Compos Sci Technol 221:109349. https://doi.org/10.1016/j.compscitech.2022.109349
37. Mathew M, Midhun Dominic CD, Neenu KV, Begum PMS, Dileep P, Kumar TGA, Sabu AA, Nagane D, Parameswaranpillai J, Badawi M (2023) Carbon black and chitin nanofibers for green tyres: preparation and property evaluation. Carbohyd Polym 310:120700. https://doi.org/10.1016/j.carbpol.2023.120700
38. Visakh PM, Monti M, Puglia D, Rallini M, Santulli C, Sarasini F, Thomas S, Kenny JM (2012) Mechanical and thermal properties of crab chitin reinforced carboxylated SBR composites. Expr Polym Lett 6:396–409
39. Midhun. Dominic CD, Joseph R, Sabura Begum PM, Raghunandanan A, Vackkachan NT, Padmanabhan D, Formela K (2020) Chitin nanowhiskers from shrimp shell waste as green filler in acrylonitrile-butadiene rubber: processing and performance properties. Carbohyd Polym 245:116505
40. Alkhursani SA et al (2022) Application of nano-inspired scaffolds-based biopolymer hydrogel for bone and periodontal tissue regeneration. Polymers 14(18):3791
41. Ahmad SI et al (2020) Chitin and its derivatives: structural properties and biomedical applications. 164:526–539
42. Kavimani V et al (2023) A CRITIC integrated WASPAS approach for selection of natural and synthetic fibers embedded hybrid polymer composite configuration. 1–16
43. Ismail A et al (2023) Microencapsulated phase change materials for enhanced thermal energy storage performance in construction materials: a critical review. 401:132877
44. Qiu J, Wang J (2017) Chitin in rubber based blends and micro composites. Rubber Based Bionanocompos Prep 71–107
45. Gaona CGC et al (2023) Composites and novel applications in the biomedical field. 129
46. Liu C et al (2023) Polymer engineering in phase change thermal storage materials. 188:113814
47. Chakraborty I, Pallen S, Shetty Y, et al (2020) Advanced microscopy techniques for revealing molecular structure of starch granules. 105–122
48. Mahmoudi Najafi SH, Baghaie M, Ashori A (2016) Preparation and characterization of acetylated starch nanoparticles as drug carrier: ciprofloxacin as a model. Int J Biol Macromol 87:48–54. https://doi.org/10.1016/j.ijbiomac.2016.02.030
49. Lecorre DS, Bras J, Dufresne A (2012) Influence of the botanic origin of starch nanocrystals on the morphological and mechanical properties of natural rubber nanocomposites. Macromol Mater Eng 297:969–978. https://doi.org/10.1002/mame.201100317
50. Vudjung C, Saengsuwan S (2017) Synthesis and properties of biodegradable hydrogels based on cross-linked natural rubber and cassava starch. J Elastomers Plast 49:574–594. https://doi.org/10.1177/0095244316676868

51. Du X, Zhang Y, Pan X et al (2019) Preparation and properties of modified porous starch/carbon black/natural rubber composites. Compos Part B Eng 156:1–7. https://doi.org/10.1016/j.compositesb.2018.08.033
52. Portela R, Leal CR, Almeida PL, Sobral RG (2019) Bacterial cellulose: a versatile biopolymer for wound dressing applications. Microb Biotechnol 12(4):586–610
53. Azizi Samir MAS, Alloin F, Dufresne A (2005) Review of recent research into cellulosic whiskers, their properties and their application in nanocomposite field. Biomacromolecules 6(2):612–626
54. Hirai A, Inui O, Horii F, Tsuji M (2009) Phase separation behavior in aqueous suspensions of bacterial cellulose nanocrystals prepared by sulfuric acid treatment. Langmuir 25(1):497–502
55. Habibi Y, Lucia LA, Rojas OJ (2010) Cellulose nanocrystals: chemistry, self-assembly, and applications. Chem Rev 110(6):3479–3500
56. George J, Ramana KV, Bawa AS (2011) Bacterial cellulose nanocrystals exhibiting high thermal stability and their polymer nanocomposites. Int J Biol Macromol 48(1):50–57
57. George J (2012) High performance edible nanocomposite films containing bacterial cellulose nanocrystals. Carbohyd Polym 87(3):2031–2037
58. Liu W, Du H, Zhang M, Liu K, Liu H, Xie H et al (2020) Bacterial cellulose-based composite scaffolds for biomedical applications: a review. ACS Sustain Chem Eng 8(20):7536–7562
59. Mohite BV, Patil SV (2014) Physical, structural, mechanical and thermal characterization of bacterial cellulose by G. hansenii NCIM 2529. Carbohyd Polym 106:132–141
60. Trovatti E, Carvalho AJ, Ribeiro SJ, Gandini A (2013) Simple green approach to reinforce natural rubber with bacterial cellulose nanofibers. Biomacromolecules 14(8):2667–2674
61. Chen Y, Li G, Yin Q, Jia H, Ji Q, Wang L et al (2018) Stimuli-responsive polymer nanocomposites based on styrene-butadiene rubber and bacterial cellulose whiskers. Polym Adv Technol 29(5):1507–1517
62. Zeng M, Laromaine A, Roig A (2014) Bacterial cellulose films: influence of bacterial strain and drying route on film properties. Cellulose 21:4455–4469
63. Ogrizek L, Lamovšek J, Čuš F, Leskovšek M, Gorjanc M (2021) Properties of bacterial cellulose produced using white and red grape bagasse as a nutrient source. Processes 9(7):1088
64. Saleh AK, El-Gendi H, Ray JB, Taha TH (2021) A low-cost effective media from starch kitchen waste for bacterial cellulose production and its application as simultaneous absorbance for methylene blue dye removal. Biomass Convers Biorefin 1–13
65. Singh O, Panesar PS, Chopra HK (2017) Response surface optimization for cellulose production from agro industrial waste by using new bacterial isolate Gluconacetobacter xylinus C18. Food Sci Biotechnol 26:1019–1028
66. Carreira P, Mendes JA, Trovatti E, Serafim LS, Freire CS, Silvestre AJ, Neto CP (2011) Utilization of residues from agro-forest industries in the production of high value bacterial cellulose. Biores Technol 102(15):7354–7360
67. Fontana JD, Franco VC, De Souza SJ, Lyra IN, De Souza AM (1991) Nature of plant stimulators in the production of Acetobacter xylinum ("tea fungus") biofilm used in skin therapy. Appl Biochem Biotechnol 28:341–351
68. Toyosaki H, Naritomi T, Seto A, Matsuoka M, Tsuchida T, Yoshinaga F (1995) Screening of bacterial cellulose-producing Acetobacter strains suitable for agitated culture. Biosci Biotechnol Biochem 59(8):1498–1502
69. Kurosumi A, Sasaki C, Yamashita Y, Nakamura Y (2009) Utilization of various fruit juices as carbon source for production of bacterial cellulose by Acetobacter xylinum NBRC 13693. Carbohyd Polym 76(2):333–335
70. Keshk S, Sameshima K (2006) The utilization of sugar cane molasses with/without the presence of lignosulfonate for the production of bacterial cellulose. Appl Microbiol Biotechnol 72:291–296
71. Bae S, Shoda M (2004) Bacterial cellulose production by fed-batch fermentation in molasses medium. Biotechnol Prog 20(5):1366–1371
72. Hong F, Zhu YX, Yang G, Yang XX (2011) Wheat straw acid hydrolysate as a potential cost-effective feedstock for production of bacterial cellulose. J Chem Technol Biotechnol 86(5):675–680

73. Hong F, Guo X, Zhang S, Han SF, Yang G, Jönsson LJ (2012) Bacterial cellulose production from cotton-based waste textiles: enzymatic saccharification enhanced by ionic liquid pretreatment. Biores Technol 104:503–508
74. Goelzer FDE, Faria-Tischer PCS, Vitorino JC, Sierakowski MR, Tischer CA (2009) Production and characterization of nanospheres of bacterial cellulose from Acetobacter xylinum from processed rice bark. Mater Sci Eng C 29(2):546–551
75. Hong F, Qiu K (2008) An alternative carbon source from konjac powder for enhancing production of bacterial cellulose in static cultures by a model strain Acetobacter aceti subsp. xylinus ATCC 23770. Carbohydr Polym 72(3):545–549
76. Gunasekaran S, Natarajan RK, Kala A (2007) FTIR spectra and mechanical strength analysis of some selected rubber derivatives. Spectrochim Acta Part A Mol Biomol Spectrosc 68(2):323–330
77. Zevallos Torres LA, Lorenci Woiciechowski A, de Andrade Tanobe VO, Karp SG, Guimarães Lorenci LC, Faulds C et al (2020) Lignin as a potential source of high-added value compounds: a review. J Clean Prod 263:121499
78. Tardy BL, Lizundia E, Guizani C, Hakkarainen M, Sipponen MH (2023) Prospects for the integration of lignin materials into the circular economy. Mater Today 65:122–132
79. Hosseinmardi A, Amiralian N, Hayati AN, Martin DJ, Annamalai PK (2021) Toughening of natural rubber nanocomposites by the incorporation of nanoscale lignin combined with an industrially relevant leaching process. Ind Crops Prod 159:113063
80. Liu R, Li J, Lu T, Han X, Yan Z, Zhao S et al (2022) Comparative study on the synergistic reinforcement of lignin between carbon black/lignin and silica/lignin hybrid filled natural rubber composites. Ind Crop Prod 187:115378
81. Nguyen NA, Barnes SH, Bowland CC, Meek KM, Littrell KC, Keum JK et al (2018) A path for lignin valorization via additive manufacturing of high-performance sustainable composites with enhanced 3D printability. Sci Adv 4:1–15
82. Bova T, Tran CD, Balakshin MY, Chen J, Capanema EA, Naskar AK (2016) An approach towards tailoring interfacial structures and properties of multiphase renewable thermoplastics from lignin-nitrile rubber. Green Chem 18:5423–5437
83. Feng X, Yang Z, Chmely S, Wang Q, Wang S, Xie Y (2017) Lignin-coated cellulose nanocrystal filled methacrylate composites prepared via 3D stereolithography printing: mechanical reinforcement and thermal stabilization. Carbohydr Polym [Internet]. 169:272–281. https://doi.org/10.1016/j.carbpol.2017.04.001
84. Tran CD, Chen J, Keum JK, Naskar AK (2016) A new class of renewable thermoplastics with extraordinary performance from nanostructured lignin-elastomers. Adv Funct Mater 26:2677–2685
85. Naeimi A et al (2022) First and efficient bio-nano composite, SnO_2/calcite based on Cypress leaves and eggshell wastes, for cytotoxic effects on HepG2 liver cancer cell lines and its antioxidant and antimicrobial activity. J Mol Struct 1259:132690. https://doi.org/10.1016/j.molstruc.2022.132690
86. Kumar P, Roopa CP et al (2022) Natural nano-fillers materials for the Bio-composites: a review. J Indian Chem Soc 99(10):100715. https://doi.org/10.1016/j.jics.2022.100715
87. Kasbaji M et al (2023) Future trends in dye removal by metal oxides and their nano/composites: a comprehensive review. Inorg Chem Commun 158:111546. https://doi.org/10.1016/j.inoche.2023.111546
88. Verma J, Petru M, Goel S (2024) Cellulose based materials to accelerate the transition towards sustainability. Ind Crops Prod 210:118078. https://doi.org/10.1016/j.indcrop.2024.118078
89. Li X et al (2024) Promising cellulose–based functional gels for advanced biomedical applications: a review. Int J Biol Macromol 260:129600. https://doi.org/10.1016/j.ijbiomac.2024.129600
90. Al Mahmud MZ (2023) Exploring the versatile applications of biocomposites in the medical field. Bioprinting 36:e00319. https://doi.org/10.1016/j.bprint.2023.e00319
91. Okwuwa CC et al (2023) Cellulose dissolution for edible biocomposites in deep eutectic solvents: a review. J Clean Prod 427:139166. https://doi.org/10.1016/j.jclepro.2023.139166

92. Agbakoba VC et al (2023) Preparation of cellulose nanocrystal (CNCs) reinforced polylactic acid (PLA) bionanocomposites filaments using biobased additives for 3D printing applications. Nanoscale Adv 5(17):4447–4463. https://doi.org/10.1039/D3NA00281K
93. Leite LSF et al (2020) Scaled-up production of gelatin-cellulose nanocrystal bionanocomposite films by continuous casting. Carbohydr Polym 238:116198. https://doi.org/10.1016/j.carbpol.2020.116198
94. Ghosal K et al (2024) Polyvinyl alcohol-based bionanocomposites: synthesis, properties, and applications. In: Advances in bionanocomposites. Elsevier, pp 117–132. https://doi.org/10.1016/B978-0-323-91764-3.00010-3
95. Wu Y, He J, Xiong L (2024) Price discovery in Chinese PVC futures and spot markets: impacts of COVID-19 and benchmark analysis. Heliyon 10(2):e24138. https://doi.org/10.1016/j.heliyon.2024.e24138
96. Dhal MK et al (2023) Polylactic acid/polycaprolactone/sawdust based biocomposites trays with enhanced compostability. Int J Biol Macromol 253:126977. https://doi.org/10.1016/j.ijbiomac.2023.126977
97. Vijayaram TR (2009) A technical review of natural rubber. Int J Desa Manuf Technol 3(1):25–37
98. Vunovic X (2009) Tapping the amazon for victory: Brazil's "battle for rubber" of world war II. Doctoral dissertation. Dec. 2, Faculty of the Graduate School of Arts and Sciences Of Georgetown University, Washington D.C.
99. Prabhakaran Nair KP (2010) The agronomy and economy of important tree crops of the developing world, pp 237–273. ISBN 978-0-12-384677-8. https://www.quimitube.com, 2017
100. Bruins J (2003) World agriculture: towards 2015/2030. A FAO perspective. In: Bruins J (ed) Earthscan Publications Ltd., London, 432 pp
101. Cornish K (1996) Hypoallergenic natural rubber products from *Parthenium argentatum* (Gray) and other non *Hevea brasiliensis* species. United States Patent. Patent number: 5.580.942; Dec. 3
102. Whaley WG (1948) Rubber-the primary source for American production. Econ Bot 2(2):198–216. https://doi.org/10.1007/BF02859004. Online ISSN: 0013-0001
103. Franck A, De Leenheer L (2005) Inulin. Biopolymers Online 6:439–448. https://doi.org/10.1002/3527600035.bpol6014
104. Coussement P (1999) Inulin and oligofructose as dietary fiber: analytical, nutrition and legal aspects. In: Cho SS, Prosky L, Dreher M (eds) Complex carbohydrates in foods. New York, pp 203–212
105. De Leenheer L, Hoebregs H (1994) Progress in the elucidation of the composition of chicory inulin. Starch 46:193–196
106. Flamm G, Glinsmann W, Kritchevsky D, Prosky L, Roberfroid M (2001) Inulin and oligofructose as dietary fiber: a review of the evidence. Crit Rev Food Sci Nutr 41:353–362
107. van Laere A, Van den Ende W (2002) Inulin metabolism in dicots: chicory as a model system. Plant Cell Environ 25:803–813
108. Schütz K, Muks E, Carle R, Schieber A (2006) Separation and quantification of inulin in selected artichoke (Cynara scolymus L.) cultivars and dandelion (Taraxacum officinale WEB. Ex WIGG.) roots by high performance anion exchange chromatography with pulsed amperometric detection

Chapter 2
Cellulose-Based Rubber Bionanocomposites

Denisse Ochoa Torres and María Inés Errea

Abstract Rubber composites are part of more than 50,000 products that are used daily all over the world. However, traditional fillers in the rubber industry often pose environmental and health concerns. To address this, cellulose has emerged as a sustainable alternative due to its low cost, biodegradability, and renewable origin. Particularly noteworthy is nanocellulose, which, due to its small size, exhibits superior reinforcing capabilities in rubber composites. In this line, cellulose-based rubber bionanocomposites are high-performance environmentally friendly materials with potential in different fields. Among them, the following stand out: (i) biomedical, with a particular focus on tissue engineering (ii) aerospace industry, through the development of ultralight aerogels with low thermal conductivity, (iii) separation process technology, through the preparation of membranes, and (iv) food industry, due to the feasibility of preparing coating materials for food preservation. Besides, studies show that depending on the final desired properties of the composite the most appropriate form of nanocellulose, either nanocrystals (CNCs) or nanofibrils (CNFs), should be carefully chosen. Bacterial nanocellulose (BC) remains unexplored for this type of material. There is still a long road to be explored but results obtained so far serve as encouragement to continue investigating new novel applications for rubber composites with nanocellulose.

Keywords Nanocellulose · Nanocrystals (CNCs) · Nanofibrils (CNFs) · Bacterial nanocellulose (BC) · Rubber composites · Environmentally friendly materials · Natural rubber nanocomposites · Synthetic rubber nanocomposites and biomedical applications

---

D. Ochoa Torres · M. I. Errea (✉)
Instituto Tecnológico de Buenos Aires, Buenos Aires (ITBA), Ciudad Autónoma de Buenos Aires, Argentina
e-mail: merrea@itba.edu.ar

CONICET (Consejo de Investigaciones Científicas y Técnicas), Ciudad Autónoma de Buenos Aires, Argentina

Visakh P. M. *Rubber Based Bionanocomposites*, Advanced Structured Materials 210,
https://doi.org/10.1007/978-981-10-2978-3_2

2.1 Introduction

Rubber is a polymer, either of natural or synthetic origin, widely used around the world. Natural rubber latex (NRL) is produced by more than 2000 plant species (rubber plants), but only NRL from *Hevea brasiliensis* and *Parthenium argentatum* are used commercially with 99% and 1% of the world market, respectively [1]. Among the countries that cultivate these plant species, Indonesia and Thailand have the largest cultivated areas [2]. NRL is a colloidal system containing approximately 50 wt.% water, 30–45 wt.% rubber particles, and 4–5 wt.% of proteins, lipids, and carbohydrates [3]. Natural Rubber (NR), obtained from NRL is mainly constituted by a polymer formed by repeating units of cis-1,4-isoprene (Fig. 2.1) [4], having molecular masses of about 10^6 Da [5].

NRL was first obtained from *Hevea brasiliensis* in South America [6] and it has been used for the manufacture of a wide variety of products for long time ago. In fact, first records of natives from the Island of Haiti playing with "balls made of gum from a tree" are traced back to the sixteenth century during Colombus' second voyage [7]. Even though this unique material caught the attention of people, it had no commercial use for approximately 300 years after Columbus. With the development of new technologies, the demand for NR started to increase, also promoting the development of synthetic rubber (SR). Fabrication of SR peaked during the period of the Second World War when the supply of NR resulted insufficient [8, 9]. Since then, NR and SR started a market competition that was initially won by SR (with 70% of the market) but has evolved to an almost 50/50 share in 2022 [10].

Although rubber possesses distinctive properties like great elasticity, flexibility, and versatility, it also has some drawbacks such as low strength and poor stability (it is susceptible to UV light, becomes sticky in high temperatures and brittle in low temperatures) [9, 11, 12]. To improve physical, chemical, and mechanical properties of rubber, the polymer chains are subjected to crosslinking processes [13]. The crosslinker may be sulfur atoms in short chains, carbon-to-carbon bond, polyvalent organic radical, ionic cluster, or polyvalent metal ion, being sulfur chains the most common crosslinkers (Fig. 2.2). The process involving the latter is the most widely used and is referred to as vulcanization [14, 15].

Nowadays, vulcanized rubber is used as a raw material for the manufacture of more than 50,000 different products [16], i.e., balloons, gloves, pillows, footwear, catheters, and tires [17]. The very spread use of materials obtained from rubber all

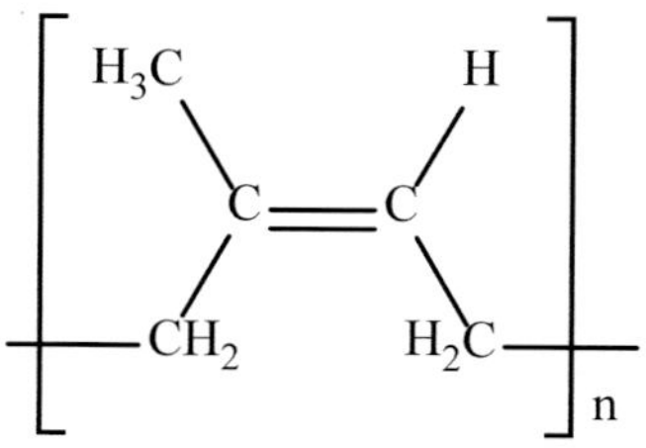

$$\left[\begin{array}{c} H_3C \quad\quad\quad H \\ C{=}C \\ -CH_2 \quad\quad H_2C- \end{array} \right]_n$$

Fig. 2.1 Chemical structure of natural rubber

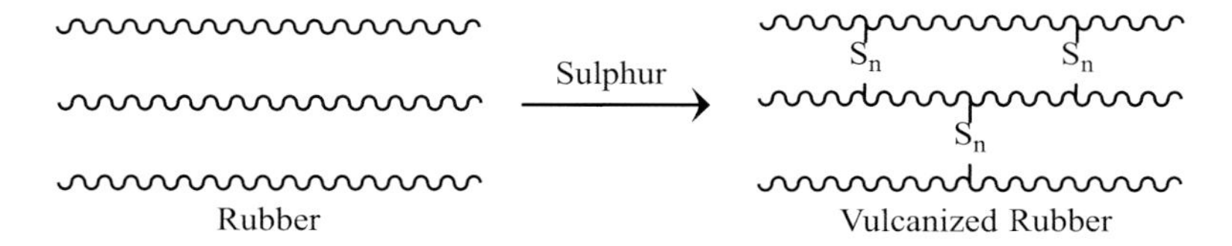

Fig. 2.2 Vulcanization of rubber

over the world explains why it is of special importance to record the recent advances and new technologies developed around it.

In order to impart to the vulcanized rubber the final properties required for the different applications, additives (such as stabilizers) are usually added and/or composite materials are prepared [18, 19].

Composites are materials that have two or more phases, involving a matrix (in this case rubber) and other compounds, which are commonly referred to as fillers, that are introduced to reinforce mechanical and thermal properties of the matrix [20]. It is important to note that the final properties of the composite are highly influenced by the shape, size, and orientation of the fillers [21]. In this line, it has been demonstrated the significant impact of the size of the fillers in the mechanical properties of the rubber composites. Sizes greater than 10,000 nm have degradant mechanical effects, from 100 to 1000 nm there is a semi-reinforcing effect and only the ones smaller than 100 nm, which are called nanofillers, have a significant reinforcing effect [22]. Given this, the current trend is to use nanocomposites with nanoscale fillers that result in a high performance in the desired applications [23].

Currently, in the rubber industry, carbon black (CB) is the most used reinforcing filler with 90% of its production being used in the manufacturing of tires [24]. However, given that CB is derived from petroleum, its use has serious drawbacks such as contamination and the increase in GHG emissions. Moreover, its utilization implies a major health concern in workplaces because it is listed by the International Agency for Research on Cancer (IARC) as possibly carcinogenic to humans [25]. This situation becomes even more serious when considering that the content of CB in some rubber products can be as high as 50% (m/m) [11].

Taking into account the above discussion, sustainable solutions to enhance rubber properties are now demanded. A feasible solution is the preparation of biocomposites, which involves the use of natural, biocompatible, and biodegradable materials as fillers. In this context, cellulose, a natural abundant polymer consisting of D-glucopyranose units linked by $\beta(1 \rightarrow 4)$ bonds [26] (Fig. 2.3) has aroused great interest as a filler because of its low cost, biodegradability, and renewable origin. Furthermore, it has outstanding mechanical properties, such as great mechanical stiffness and strength which make it very attractive as a reinforcement for rubber matrices [27].

Cellulose has been traditionally obtained from wood, but there are other alternative sources. Among them, plants such as cotton, ramie, sisal, flax, wheat, potato tubers, beet pulp, soybeans, as well as marine algae and even some marine organisms, such as

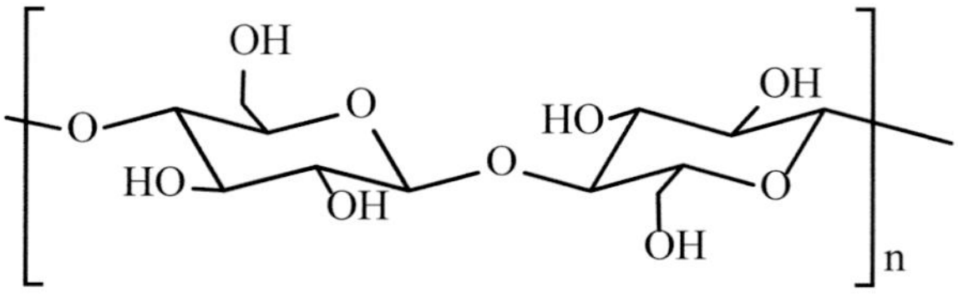

Fig. 2.3 Chemical structure of cellulose

tunicates, can be mentioned. Moreover, there are also specific bacteria that synthesize pure nanofibrilated cellulose [28].

Considering both, sustainability, and the impact of the filler's size on the rubber properties, mentioned above, the preparation of bionanocomposites have attracted the attention of the rubber industry, leading to the development of composites in which the filler is a material of biological origin with sizes that have at least one dimension on the nanometer scale [29]. Among them, nanocelluloses stand out. Nanocelluloses can be categorized in three forms, (1) cellulose nanofibers (CNFs), (2) cellulose nanocrystals (CNCs), and (3) bacterial nanocellulose (BNC) [30]. CNCs have rod-like shapes or whiskers shapes with 3–50 nm in diameter and 100–500 nm in length which are obtained by acid or enzymatic treatment of cellulose of plant origin. CNFs are constituted by alternated amorphous and crystalline regions, with the diameter in the nanometric scale (1–100 nm) and length between 500 and 2000 nm [31]. CNFs are obtained by mechanical processing of macrofibers of wood or plants, sometimes subjected to chemical or enzymatic pretreatment. BNC has unique characteristics such as high chemical purity, crystallinity, and polymerization degree, and it is obtained as a nanostructured material avoiding the high-energy demanded process involved in the obtention of CNFs.

The high aspect ratio and large surface area of CNFs, CNCs, and BNC can be used to create strong interfacial bonding to the polymer matrix when they are used as reinforcement fillers. However, nanocellulose hydrophilicity nature becomes a challenge when trying to introduce it to a rubber hydrophobic polymer matrix. In this case, surface functionalization is required to accomplish the interaction between the filler and the matrix [11].

From all discussed above, cellulose-based rubber bionanocomposites are of special interest in the line of producing high-performance environmentally friendly materials. These biocomposites have the potential to be used in various engineering applications in several different industries such as medical, automotive, military, aerospace, packaging, coating, structural, and membranes (Fig. 2.4) which are currently being investigated and will be further described in the next sections.

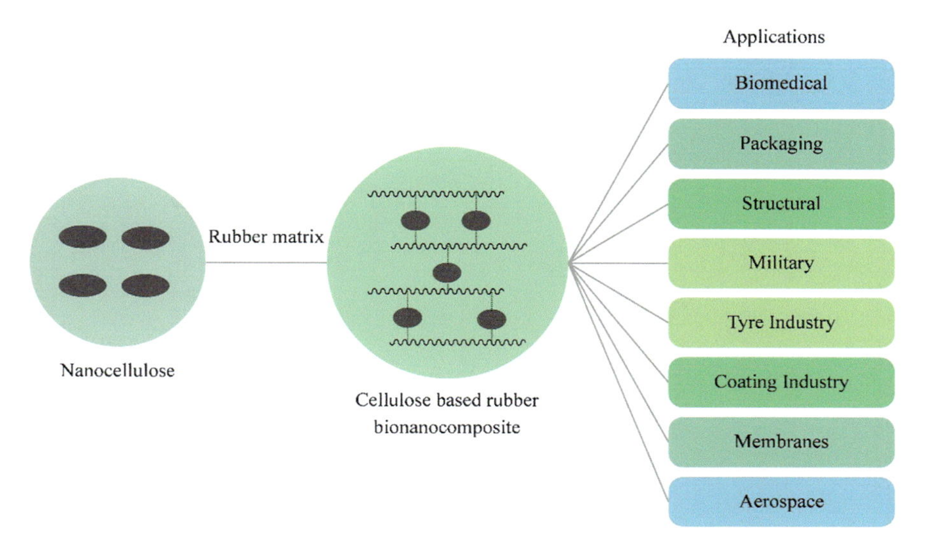

Fig. 2.4 Applications of cellulose-based rubber bionanocomposites

2.2 Applications of Cellulose/Natural Rubber Nanocomposites

Natural rubber, due to its renewable origin, is undoubtedly more sustainable than the synthetic one, but its production remains insufficient to meet global demand. Therefore, the replacement of NR, at least in some industrial sectors, by SR is inevitable. In addition, the structural diversity of synthetic rubber makes it possible to further adjust the properties of the final product according to the requirements of different specific applications.

In the case of biomedical applications, as described in Sect. 2.4, only natural rubber is used, due to its high biocompatibility.

2.3 Applications of Cellulose/Synthetic Rubber Nanocomposites

Since the nineteenth century, the preparation of synthetic rubber with outstanding mechanical properties has been an important topic in the development of an alternative source of natural rubber [32].

In order to achieve the specific properties required for the different applications, synthetic rubbers have been prepared with various chemical structures, among which, due to their widespread use, the following are worth to be mentioned: (i) Styrene butadiene styrene (SBS), (ii) Styrene butadiene (SBR), (iii) Polybutadiene (PB), (iv)

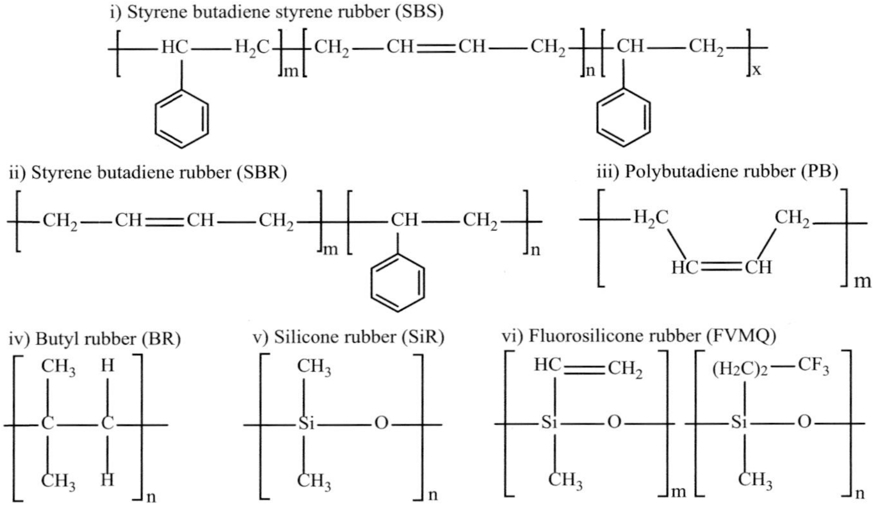

i) Styrene butadiene styrene rubber (SBS)

ii) Styrene butadiene rubber (SBR)

iii) Polybutadiene rubber (PB)

iv) Butyl rubber (BR)

v) Silicone rubber (SiR)

vi) Fluorosilicone rubber (FVMQ)

Fig. 2.5 Synthetic rubber structures

Butyl rubber (BR), (v) Silicone rubber (SiR), and (vi) Fluorosilicone rubber (FVMQ) (Fig. 2.5).

As mentioned above, despite the environmental problems generated using materials from the petroleum industry, there are applications such as tire production where the properties of synthetic rubber can be required.

2.4 Biomedical Applications

Biomaterials are substances that interact with biological systems to evaluate, treat, augment, or replace any tissue, organ, or function of the body. Biomaterials have revolutionized the biomedical field by providing innovative solutions for medical implants, tissue engineering, drug delivery, and diagnostic technologies. The ongoing research in biomaterial science continues to expand the possibilities for improving healthcare outcomes and patient well-being. Materials need to meet certain criteria to be suitable for biomedical uses. Among them, it can be highlighted the biocompatibility, stimuli-responsive to external/internal inputs [33] and specific surface properties for the correct interaction between the interface of the biological tissue and the biomaterial [34].

Most materials used in the medical field are still of synthetic origin [35], but natural polymers have gained popularity in recent years. Specifically, NRL due to its low cost and excellent biocompatibility appears as a promising biomaterial to be applied in this field. Furthermore, it has been proved that NRL has angiogenic properties, meaning that it promotes the formation of new blood vessels, which is

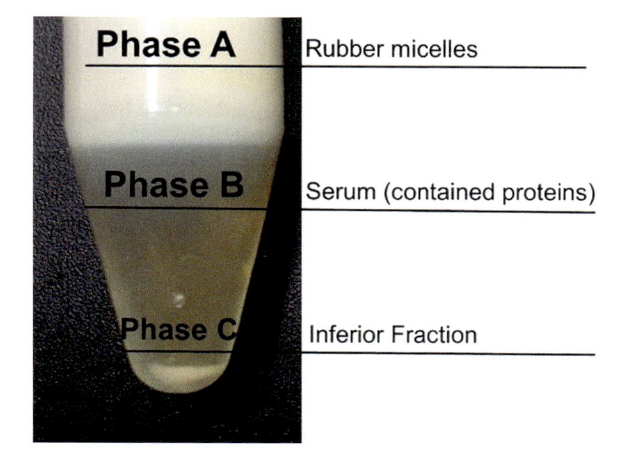

Fig. 2.6 Processing of NRL by centrifugation. Reprinted from [45] Copyright (2023) with permission from Elsevier

of special interest in wound healing and tissue engineering [36, 37]. Respect to the use of NRL for biomedical application, although they have proven to possess low cytotoxicity/genotoxicity [38] the processing of the NRL previous to the fabrication of the membranes is a key point to consider. After centrifugation of NRL three phases are obtained: (i) an upper phase which contains mainly the rubber particles with minor proportion of proteins and lipids; (ii) a serum as intermediate phase, which contains amino acids and proteins; and (iii) a bottom phase containing mainly large molecular weight proteins (Fig. 2.6) [37]. Taking into account that the angiogenic behavior is attributed to the proteins remaining in the rubber fraction [39] and the non-rubber fractions of NRL negatively affected the cytotoxicity [40], in general, materials for biomedical applications are prepared using only the upper fraction obtained after the NRL centrifugation process, to ensure both, angiogenesis and biocompatibility [39, 41–43].

On the other hand, nanocellulose also exhibits good biocompatibility and low cytotoxicity/genotoxicity as was demonstrated in both in vitro and in vivo studies [44].

Taking into account the above discussion, the preparation of composites of rubber and cellulose is particularly interesting in the biomedical field and some of the advances in this field are described below.

Tissue Engineering and Bone Regeneration

The increase in life expectancy has been a direct result of the different challenges overcome by modern medicine. At the same time, new age-related diseases need to now be addressed in order to guarantee longevity with a good quality of life. Tissue engineering and regenerative medicine aim to find more reliable solutions for the loss of skeletal tissue than the current approaches used [46]. In this line, occlusive membranes for guided bone regeneration prepared from NRL are being studied. These membranes act as barriers for unwanted epithelial cells and also provide the osteogenic cells with better conditions for bone formation [39, 43], due to the capacity of NRL to promote angiogenesis. Despite the recognized improvement of the mechanical properties of NRL when nanocellulose is added as reinforcement, there is still a lack of reported studies using cellulose–rubber bionanocomposites as occlusive membranes. However, composites including CNCs and bioglass particles as reinforce fillers of NRL matrices for tissue engineering were recently published with promising results [34].

Another interesting biomedical application of bionanocomposites is the development of neuromotor prostheses (NMPs). Several people across the globe experience various types of motor impairments, which reduce their life quality by limiting their independence and mobility. NMPs look for a solution to restore the lost motor functions by recording and guiding the brain signals related to the movement to external aids (i.e., wheelchairs, robotic arms) [47]. One of the challenges of NMPs is to match the mechanical properties of the hard rigid electrodes with the soft cortical tissue. This disparity leads to scar formation and death tissue in the zone of the implant, which results in the malfunction of the device [48]. A feasible solution is the development of mechanically adaptive materials to create a sufficiently rigid material to be able to insert it, but that softens once it is in contact with the brain tissue [49]. Specifically, water-responsive CNC-based rubber bionanocomposites with mechanically adaptive behavior have been developed with both NRL and epoxidized NRL (ENRL) [50, 51]. Significantly higher water sensitivity was achieved in the case of the epoxidized NRL which can be attributed to the strong hydrogen bond interactions established between the oxygens of the epoxide groups of the ENRL and the CNCs' hydroxyl groups. This interaction allowed a better filler dispersion due to the formation of a network CNC-ENRL, in addition to the CNC-CNC polymer network presented when the matrix was the non-epoxidized NR [50]. These studies evidence the possibility of tuning the water responsivity as well as the mechanical properties of the CNC–rubber biocomposite taking advantage of the versatility of CNC towards chemical modifications.

Antimicrobial

Central venous catheters are an integral part of modern medicine and catheter-related bloodstream infection is the third most frequent nosocomial infection [52]. In the same line, the formation of bacterial biofilms on urinary catheters is a leading cause of urinary tract infections in intensive care units. Biofilms enable bacteria to survive in challenging environments and enhance their resistance to immune defenses as well as commonly applied antibiotics. Consequently, infections due to biofilms are extremely difficult to treat [53]. Therefore, the development of catheters prepared with materials that prevent the formation of biofilms would be a great contribution to public health, considering the extremely high costs associated with nosocomial diseases.

In this context, given the capacity of CNC to induce flocculation and phase separation of bacteria by depletion, which reduces their diffusion from the urinary collector to the catheter, a composite material of CNCs and NRL was prepared and the release of CNCs from the composites to aqueous solutions was investigated. Findings indicated that minimal amounts of released CNCs were sufficient to control both bacterial adhesion and biofilm formation [54].

On the other hand, the versatility of nanocellulose towards chemical modifications can be advantageously used for introducing new functional groups with antimicrobial properties (aldehydes, quaternary ammonium). There are studies which indicate that the addition of antimicrobial nanofillers can confer antimicrobial attributes to the final rubber nanocomposites [55]. Furthermore, antimicrobial properties can also be imparted to the bionanocomposites by incorporating antibiotics into their composition (i.e., ampicillin, gentamycin, tetracycline, chloromycetin) [56].

However, there is still a lot to be explored in the development of antimicrobial materials from bionanocomposites of rubber and nanocellulose.

2.5 Packaging Applications

Worldwide awareness of the environmental problems generated by packaging waste has led to great efforts to develop packaging involving environmentally friendly materials. In this line, cellulose due to its renewable origin and its biodegradability, has received increasing interest in recent times. Furthermore, cellulose is odorless and tasteless, which are highly desired properties for food packaging. In particular, nanocelluloses (NC), either in nanocrystal or fibrillated form, as was mentioned in the introduction section, have outstanding properties such as low toxicity and density, high-tensile strength, and high aspect ratio. However, its hydroxyl groups impart a high hydrophilicity as well as porosity, which represent limitations for its use in packaging. In this context, the development of new packaging materials using NC as a reinforcing phase in synthetic polymeric matrices could enhance the properties of the matrices avoiding the limitations associated with cellulose hydrophilicity [57].

Moreover, the elasticity of NR is useful in packaging, but its hardness, Young's modulus, and abrasion resistance need to be improved for its use in this application [58]. In this context, films of NR reinforced with bacterial cellulose (BC) nanofibers were developed. By loading BC at 80 wt.%, the mechanical properties of the composites, such as Young's modulus and tensile strength, were approximately 2,580 times and 94 times higher, respectively, than those of pure NR films, whereas the elongation at break was decreased to 0.04 of that of the NR film. These results demonstrate the potential of NR–BC composite as biofilms with elastic properties for packaging, including food and medicinal industries [59].

2.6 Structural Applications

The disposal of waste rubber products has become one of the most serious pollution problems since the beginning of the twenty-first century, due to the low degradation rate of rubber and the lack of effective treatment methods [60]. In this context, in an attempt to mitigate the environmental deterioration caused by the continuous increase of car tires discarded annually, the reuse of rubber obtained from discarded car tires as a substitute for traditional aggregates in structural concrete has recently gained attention. Rubber concrete is a cement-based composite material that integrates waste-tire rubber particles in a specific proportion to modify its internal structure. The substitution of a portion of the fine aggregates in conventional concrete by rubber, could be an option to diminish the environmental damage caused by the discarded rubber tires and contribute to sustainable and eco-friendly constructions. However, the incorporation of rubber particles into concrete reduces its compressive strength, tensile strength, shear strength, and elastic modulus [61]. The reduction of concrete working performance due to the addition of rubber particles is attributed to the high hydrophobicity of rubber (mainly SBR or NR). Since highly hydrophobic rubber particulates cannot be dispersed effectively in the cement slurry, cavities and large cracks are formed when the two materials are combined [60].

Taking into account that the hydrophobicity of rubber itself is enhanced by the hydrophobicity of the fillers commonly used in the tire industry, such as carbon black, the replacement (at least partial) of CB by nanocellulose (Sect. 2.8), could improve the compatibility between the cement slurry and the rubber particles by enhancing the hydrophilicity of the rubber composite due to the polyhydroxylated structure of cellulose. This could be a promising strategy, bearing in mind environmental conservation and lower energy consumption. In this line, there are early reports about partial replacement of the cement Portland with biocomposites of rubber, cellulose fibers, and silica. The results showed an enhancement of the cement properties after the addition of the composite [62].

2.7 Military Applications

As mentioned above, the Second World War was the incentive to produce synthetic rubber, given the high demand that was suddenly experienced and could not be fulfilled by natural rubber. In general, the requirements of the armed forces have been strong drivers for the development of the rubber industry. When designing materials for military applications, many factors must be taken into account. These include the need to be lightweight, flexible, highly protective (such as bulletproof vests), and resistant to the harsh conditions that can be encountered in military operations, such as extreme temperature changes, rain, and sandstorms [63, 64]. In this context, polymer nanocomposites have been shown to possess many of the properties required for this industry, and rubber as a matrix reinforced by bionanocompounds such as CNCs and CNFs are targets worth to investigate [65].

On the other hand, transportation is inherently part of military defense fields which use vehicles, aircrafts, ships, and drones [65]. Given this, is important to take into account that the development of the tire industry (Sect. 2.8) has a significant impact on the quality of the army's means of transport.

Besides, rubber bullets are part of non-lethal weaponry where various types of rubber (NR and SR, including PB, SBR, BR) highly filled with CB, clay, calcium carbonate, and other metal fillers [66] are used. Considering all previous discussion about the advantages of NC as replacement of CB, CNCs, or CNFs are worth to assay as fillers to create rubber bullets, enhancing the sustainability of military industry.

Furthermore, high-tech armament can be developed with cellulose-based rubber bionanocomposites, which, with the incorporation of graphene, have proven to have high electric conductivity. This property can be used as a parameter to sense various external outputs [67].

2.8 Tire Industry

In addition to the environmental deterioration associated with the use of synthetic rubber, derived from the petroleum industry, both natural and synthetic rubber are facing resource scarcity issues. This is an important topic considering that the tire industry is highly raw material demanded [68]. For this reason, there is a high interest in developing green processing technologies that implement sustainable solutions for this industry [69].

In the tire production, the compound formulations for tire tread can include NR, SBR, PB, fillers (CB and silica), crosslinkers, and other additives (activators, plasticizers, and accelerators) [70], according to the final properties required, which depend on the type of vehicle the tire is designed for. As was mentioned above, there are many studies aiming to replace at least part of CB with more sustainable fillers, like nanocellulose. In this line, there are studies that showed that it is possible to replace up to 34% of CB by CNCs, without any evidence of a negative impact in the composite properties as can be seen in Table 2.1. Higher percentage of replacement of CB led to a deterioration of the tensile and tear strength of the composite (Table 2.1) [71, 72]. However, the elongation at break continuously increased with the increase in the percentage of replacement of CB (Table 2.1). Regarding the heat build-up temperature, its values decreased with the increase of CNCs up to 44%, and at higher values, a slight increase was observed.

These results can be explained in terms of dispersibility of the fillers in the rubber matrix. The addition of CNCs contributes to the dispersion of CB in the matrix, reducing the number of CB aggregates (Fig. 2.7), which consequently enhanced the rubber elastic deformation while maintaining similar tensile and tear strength.

In the same line, the generation of heat during dynamic compression results from the delay caused by the deformation and rearrangement of filler aggregates

Table 2.1 Mechanical properties of NR/CB/CNC composites with various CB/CNC ratios. The total filler loading of CB/CNC was fixed at 45 phr

Percentage of CB replaced with CNCs (%)	Tensile strength (MPa)	Tear strength (kNm^{-1})	Elongation at break (%)	Heat build-up (°C)
0	28.56	75.74	463.2	23.4
11	29.55	81.33	477.5	21.7
22	29.34	82.15	514.2	18.6
34	27.67	74.64	536.3	17.8
44	27.31	70.18	554.6	17.7
55	26.28	62.55	566.1	19.8

Note This table was adapted from [72]

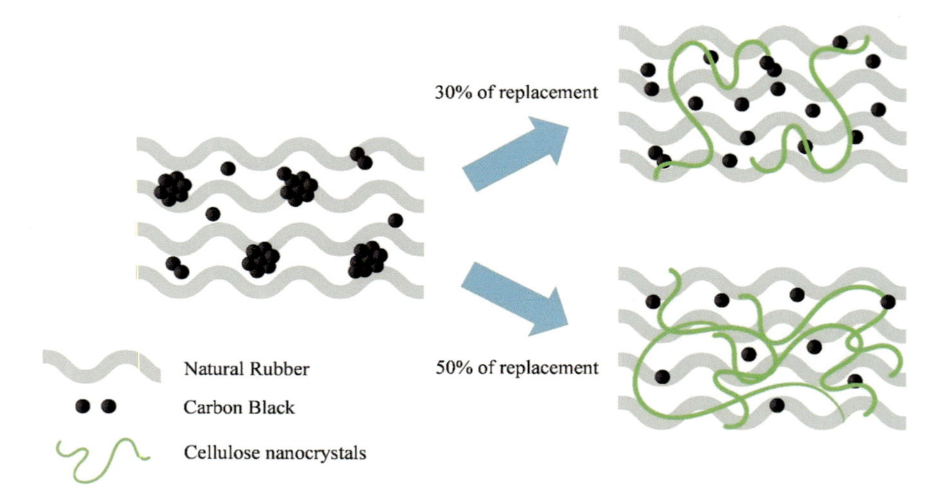

Fig. 2.7 Interaction between carbon black (CB) and cellulose nanocrystals (CNCs) in a natural rubber (NR) matrix. Percentages refer to the mass of CB replaced by CNCs

in composites under high dynamic frequency conditions, as detailed in the literature [73, 74]. As CNCs improve the CB dispersion in the rubber matrix, the lag effect is less, and consequently, the heat build-up is reduced.

On the other hand, relevant parameters for the vulcanization process were analyzed for rubber composites in which different percentages of CB were replaced by CNCs (Table 2.2) [75].

The studies showed that the cure time of the composites increased while replacing carbon black with nanocellulose and this was attributed to an increase of the composites' active surface due to the addition of CNCs. Besides, as it can be seen in Table 2.2, the analysis also showed an increase in the scorch time of the composites containing CNCs. Moreover, the cure rate index increased when nanocellulose was added,

Table 2.2 Curing parameters for NR/CB/CNC composites with various CB/CNC ratios. The total filler loading of CB/CNC was fixed at 30 phr

Percentage of CB replaced with CNCs (%)	Scorch time t_{10} (min)	Optimum cure time t_{90} (min)	Cure rate index CRI (min^{-1})	Minimum torque M_L (dNm)	Maximum torque M_H (dNm)	M_H-M_L (dNm)
0	2.07	8.15	16.44	0.35	11.12	10.77
16	2.11	8.17	16.5	0.39	11.55	11.16
33	2.23	8.26	16.58	0.27	10.37	10.10
50	2.34	8.36	16.61	0.30	9.69	9.39

Note This table was adapted from [75]

which means that the processing of the composite became easier by CB replacement. Furthermore, the curing index rate was also increased respect of rubber, when nanocomposites were prepared using only CNCs as filler, avoiding completely the use of CB [76]. In addition, the minimum torque value (which shows that viscosity was affected by the addition of nanocellulose [77]), was found to be lower for the composite in which 16% of CB was replaced with CNCs. Besides, in this composite, the highest value of differential torque was achieved (Table 2.2). The increase in the differential torque value was related to a higher restriction in the mobility of the rubber chains caused by the entanglement with the cellulose nanofibers, which provides more stiffness and rigidity to the composite.

In conclusion, the studies agree with the feasibility of replacing at least a part of CB in the tyre production with nanocellulose, which leads to composites with improved mechanical and processing properties. These results are promising from both environmental and economic points of view. However, great efforts need to still be made in order to find sustainable solutions that allow for replacing the entirety of CB in the rubber composites.

2.9 Coating Industry

The outstanding oxygen and gas barrier properties of CNFs as well as their excellent mechanical properties have promoted studies aiming to use them in the development of biodegradable films for coating applications. However, there are some drawbacks to overcome before they can be used efficiently for this purpose, such as their high sensitivity to moisture and their poor water vapor barrier properties. Taking into account that these drawbacks are mainly attributed to the high hydrophilicity of cellulose, which can establish strong hydrogen bond interactions with the water molecules, the preparation of composites with less hydrophilic materials could turn more tortuous the path of water through the composite, diminishing its sensitivity to moisture and enhancing its vapor barrier properties. In this line, a hydrophobic

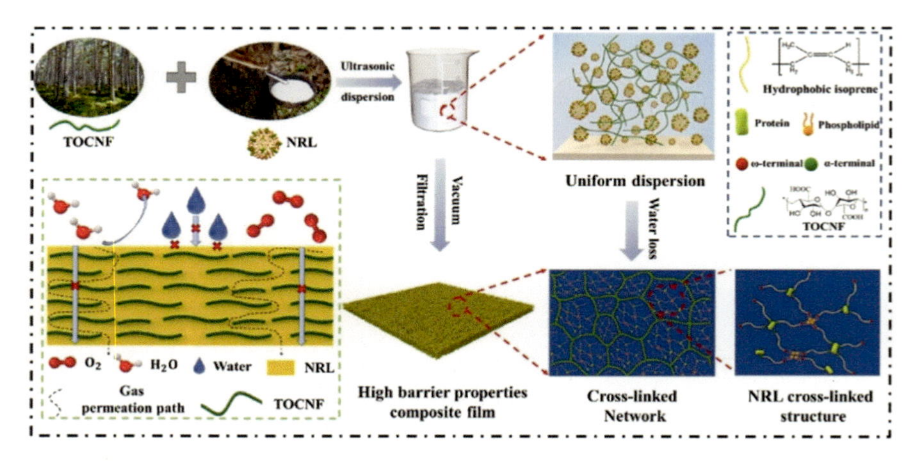

Fig. 2.8 Preparation of oxidated bacterial cellulose nanofibers (TOCNFs) and NRL bionanocomposites. Reprinted from [78] Copyright 2024 with permission from Elsevier

component such as NRL could be added to improve the properties of CNFs for coating applications. Moreover, NRL has additional advantages, such as low cost, good elasticity, and good film-forming properties are worth to mention [78].

In this context, oxidated bacterial cellulose nanofibers (TOCNFs) prepared by treatment of bacterial cellulose nanofibers with 2,2,6,6-tetramethylpiperidin-1-yloxy radical (TEMPO) were composited with NRL (Fig. 2.8). Various composites were prepared by mixing different ratios of TOCNFs and NRL. The findings exhibited that the composite formed by mixing equal proportion of TOCNFs and NRL (C50N50) had the best results. In these conditions, a film which provides a good barrier to water vapor and oxygen was obtained.

Strawberry preservation experiments were conducted to investigate the performance of this new film [78]. Three experiments were carried out in paralell: in one of them, the strawberries were encapsulated with C50N50, while, in a second experiment, they were encapsulated with polyethylene (PE) and finally, in a third experiment, the fruits were left exposed to the air without any kind of protection (CK) (Fig. 2.9).

Figure 2.9a–c represent the variation of firmness, color, and total soluble solids content of the strawberries. As it can observed in the figures, all three parameters decayed during the study period, but the decay was less pronounced in the case of strawberries covered with C50N50, and significantly faster in the group of strawberries exposed to air.

Figure 2.9d–f represent the variation of decay index, relative conductivity, and weight loss of strawberries. The decay index is a weighted average of the fruit decay grades. The higher the decay index, the more severe the fruit decay. Relative conductivity is an indicator of the fruit cell membrane damage: when cytoplasmic membrane decreased, membrane permeability increased, leakage of intracellular electrolytes to the outside occurred, and the relative conductivity became large [79].

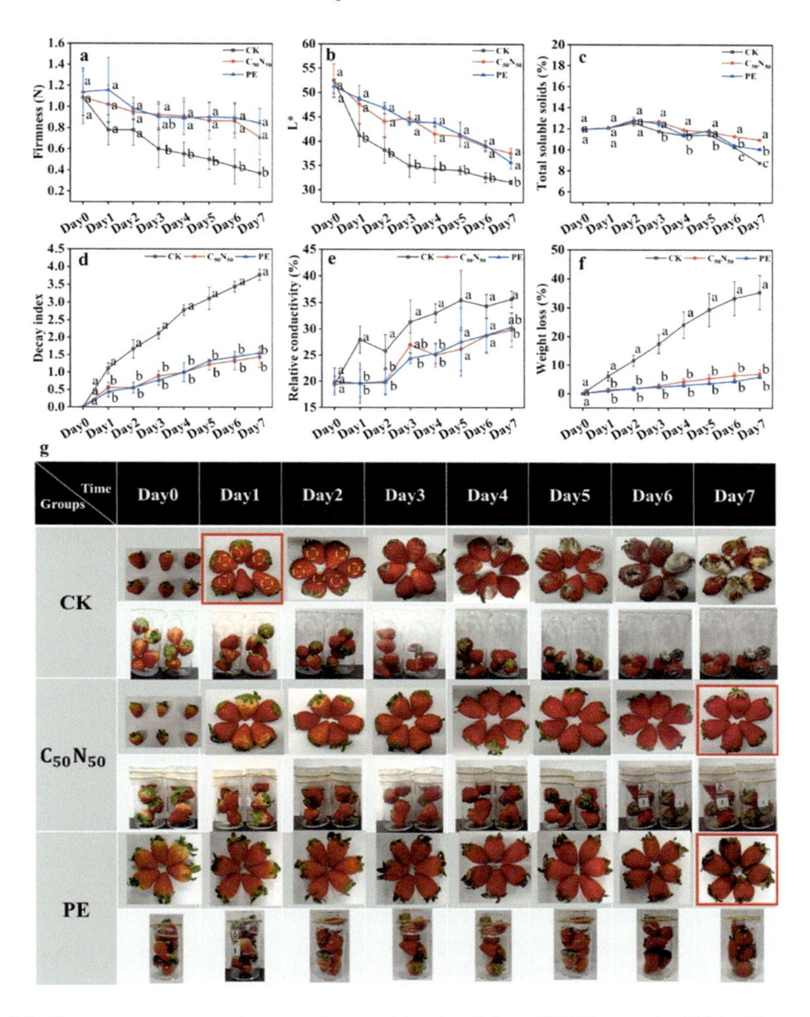

Fig. 2.9 Strawberry preservation experiments. Reprinted from [78] Copyright 2024 with permission from Elsevier

The decay index, relative conductivity, and weight loss of strawberries gradually increased in all groups during storage. However, the increasing trend in the CK group was significantly higher than those in the C50N50 and PE groups, which is consistent with the fact that both films represent a barrier to water and oxygen transmission, thus contributing to strawberry preservation.

Figure 2.9g represents the changes of strawberries during the study period that can be observed with the naked eye. On Day 0, each group of strawberries shows a fresh appearance, and no wrinkles are observed on the surface of none of them. As the time passed, the strawberries in each group showed gradual ripening until they started to rot. As it can be observed in the figure, the decay in the group of strawberries that was no protected, started at the beginning of Day 1, and on Day 3, they were not fit anymore for consumption. On the contrary, the C50N50 and PE group still did not show signs of extensive decay on the surface until Day 7, demonstrating the preservation effect of the C50N50 and PE film.

These results are very interesting because they show that a film obtained from renewable raw materials, such as C50N50, could have the same efficiency for food preservation given by a petroleum-based plastic like PE.

On the other hand, to address the growing social demand of materials from renewable sources, research on cellulosic packaging, such as paper and cardboard, has been carried out. Unfortunately, cellulose cannot be used for liquid-damp materials because of its great hydrophilicity given by its hydroxyl groups. Furthermore, as it was mentioned before, its porosity is another drawback which needs to be solved, to achieve the properties needed for an efficient use of cellulose for this application. Paper coatings, such as CNFs, have been proposed to improve cellulose's paper properties [80]. However, although the presence of CNFs represents an advantage on the paper's mechanical properties, their hydrophilicity remains unchanged. In this sense, it is recommended to carry out combinations of CNFs with hydrophobic synthetic polymers, aiming to reduce the CNFs hydrophilicity and enhance their flexibility. In this line, styrene-butadiene rubber (SBR) (Fig. 2.10a), and carboxylated styrene-butadiene rubber (XSBR) (Fig. 2.10b) are polymeric materials that have been studied as additives in mixtures with CNFs [81]. Among them, XSBR stands out because of its physical properties and chemical stability.

Furthermore, the presence of carboxyl groups in XSBR increases the compatibility of this polymer with CNFs, due to the hydrogen bridge-type interactions that can be established between these groups and the hydroxyl groups of CNFs. In fact, the studies evidenced the formation of homogeneous and well-distributed structures when XSBR was added to CNF suspensions. Besides, there was a reduction in the surface porosity of the paper with the addition of XSBR, and coatings with 10% XSBR exhibited a smoother surface with 24% less roughness than the original paper. Moreover, the coating with CNFs and XSBR effectively makes more hydrophobic the paper surface, diminishing by 7% its hydrophilicity. With respect to the mechanical properties, coatings with higher concentrations of XSBR led to an enhancement of 25% in the elongation in the tensile test.

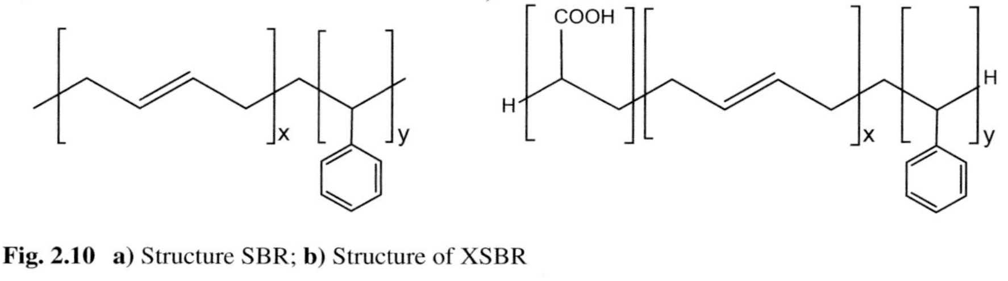

Fig. 2.10 **a)** Structure SBR; **b)** Structure of XSBR

In this context, many other chemical modifications of CNFs have been proposed in order to enhance the chemical compatibility between the hydrophilic CNFs and hydrophobic matrices. Among these modifications, it is worth to mention the work of Fein et al. [82], in which CNFs were modified with cis-5-norbornene-endo-2,3-dicarboxylic anhydride (carbic anhydride) (Fig. 2.11a) to produce norbornene functionalized CNFs that further reacted with crosslinkers possessing thiol groups in an aqueous medium, avoiding solvent exchange steps, organic solvents, or ionic liquids [83]. In this case, CNFs were utilized for thiol-ene reactions (Fig. 2.11b) to create hybrid CNF-NR particles by attaching NR to CNFs as shown in Figs. 2.11b and 2.12a.

After preparing the coating material, a vacuum filtration method for film forming was utilized to produce a coating on top of the filter paper (Fig. 2.12b). An improvement in the retention and drainage rate was observed when compared with a CNFs/NRL composite. Besides, the chemical modification of the CNFs led to a better dispersion mixtures with NR. Furthermore, NR addition to the coating materials increased the water contact angle, reduced water absorption, and decreased the water vapor transmission rate.

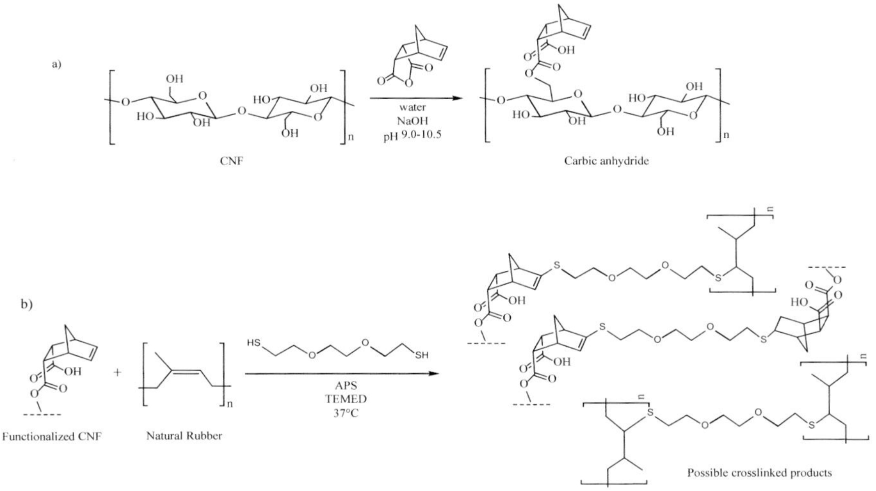

Fig. 2.11 CNF functionalization and thiol-norbornene reaction. Reprinted from [82] Copyright 2020 with permission from Elsevier

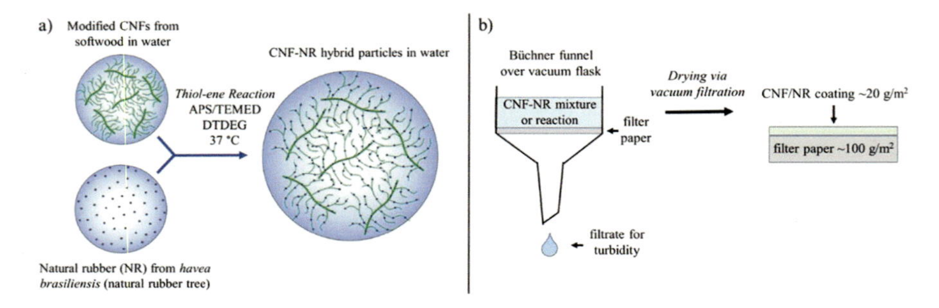

Fig. 2.12 Vacuum filtration method for film forming. Reprinted from [82] Copyright 2020 with permission from Elsevier

2.10 Membrane Technology Applications

Membranes have a relevant role in chemical technology and are currently being used increasingly in a broad range of applications. The control of the permeation of chemical species in contact with the membranes is the key property to look for when developing new materials for membrane technology applications. For instance, in drug delivery, the goal is to control the permeation rate of a specific drug from a reservoir to the body; in separation processes, membranes need to allow some components permeate while completely restricting permeation of others [84].

In this field, the challenges that are yet to be addressed are the high cost of the membranes and the high-energy demand of these systems [85]. In this context, bionanocomposites can be part of a sustainable cost-effective solution to overcome at least part of these drawbacks [86].

Diffusion is a key process in membrane technology because it is the fundamental mechanism by which mass transfer occurs. This process depends on (i) the free volume within the material, (ii) the polymer chains motion, (iii) the tortuosity of the pathway, (iv) polarity of the fillers, and (v) size of the chemical species involved [87]. Respect to the tortuosity of the pathway, it strongly depends on the size of the filler. Besides, the diffusion is also highly dependent on the crosslinking degree of this polymer.

In the case of biocomposites of rubber and nanocellulose, due the hydrophilic nature of the latter, barrier properties towards non-polar organic solvents are improved. It has been proven that small amounts of CNFs significantly reduce permeant diffusion of non-polar organic solvents by creating a tortuous path because of both, the well-known dispersibility of CNFs in the rubber matrix and their recognized high hydrophilicity. In fact, the diffusion coefficient substantially decreased after the addition of 5 wt.% of nanocelluloses in a vulcanized latex matrix [88]. There are studies that showed that CNFs were more efficient than CNCs efficient in decreasing the diffusion which is attributed to the tangling behavior of nanofibers [89].

Although much progress remains to be made in this area, the results obtained up to the moment are promising and stimulate further studies of rubber and CNC bionanocomposites as barrier membranes for separation purposes.

2.11 Aerospace Applications

Aerospace industry has a crucial role in our daily lives, since this industry is responsible for the development, not only of the airplanes that make it possible to travel around the world in reduced times but also for the design of sophisticated satellites that provide fundamental knowledge and data.

Approximately 50% of the entire advanced composites materials produced in the United States is consumed by the aerospace industry [65] and, as a result of the ongoing green technologies, this sector is also transitioning towards sustainable solutions.

In this context, fluorosilicone rubber (FVMQ) is widely used in the automotive and aerospace industry because of its chemical resistance. In particular, this material plays an important role in sealing technology, specifically when fuels and mineral oils are involved. Besides, there is evidence of the enhancement in the stiffness and oil resistance when FVMQ is reinforced with CNFs. Furthermore, the presence of CNFs improves the stress-relaxation characteristics of the composite under oil-immersion conditions. These results show the potential of using CNFs in advanced elastomer materials that have superior oil resistance and enhanced mechanical properties [90].

On the other hand, the aerospace industry is a major demander of ultralight materials. The weight of materials has a major impact on fuel consumption, speed, and airplanes' capacity, among other factors [65]. An improvement of the fuel efficiency would result in a significantly positive economic effect. For instance, an airline company with a fleet of 600 planes can save up to 11,000 gallons of fuel per year by decreasing 1 pound the weight of each aircraft [91].

Currently, in the search of ultralight materials with low thermal conductivity, aerogels made from silica are being used by NASA in astronautical missions as thermal insulators [92]. Besides, this kind of material has been proposed for insulation around battery packs in robotic rovers (i.e., Mars Sojourner Rover) and also in extravehicular activity suits for humans going out to missions in Mars [93]. However, more robust forms of aerogels are desired to extend the use of this material in aerospace [93].

In this context, aiming to develop ultralight sustainable materials, more robust than those obtained from silica, composite porous nanomaterials of CNCs and rubber, were prepared. Different materials were produced by using three different types of synthetic rubber (SBS, PB, and SiR) which were mixed with CNC surface-modified with vinyl groups (Fig. 2.13). In all cases, aerogels with enhanced mechanical performances and structural stability were obtained. Furthermore, these materials showed

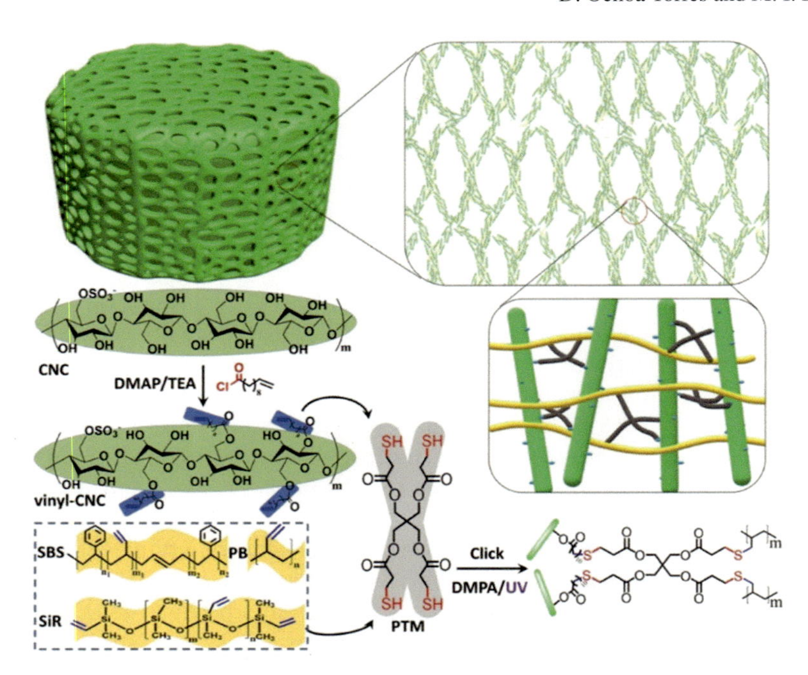

Fig. 2.13 Aerogels made from modified CNCs with silicone rubber (SiR), polybutadiene rubber (PB), or styrene butadiene styrene (SBS). Reprinted from [94] Copyright 2022 with permission from Elsevier

ultralow thermal conductivity which make them suitable as thermal insulating materials. Besides, all of them exhibited hydrophobic properties and enhanced solvent resistance [94].

All rubber–nanocellulose biocomposite had values of density, porosity, and thermal conductivity within the ranges reported for pure silica aerogels [95], but the values of surface area were lower than expected [95].

In addition to the lightness and low thermal conductivity, another property highly desirable for materials designed for aerospace is the ability to be self-healed. This kind of innovative material could reduce maintenance costs by repairing small cracks or scratches in the airplane wings or fuselages [96] which could extend the lifetime of critical structures. Besides, self-healing coatings could be used for corrosion protection.

The interest in this kind of material dates back to a couple of decades ago, with the development of self-healing epoxy resins [97]. In this context, a composite from epoxidated natural rubber (ENR) and CNCs, was prepared. The self-healing properties were achieved due to the dynamic hydrogen bond supramolecular networks between the oxygen groups of ENR and the hydroxyl groups of the CNC surface. It is worth to mention that, considering that chains interdiffusion is needed to get the self-healing properties, slightly cross-linked rubber was required for this application [98].

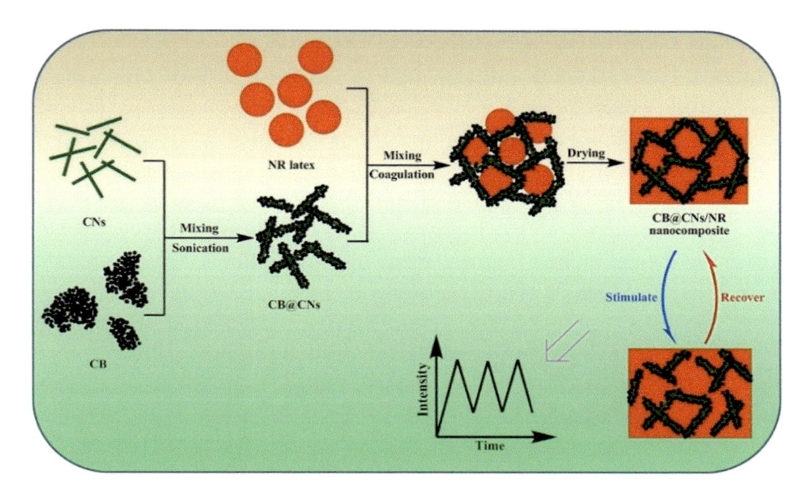

Fig. 2.14 3D hierarchical conductive structure of natural rubber latex (NR latex) with carbon black (CB) and cellulose nanocrystals (CNs). Reprinted from [100] Copyright 2016 with permission from Elsevier

On the other hand, in the aerospace industry, the sensing of strain is especially valuable because it can work as a control for structural vibration preventing possible damages associated with these vibrations [99]. In this line, smart conductive polymer composites (CPCs) have been proposed for this application. These composites are referred to as smart materials because of their stimuli-responsive properties to external inputs like strain, pressure, and temperature. The external changes could promote the destruction and reconstruction of the conductive networks in the matrix generating a change in the electric signal that can be detected by an external device [100]. Considering that that well-constructed networks yield stable output signals, but have low sensitivity to the external stimuli, the challenge in the design of this sensing material is to get networks with stable and strong output signals [100].

In this line, a brittle bionanocomposite with NRL, CB, and CNC that form a 3D hierarchical conductive structure (Fig. 2.14) was developed. The new material had high electrical conductivity, fast response, and good reproducibility. Furthermore, results showed that the sensing behavior can be tuned by changing the loadings of the CB and CNC in the composite [100].

2.12 Conclusion

In summary, the bionanocomposites of rubber and nanocellulose are highly promising for achieving significant sustainable technological advances in various fields, including (i) biomedical, with a particular focus on tissue engineering, (ii) aerospace industry, through the development of ultralight aerogels with low thermal conductivity, (iii) separation process technology, through the preparation of membranes, and (iv) food industry, due to the feasibility of preparing coating materials for food preservation.

In addition, we cannot fail to mention the importance that the replacement of carbon black by nanocellulose can have in terms of increasing the sustainability of the processes and materials involved in industries that are traditionally large consumers of rubber and CB, such as the tire industry.

Besides, there is evidence that the form of nanocellulose used (CNCs, CNFs) impacts in the properties of the biocomposite. In other words, depending on the application, the most appropriate form should be carefully chosen. Unfortunately, in the case of BC there are not enough studies to draw a conclusion.

The results obtained so far serve as encouragement to continue on the long road that still remains to be explored.

References

1. Tanaka Y, Sakdapipanich JT (2001) Chemical structure and occurrence of natural polyisoprenes. Biopolym Online. https://doi.org/10.1002/3527600035.bpol2001
2. Ali Shah A, Hasan F, Shah Z, Kanwal N, Zeb S (2013) Biodegradation of natural and synthetic rubbers: a review. Int Biodeterior Biodegrad 83:145–157
3. Andrade KL, Ramlow H, Floriano JF, Acosta ED, Faita FL, Machado RAF (2022) Latex and natural rubber: recent advances for biomedical applications. Polímeros. https://doi.org/10.1590/0104-1428.20210114
4. Kohjiya S, Ikeda Y (2021) Chemical fundamentals relevant to natural rubber. In: Chemistry, manufacture and applications of natural rubber. Elsevier, pp 3–21
5. Subramaniam A (1995) The chemistry of natural rubber latex. Immunol Allergy Clin North Am 15:1–20
6. Cornish K (2014) Biosynthesis of natural rubber (NR) in different rubber-producing species. In: Chemistry, manufacture and applications of natural rubber. Elsevier Inc., pp 3–29
7. Browne EA (2014) Rubber, Second. A. & C. BLACK LTD, London
8. Morton M (1981) History of synthetic rubber. J Macromol Sci: Part-Chem 15:1289–1302
9. Hurley PE (1981) History of natural rubber. J Macromol Sci: Part-Chem 15:1279–1287
10. Malaysian Rubber Board (2023) Consumption of natural and synthetic rubber worldwide from 1990 to 2022 (in 1,000 metric tons) [Graph]. In: Statista
11. Sinclair A, Zhou X, Tangpong S, Bajwa DS, Quadir M, Jiang L (2019) High-performance styrene-butadiene rubber nanocomposites reinforced by surface-modified cellulose nanofibers. ACS Omega 4:13189–13199
12. Hakimi NMF, Hua LS, Chen LW, Mohamad SF, Osman Al Edrus SS, Park BD, Azmi A (2021) Surface modified nanocellulose and its reinforcement in natural rubber matrix nanocomposites: a review. Polym (Basel). https://doi.org/10.3390/polym13193241
13. Thorn AD, Robinson RA (1994) Compound Design. In: Bhowmick AK, Hall MM, Benarey HA (eds) Rubber products manufacturing technology. Marcel Dekker Inc., New York, pp 1–79
14. Coran AY (2013) Vulcanization. In: The science and technology of rubber. Elsevier Inc., pp 337–381
15. Fisher HL (1939) Vulcanization of Rubber. Ind Eng Chem 31:1381 – 1389
16. Junkong P, Ikeda Y (2021) Properties of natural rubbers from guayule and rubber dandelion. In: Chemistry, manufacture and applications of natural rubber. Elsevier, pp 177–201

17. Evingür G, Pekcan Ö (2023) Application and characterization of rubber materials. Appl Charact Rubber Mater. https://doi.org/10.5772/intechopen.102183
18. Rahman MR, Khui PLN, Bakri MK Bin (2021) Cellulose-reinforced rubber composites. In: Fundamentals and recent advances in nanocomposites based on polymers and nanocellulose. Elsevier, pp 175–188
19. Popescu RC, Popescu D, Grumezescu AM (2017) Applications of rubber-based blends. Recent developments in polymer macro. Preparation and Characterisation. Elsevier Inc., Micro and Nano Blends, pp 75–109
20. Yasin S, Hussain M, Zheng Q, Song Y (2021) Effects of ionic liquid on cellulosic nanofiller filled natural rubber bionanocomposites. J Colloid Interface Sci 591:409–417
21. Youssef AM, El-Sayed SM (2018) Bionanocomposites materials for food packaging applications: concepts and future outlook. Carbohydr Polym 193:19–27
22. Low DYS, Supramaniam J, Soottitantawat A, Charinpanitkul T, Tanthapanichakoon W, Tan KW, Tang SY (2021) Recent developments in nanocellulose-reinforced rubber matrix composites: a review. Polym (Basel) 13:1–35
23. Saritha A, Malhotra SK, Thomas S, Joseph K, Goda K, Sreekala MS (2013) State of the Art–Nanomechanics. In: Polym Compos. Wiley, pp 1–12
24. Roy K, Pongwisuthiruchte A, Chandra Debnath S, Potiyaraj P (2021) Application of cellulose as green filler for the development of sustainable rubber technology. Curr Res Green Sustain Chem 4:100140
25. International Agency for Research on Cancer Agents Classified by the IARC Monographs, Volumes 1–135–IARC Monographs on the Identification of Carcinogenic Hazards to Humans
26. Díaz Bukvic GD, Rossi E, Errea MI (2023) Polysaccharides as economic and sustainable raw materials for the preparation of adsorbents for water treatment. Polysaccharides 4:219–255
27. Kataria A, Chaturvedi S, Chaudhary V, Verma A, Jain N, Sanjay MR, Siengchin S (2022) Cellulose fiber-reinforced composites—history of evolution, chemistry, and structure. In: Cellulose fibre reinforced composites: interface engineering, processing and performance. Elsevier, pp 1–22
28. Ávila Ramírez JA, Cerrutti P, Bernal C, Errea MI, Foresti ML (2019) Nanocomposites based on poly(lactic acid) and bacterial cellulose acetylated by an α-hydroxyacid catalyzed route. J Polym Environ 27:510–520
29. Puiggalí J, Katsarava R (2017) Bionanocomposites. In: Clay-polymer nanocomposites. Elsevier Inc., pp 239–272
30. Shojaeiarani J, Bajwa DS, Chanda S (2021) Cellulose nanocrystal based composites: a review. Compos Part C: Open Access 5:100164
31. Phanthong P, Reubroycharoen P, Hao X, Xu G, Abudula A, Guan G (2018) Nanocellulose: extraction and application. Carbon Resour Convers 1:32–43
32. Kawahara S, Nishioka H, Yamano M, Yamamoto Y (2022) Synthetic rubber with the tensile strength of natural rubber. ACS Appl Polym Mater 4:2323–2328
33. Lavrador P, Esteves MR, Gaspar VM, Mano JF (2021) Stimuli-responsive nanocomposite hydrogels for biomedical applications. Adv Funct Mater. https://doi.org/10.1002/adfm.202005941
34. Silva MJ, Dias YJ, Zaszczyńska A, Kołbuk D, Kowalczyk T, Sajkiewicz P, Yarin AL (2023) Three-phase bio-nanocomposite natural-rubber-based microfibers reinforced with cellulose nanowhiskers and 45S5 bioglass obtained by solution blow spinning. J Appl Polym Sci. https://doi.org/10.1002/app.54661
35. Andrade KL, Ramlow H, Floriano JF, Acosta ED, Faita FL, Machado RAF (2023) Latex and natural rubber: processing techniques for biomedical applications. Braz J Chem Eng. https://doi.org/10.1007/s43153-023-00317-y
36. Guerra NB, Sant'Ana Pegorin G, Boratto MH, de Barros NR, de Oliveira Graeff CF, Herculano RD (2021) Biomedical applications of natural rubber latex from the rubber tree Hevea brasiliensis. Mater Sci Eng, C. https://doi.org/10.1016/j.msec.2021.112126
37. Ferreira M, Mendonça RJ, Coutinho-Netto J, Mulato M (2009) Angiogenic properties of natural rubber latex biomembranes and the serum fraction of Hevea brasiliensis

38. Floriano JF, Da Mota LSLS, Furtado EL, Rossetto VJV, Graeff CFO (2014) Biocompatibility studies of natural rubber latex from different tree clones and collection methods. J Mater Sci Mater Med 25:461–470
39. Borges FA, de Barros NR, Garms BC, Miranda MCR, Gemeinder JLP, Ribeiro-Paes JT, Silva RF, de Toledo KA, Herculano RD (2017) Application of natural rubber latex as scaffold for osteoblast to guided bone regeneration. J Appl Polym Sci. https://doi.org/10.1002/app.45321
40. Furuya M, Shimono N, Yamazaki K, Domura R, Okamoto M (2017) Evaluation on cytotoxicity of natural rubber latex nanoparticles and application in bone tissue engineering
41. Miranda MCR, Borges FA, Barros NR, Santos Filho NA, Mendonça RJ, Herculano RD, Cilli EM (2018) Evaluation of peptides release using a natural rubber latex biomembrane as a carrier. Amino Acids 50:503–511
42. Mrue F, Coutinho Netto J, Ceneviva R, Lachat JJ, Thomazini JA, Tambelini H (2004) Evaluation of the biocompatibility of a new biomembrane
43. Ereno C, Guimarães SAC, Pasetto S, Herculano RD, Silva CP, Graeff CFO, Tavano O, Baffa O, Kinoshita A (2010) Latex use as an occlusive membrane for guided bone regeneration. J Biomed Mater Res A 95:932–939
44. Klemm D, Kramer F, Moritz S, Lindström T, Ankerfors M, Gray D, Dorris A (2011) Nanocelluloses: a new family of nature-based materials. Angew Chem-Int Ed 50:5438–5466
45. Asami J, Quevedo BV, Santos AR, Giorno LP, Komatsu D, de Rezende Duek EA (2023) The impact of non-deproteinization on physicochemical and biological properties of natural rubber latex for biomedical applications. Int J Biol Macromol 253:126782
46. Black CRM, Goriainov V, Gibbs D, Kanczler J, Tare RS, Oreffo ROC (2015) Bone tissue engineering. Curr Mol Biol Rep 1:132–140
47. Hochberg LR, Serruya MD, Friehs GM, Mukand JA, Saleh M, Caplan AH, Branner A, Chen D, Penn RD, Donoghue JP (2006) Neuronal ensemble control of prosthetic devices by a human with tetraplegia. Nature 442:164–171
48. Capadona JR, Tyler DJ, Zorman CA, Rowan SJ, Weder C (2012) Mechanically adaptive nanocomposites for neural interfacing. MRS Bull 37:581–589
49. Montero De Espinosa L, Meesorn W, Moatsou D, Weder C (2017) Bioinspired polymer systems with stimuli-responsive mechanical properties. Chem Rev 117:12851–12892
50. Tian M, Zhen X, Wang Z, Zou H, Zhang L, Ning N (2017) Bioderived rubber-cellulose nanocrystal composites with tunable water-responsive adaptive mechanical behavior. ACS Appl Mater Interfaces 9:6482–6487
51. Sahu P, Bhowmick AK (2020) Sustainable water responsive mechanically adaptive and self-healable polymer composites derived from biomass. Processes 8:1–16
52. Bouza E, Burillo A, Muñoz P (2002) Catheter-related infections: diagnosis and intravascular treatment. Clin Microbiol Infect 8:265–274
53. Djeribi R, Bouchloukh W, Jouenne T, Menaa B (2012) Characterization of bacterial biofilms formed on urinary catheters. Am J Infect Control 40:854–859
54. Gong X, Liu T, Zhang H, Liu Y, Boluk Y (2021) Release of cellulose nanocrystal particles from natural rubber latex composites into immersed aqueous media. ACS Appl Bio Mater 4:1413–1423
55. Sethulekshmi AS, Jayan JS, Saritha A, Joseph K, Aprem AS, Sisupal SB (2022) Antimicrobial studies in rubber nanocomposites—a mini review. Ind Crops Prod. https://doi.org/10.1016/j.indcrop.2022.115374
56. Li J, Cha R, Mou K, Zhao X, Long K, Luo H, Zhou F, Jiang X (2018) Nanocellulose-based antibacterial materials. Adv Healthc Mater. https://doi.org/10.1002/adhm.201800334
57. Amara C, El Mahdi A, Medimagh R, Khwaldia K (2021) Nanocellulose-based composites for packaging applications. Curr Opin Green Sustain Chem. https://doi.org/10.1016/j.cogsc.2021.100512
58. Potivara K, Phisalaphong M (2019) Development and characterization of bacterial cellulose reinforced with natural rubber. Materials. https://doi.org/10.3390/ma12142323
59. Phomrak S, Phisalaphong M (2017) Reinforcement of natural rubber with bacterial cellulose via a latex aqueous microdispersion process. J Nanomater. https://doi.org/10.1155/2017/4739793

60. Bu C, Zhu D, Liu L, Lu X, Sun Y, Yu L, OuYang Y, Cao X, Wang F (2022) Research progress on rubber concrete properties: a review. J Rubber Res 25:105–125
61. He S, Jiang Z, Chen H, Chen Z, Ding J, Deng H, Mosallam AS (2023) Mechanical properties, durability, and structural applications of rubber concrete: a state-of-the-art-review. Sustainability 15:8541
62. Júnior Carvalho Machado P, dos Reis A, Ferreira R, de Castro A, Motta L, Pasquini D (2020) Characterization and properties of cementitious composites with cellulose fiber, silica fume and latex. Constr Build Mater. https://doi.org/10.1016/j.conbuildmat.2020.119602
63. Rashid AB, Hoque ME (2022) Polymer nanocomposites for defense applications. In: Advanced polymer nanocomposites: science, technology and applications. Elsevier, pp 373–414
64. Jasni AH, Salleh EM, Zakaria FZ, Razali NAM, Ya'Acob WHW, Aziz FHA (2019) Fabrication and preliminary characterization of rubber cellulose/natural rubber latex (C-NRL) nanocomposite fibers. AIP Conf Proc. https://doi.org/10.1063/1.5089362
65. Oladele IO, Omotosho TF, Adediran AA (2020) Polymer-based composites: an indispensable material for present and future applications. Int J Polym Sci. https://doi.org/10.1155/2020/8834518
66. Roland CM (2009) Naval and space applications of rubber
67. Cao J, Zhang X, Wu X, Wang S, Lu C (2016) Cellulose nanocrystals mediated assembly of graphene in rubber composites for chemical sensing applications. Carbohydr Polym 140:88–95
68. Rawat NK, Volova TG, Haghi AK (2021) Applied biopolymer technology and bioplastics: sustainable development by green engineering materials. CRC Press
69. Deng S, Chen R, Duan S, Jia Q, Hao X, Zhang L (2023) Research progress on sustainability of key tire materials. SusMat 3:581–608
70. Srivastava SK, Bhuyan B (2019) Rubber nanocomposites for tyre tread applications. In: Rubber nanocomposites and nanotextiles: perspectives in automobile technologies. De Gruyter, pp 31–74
71. Jiang W, Cheng Z, Wang J, Gu J (2022) Modified nanocrystalline cellulose partially replaced carbon black to reinforce natural rubber composites. J Appl Polym Sci. https://doi.org/10.1002/app.52057
72. Li C, Huang F, Wang J, Liang X, Huang S, Gu J (2018) Effects of partial replacement of carbon black with nanocrystalline cellulose on properties of natural rubber nanocomposites. J Polym Eng 38:137–146
73. Xu SH, Gu J, Luo YF, Jia DM (2012) Effects of partial replacement of silica with surface modified nanocrystalline cellulose on properties of natural rubber nanocomposites. Express Polym Lett 6:14–25
74. Liu Y, Chen W, Jiang D (2022) Review on heat generation of rubber composites. Polymers (Basel) 15:2
75. Dominic M, Joseph R, Sabura Begum PM, Kanoth BP, Chandra J, Thomas S (2020) Green tire technology: effect of rice husk derived nanocellulose (RHNC) in replacing carbon black (CB) in natural rubber (NR) compounding. Carbohydr Polym. https://doi.org/10.1016/j.carbpol.2019.115620
76. Correia CA, Valera TS (2019) Cellulose nanocrystals and jute fiber-reinforced natural rubber composites: cure characteristics and mechanical properties. Mater Res. https://doi.org/10.1590/1980-5373-MR-2019-0192
77. Dominic MCD, Joseph R, Begum PMS, Joseph M, Padmanabhan D, Morris LA, Kumar AS, Formela K (2020) Cellulose nanofibers isolated from the Cuscuta Reflexa plant as a green reinforcement of natural rubber. Polym (Basel). https://doi.org/10.3390/POLYM12040814
78. Meng L, Xi J, Bian H, Xiao H, Wu W (2024) Nanocellulose/natural rubber latex composite film with high barrier and preservation properties. Sustain Chem Pharm. https://doi.org/10.1016/j.scp.2023.101399
79. Xu C, Zhang X, Liang J, Fu Y, Wang J, Jiang M, Pan L (2022) Cell wall and reactive oxygen metabolism responses of strawberry fruit during storage to low voltage electrostatic field treatment. Postharvest Biol Technol 192:112017

80. Hugen LN, de Miranda EHN, Santos ADAD, Lago RCD, Silva LE, Tonoli GHD, Ferreira SR (2023) Effect of cellulose micro/nanofibrils and carboxylated styrene butadiene rubber coating on sack kraft paper. Nord Pulp Paper Res J 38:481–489
81. Annamalai PK, Dagnon KL, Monemian S, Foster EJ, Rowan SJ, Weder C (2014) Water-responsive mechanically adaptive nanocomposites based on styrene-butadiene rubber and cellulose nanocrystals-Processing matters. ACS Appl Mater Interfaces 6:967–976
82. Fein K, Bousfield DW, Gramlich WM (2020) Thiol-norbornene reactions to improve natural rubber dispersion in cellulose nanofiber coatings. Carbohydr Polym. https://doi.org/10.1016/j.carbpol.2020.117001
83. Fein K, Bousfield DW, Gramlich WM (2020) The influence of versatile thiol-norbornene modifications to cellulose nanofibers on rheology and film properties. Carbohydr Polym 230:115672
84. Baker RW (2005) Membrane technology. Kirk-Othmer Encycl Chem Technol. https://doi.org/10.1002/0471238961.1305130202011105.a01.pub2
85. Muralidhara HS (2010) Challenges of membrane technology in the XXI Century. In: Membrane technology: a practical guide to membrane technology and applications in food and bioprocessing. Elsevier, pp 19–32
86. Buonomenna MG (2016) Smart composite membranes for advanced wastewater treatments. In: Smart composite coatings and membranes. Elsevier, pp 371–419
87. Mathew AP, Packirisamy S, Stephen R, Thomas S (2002) Transport of aromatic solvents through natural rubber/polystyrene (NR/PS) interpenetrating polymer network membranes
88. Abraham E, Thomas MS, John C, Pothen LA, Shoseyov O, Thomas S (2013) Green nanocomposites of natural rubber/nanocellulose: membrane transport, rheological and thermal degradation characterisations. Ind Crops Prod 51:415–424
89. Visakh PM, Thomas S, Oksman K, Mathew AP (2012) Cellulose nanofibres and cellulose nanowhiskers based natural rubber composites: diffusion, sorption, and permeation of aromatic organic solvents. J Appl Polym Sci 124:1614–1623
90. Park YW, Yoon JH, Shin KH, Cho YJ, Yun JH, Han WH, Hong MH, Kang DG, Kim HY (2023) Enhancing stiffness and oil resistance of fluorosilicone rubber composites through untreated cellulose reinforcement. Polymers (Basel). https://doi.org/10.3390/polym15234489
91. Koniuszewska AG, Kaczmar JW (2016) Application of polymer based composite materials in transportation
92. Bheekhun N, Abu Talib AR, Hassan MR (2013) Aerogels in aerospace: an overview. Adv Mater Sci Eng. https://doi.org/10.1155/2013/406065
93. Randall JP, Meador MAB, Jana SC (2011) Tailoring mechanical properties of aerogels for aerospace applications. ACS Appl Mater Interfaces 3:613–626
94. Chen Z, Li Z, Lan P, Xu H, Lin N (2022) Hydrophobic and thermal-insulating aerogels based on rigid cellulose nanocrystal and elastic rubber. Carbohydr Polym. https://doi.org/10.1016/j.carbpol.2021.118708
95. Khan NR, Sharmin T, Bin Rashid A (2024) Exploring the versatility of aerogels: broad applications in biomedical engineering, astronautics, energy storage, biosensing, and current progress. Heliyon 10:e23102
96. Jadoun S (2023) Synthesis, mechanism, and applications of self-healing materials. Biomed Mater & Devices. https://doi.org/10.1007/s44174-023-00107-7
97. Das R, Melchior C, Karumbaiah KM (2016) Self-healing composites for aerospace applications. In: Advanced composite materials for aerospace engineering. Elsevier, pp 333–364
98. Cao L, Yuan D, Xu C, Chen Y (2017) Biobased, self-healable, high strength rubber with tunicate cellulose nanocrystals. Nanoscale 9:15696–15706
99. Chung DDL (2016) Self-sensing structural composites in aerospace engineering. In: Advanced composite materials for aerospace engineering. Elsevier, pp 295–331
100. Wu X, Lu C, Han Y, Zhou Z, Yuan G, Zhang X (2016) Cellulose nanowhisker modulated 3D hierarchical conductive structure of carbon black/natural rubber nanocomposites for liquid and strain sensing application. Compos Sci Technol 124:44–51

Chapter 3
Cellulose Based Rubber Blends and Microcomposites

S. Sahila, L. S. Jayakumari, P. S. Sampath, Sivasubramanian Palanisamy, and Murugesan Palaniappan

Abstract Cellulose-based rubber blends and microcomposites have attracted the attention of many researchers due to their unique mechanical properties. This chapter provides a thorough summary of the mechanical properties of cellulose-reinforced natural rubber (NR) microcomposite blends and an outline of the isolation of cellulose and its derivatives. A succinct overview of the literature is presented including types of composites, blends, isolation of cellulose and its derivatives, and characteristics of cellulose-based rubber blends and microcomposites. Besides the challenges and future prospects on cellulose-based rubber blends and microcomposites are also discussed in this chapter.

Keywords Cellulose-based rubber blends · Microcomposites characterization techniques · Tensile testing · Differential scanning calorimeter · Dynamical mechanical analysis · Thermogravimetric analysis · Scanning electron microscopy · Transmission electron microscopy · Atomic force microscopy

---

S. Sahila
Department of Chemistry (SFS), Madras Christian College (Autonomous), University of Madras, Tambaram East, Chennai, Tamilnadu 600059, India

L. S. Jayakumari
Department of Rubber and Plastics Technology, Anna University, MIT Campus, Chrompet, Chennai, Tamilnadu 600044, India

P. S. Sampath
Department of Mechanical Engineering, K.S. Rangasamy College of Technology, KSR Kalvi Nagar, Tiruchengode 637215, Tamilnadu, India

S. Palanisamy (✉)
Department of Mechanical Engineering, P.T.R. College of Engineering and Technology, Thanapandiyan Nagar, Austinpatti, Tirumangalam Road, Madurai, Tamilnadu 625008, India
e-mail: sivaresearch948@gmail.com

M. Palaniappan
Department of Mechanical Engineering, College of Engineering, Imam Mohammad Ibn Saud Islamic University, Riyadh 11432, Saudi Arabia

Visakh P. M. *Rubber Based Bionanocomposites*, Advanced Structured Materials 210,
https://doi.org/10.1007/978-981-10-2978-3_3

3.1 Introduction

Nowadays, a variety of renewable resource-based biodegradable and environmentally friendly polymers play a crucial role in a wide spectrum of applications [1]. Eco-friendly materials lends a hand to maintain a cleaner, greener environment because they break down quickly in the environmental conditions. Consequently, the amount of goods that are produced from renewable resources is obviously on the rise in the recent past. Eco-friendly products, for instance, are needed in the current situation in the areas of biodegradable and edible packaging, biodegradable drug carriers, biocompatible scaffolds, medical implants, composite technology, and eco-friendly sorbents [2]. The source and applications of cellulose based rubber blends and microcomposites are represented in Scheme 3.1.

Microcomposites are composite materials in which filler (1–100 μm) is used to reinforce a larger-scale matrix component. These microscopic fillers dramatically increase the matrix's mechanical properties, producing materials with greater stiffness and strength, better wear resistance, increased thermal stability, decreased thermal expansion, improved electrical conductivity, light weight, durability, cost-effectiveness, and design flexibility. Microcomposites are used in construction, sports equipment, automotive, aerospace, and marine industries [3].

Types of Microcomposites

- **Polymer matrix microcomposites**: It is the most popular type, consisting of polymers such as epoxies, polyesters, and nylons reinforced with crystalline fibers, glass, carbon, or aramid fibers, or ceramic or metal particles.

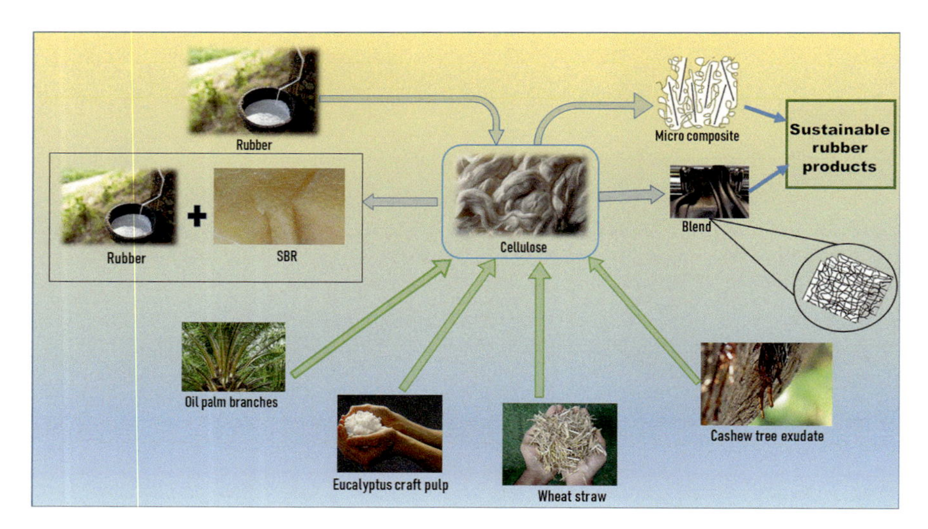

Scheme 3.1 A schematic depiction of rubber blends and composites based on cellulose

- **Metal matrix microcomposites**: It is a more rigid and strong at high temperatures microcomposites. Aluminum strengthened with silicon carbide particles is an example of this type.
- **Ceramic matrix microcomposites**: It provides excellent wear resistance, stiffness, and strength at extremely high temperatures. Zirconium oxide particle-reinforced silicon carbide is an example of this type.

Blends

Blends are molecular mixtures of two or more distinct polymers that combine the advantages and disadvantages of the original polymers to form a new substance with hybrid qualities. It possesses intriguing qualities like processability, cost-effectiveness, and customized characteristics [4]. In addition to providing a more affordable method of achieving the needed qualities than pure polymers, mixing easily available, less expensive polymers with high-performance ones can frequently lead to easier blending and production.

Types of Polymer Blends

Miscible Blends

These blends consist of molecularly blending polymers that combine to form a single homogeneous phase. Miscible blends combine the strengths of distinct polymers by tailoring the best possible property [5]. For enhanced transparency and impact resistance, polystyrene and poly (methyl methacrylate) are two examples.

Immiscible Blends: These blends are made up of polymers that don't easily mix due to differences in chemical compatibility. They form multiple distinct phases within the material, each having unique characteristics. Immiscible blends provide advantages such as enhanced toughness and flexibility [6]. Examples include polypropylene/polystyrene blends for shock absorption in packaging materials. Blends are utilized in various industries, including medical devices, automotive, building and construction, and packaging [7]. Due to its high flexibility, flexible film-forming capacity, resilience, abrasion resistance, and water repellent, natural rubber (NR) is one of the most significant elastomers. It is a renewable polymer with notable properties that are used globally in a variety of applications, with tires being the most significant NR-based material. Some NR features are designed to improve products with more notable features [8, 9].

The oldest method of reinforcing NR is the addition of nanoscopic additives to the NR matrix. The most popular fillers among these are silica nanoparticles and carbon black, nevertheless, in recent decades, discontinuous fibers have been used in the hunt for innovative fillers that offer a viable replacement for the traditional fillers, especially when it comes to mechanical reinforcement. Aramid, polyester, nylon, Kenaf, glass fibers, and regular cellulose fibers are a few effective types of fibers that have been used to reinforce NR [10].

The science of employing cellulose nanofibers as fillers for reinforcing natural or synthetic polymeric matrices is developing swiftly due to the increasing interest in their processing and characterization [11]. In addition to vegetal nanofibers and cellulose nanowhiskers (CNW), bacterial cellulose fibers have also been utilized as reinforcement for synthetic and natural polymers. These nanocelluloses are all more widely available, less expensive, biodegradable, renewable, and have a low density. Their use as feasible substitutes for traditional materials has been aided by their nanoscale characteristics and dimensions [12].

Products manufactured from sustainable and renewable non-petroleum-based resources are in greater demand. Cellulose, the most common polymer in the world, is nontoxic, renewable, and biodegradable. Cellulose or its derivatives, particularly cellulose nanocrystals (CNC) [13] or cellulose nanofibrils (CNF) [2], cellulose nanowhiskers (CNs) and microfibrillated cellulose [8], have been used widely as model filler in various polymeric matrices since the first report of utilizing nanocellulose as a reinforcing agent in a polymeric matrix [14]. These are particularly intriguing because they could increase the mechanical or functional performances of the resulting composite as well as its "green" attribute that is favorable for a sustainable world [2]. Because these biofillers are abundant, inexpensive, and biodegradable in comparison to synthetic fillers (carbon, glass, and aramid), they have a wide range of applications. While biocomposites have many advantages, they also have certain disadvantages, such as low heat resistance, poor dimensional stability, and moisture absorption qualities [15–17].

As the filler and the NR matrix have distinct structural characteristics, it is difficult to mix the filler and rubber matrix uniformly. The boundary region of the composite experiences stress because of the filler's non-uniform distribution. Hence, it is necessary to modify the filler or matrix with any functional groups to lower the stress in composites. This could significantly alter the properties of the composites and improve the compatibility of the filler and matrix [18]. The filler's functional groups and particle size/aspect ratio affect the thermo-mechanical properties of the composites regardless of the type of filler used. Greater surface area presented by smaller particle sizes leads to improved physical interaction with the NR matrix. The presence of hydroxyl groups in the biofillers can reduce their compatibility with NR. The chemical modification of the cellulose, which lowers the hydroxyl groups found in the biofillers, is expected to result in better strength and high crystallinity of the composites [2, 19].

Cellulose is a frequently used organic filler due to its low density, renewability, biodegradability, and better thermomechanical properties. Furthermore, cellulose-based materials are incredibly nontoxic due to their natural origins. The cellulose's structure is made up of repeating units of β-D-glucopyranose connected by β − 1,4-linkages [20]. In addition to giving stability, the hydrogen bonds that bind the repeating units together render a linear structure in cellulose. The aspect ratio, morphology, and particle size of these fillers vary, which gives rise to their distinct characteristics [21].

Cellulose fibers are incompatible with many polymer matrices and have a tendency to absorb moisture, despite their many benefits. However, applying surface treatments could enhance the compatibility between the fiber and matrix [22]. Through the use of sulfuric acid hydrolysis, Thomas et al. [2] recently extracted cellulose nanowhiskers from the Acacia Caesia plant. The resulting nanowhiskers displayed a high value of 79.65% the crystallinity index. Cellulose is used in a wide range of products, including flexible displays, textiles, sensors, composites, membranes, light-emitting diodes and batteries. The only elements that determine whether an application is suitable for them are their source, synthesis method, size, and processing conditions [2] (Scheme 3.1).

Isolation of Cellulose

One can extract cellulose from a wide range of sources along with bacteria and plants [23]. It is extracted from plant sources using a variety of physical and chemical techniques, including acid hydrolysis, steam explosion, and enzymatic pre-treatment. For example, Bras et al. [24] used bleached kraft bagasse pulp to separate cellulose whiskers by acid hydrolysis. The bagasse pulp was treated with 65 weight % H_2SO_4 for 45 min at 45 °C. After treatment, the pulp underwent centrifugation and dialyzation using distilled water. A few drops of $CHCl_3$ were added after 2 min of sonication, and the resultant cellulose whiskers were stored at 4 °C. This technique was used to create the 10 nm-diameter nanowhiskers [25].

Fruit bunches of oil palm (FBOP) were treated by acid and alkaline hydrolysis to extract nanocellulose (NC). The OPFB was first ground, pulverized, and rinsed with deionized water before being immersed in ethanol. Moreover, 4% NaOH was applied to it in order to eliminate lignin and hemicellulose. 30% HCl was used for acid hydrolysis in order to extract the leftover hemicellulose [26]. After that, 3.5 weight % of NaClO was used to bleach the fibers until they turned white. After being cleaned with deionized water, the resulting nanofibers were left to dry for 48 h. The cellulose's crystallinity increased from about 36–53% after the chemical treatment [27].

Bleached eucalyptus Kraft pulp was used to extract cellulose nanofibers (CNFs). After an acid wash, the pulp was immersed in $NaClO_2$. Using TEMPO (2,2,6,6-tetramethylpiperidine-1-oxyl), NaClO, and $NaClO_2$, the treated fiber was carboxylated for 3 days at 60 °C. After washing, homogenizing, diluting, and centrifuging the fibers, each individual fiber was separated. The resulting CNF fiber morphology revealed that the fibrils have a diameter of 5–7 nm and a length of micrometers [28–30].

Jute, Banana, and pineapple fibers were treated with a mild chemical treatment and a steam explosion to separate NC. The morphological analysis revealed that the NC with sizes ranging from 5 to 40 nm was isolated. Compared to the raw fibers, the NC showed better thermal stability and crystallinity. The results demonstrated that low-strength chemical treatment methods combined with the steam explosion were an effective way to isolate NC [31].

Cellulose was extracted from wheat straw using a chemo-mechanical process. FTIR and field emission scanning electron microscope (FESEM) results confirm the formation of nanocellulosic fibers with a diameter ranging from 5 to 90 nm. 51% crystallinity was revealed by the XRD data of the extracted CNFs. The study reveals that the chemo-mechanical method is effective at isolating NC [2, 32].

Compared to cellulose derived from plant fiber, bacterial cellulose (BC) offers a number of benefits, including superior biocompatibility, nontoxicity, and high purity. Two powerful bacterial species that produce cellulose are Gluconacetobacter and Komagataeibacter [33, 34]. However, their manufacturing is expensive due to the high expense of their synthetic medium and nutrition source. Several studies on low-cost BC nutrient sources have been conducted recently. Abol-Fotouh et al. [32] employed agro-waste extracts, such as date fruit waste, figure fruit waste, and sugar-cane molasses, in place of glucose as the carbon source in Hestrin-Schramm (HS) medium for the production of BC utilizing Komagataeibacter saccharivorans MD1. The sugarcane molasses extract yields the highest amount of BC, and high-quality BC was obtained [32]. Using Komagataeibacter rhaeticus, cashew gum and cashew tree exudate (CTE) were used to prepare BC in HS medium. Good thermal stability and crystallinity were demonstrated by the obtained BC [35].

3.2 Characterization Techniques

3.2.1 Tensile Testing

Tensile testing is an essential tool for evaluating the mechanical characteristics of natural rubber and its blends. The insights gained from tensile testing are crucial for material characterization, quality control, product development, and failure analysis, attributing to the manufacture of high-performance NR-based products for various applications [36].

In comparison to cellulose, natural rubber exhibits greater tensile strength and elongation at break, but cellulose has a higher Young's modulus. The properties of the natural rubber-cellulose composite are halfway between the two. By supporting the natural rubber matrix, the addition of cellulose to natural rubber can improve the composite's tensile characteristics [37]. As a result, the composite's tensile strength and Young's modulus may rise. It has the ability to swell and absorb water, which can increase the composite's elongation at break. Cellulose can interact with the natural rubber chains, which can improve the compatibility of the two materials. This can reduce the tendency of the composite to crack or delaminate. Hence, the tensile properties of natural rubber and cellulose can be tailored by adjusting the processing conditions, filler content, and crosslinking density. Because of its excellent tensile qualities, durability, and environmental friendliness, the natural rubber-cellulose composite is a material that may be utilized for a wide range of applications [38].

A room-temperature tensile test was used to characterize the non-linear mechanical behavior of NR/cellulose whiskers nanocomposites reinforced with varying contents of bagasse whiskers. Figure 3.1 displays typical stress–strain curves and the tensile properties of nanocomposite films. The increased initial slopes of the stress-stain curves clearly illustrate the stiffening effect of the whiskers on NR

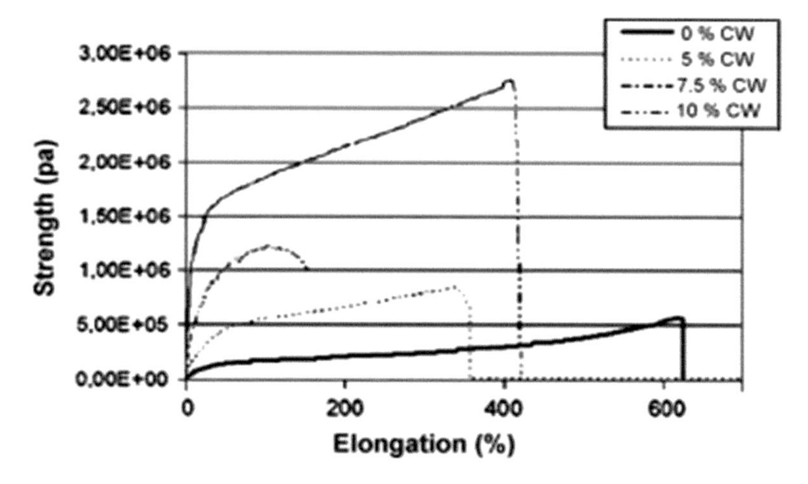

Fig. 3.1 Stress–strain curve from tensile test for rubber/cellulose whiskers nanocomposite films

nanocomposite films. NR nanocomposites exhibited significantly different stress–strain behavior from neat NR. At temperatures below their glass transition temperature, the latter showed an elastic nonlinear behavior characteristic of amorphous polymers, i.e., a regular and slight increase in stress that remained relatively constant until the fracture. When whiskers were added to rubber, the mechanical properties of NR nanocomposites, as shown in Fig. 3.2, demonstrated a significant increase in both strength and Young's modulus, but a decrease in strain at break. Tensile stress and modulus increased to their maximums of 374 and 530%, respectively, at 10 weight % whiskers. This strong reinforcement effect may be due to the mechanical percolation phenomenon of cellulose whiskers, which create a rigid, continuous network of hydrogen-bonded cellulosic nanoparticles. It is commonly known that the adhesion between the components and the matrix's composition has a direct impact on the blend properties [24].

3.2.2 *Differential Scanning Calorimeter*

Thermal behavior of natural rubber (NR) can be effectively characterized using differential scanning calorimetry (DSC). DSC can offer important insights into the crystallization behavior, melting temperature (Tm), and glass transition temperature (Tg) of NR by detecting the heat flow connected to transitions in the NR structure [39].

The Tg of NR is typically around −70 °C, indicating that NR is a glassy material at room temperature. NR exhibits a melting peak in the DSC thermogram at around −50 °C. This peak indicates the melting point of the crystalline phase in NR. The

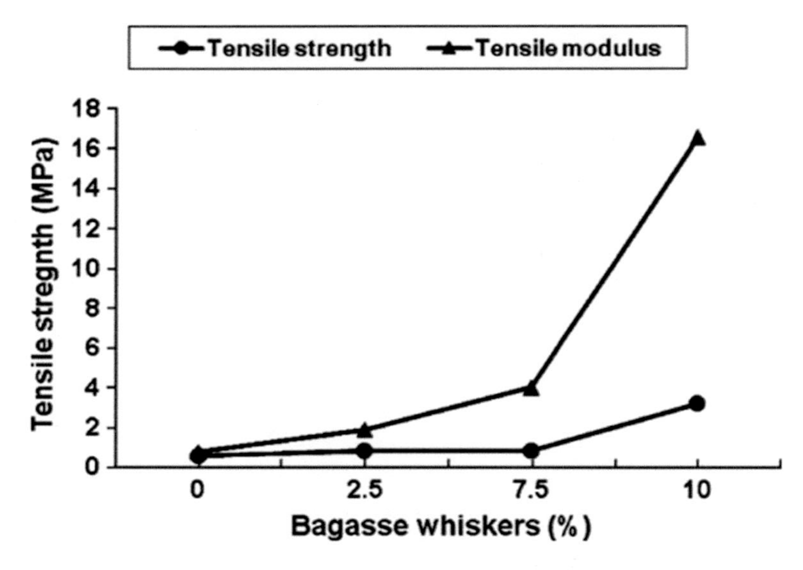

Fig. 3.2 Rubber/cellulose whiskers nanocomposite films: tensile strength and modulus as a function of composition

Tg of cellulose depends on its crystallinity and morphology [40]. For crystalline cellulose, the Tg is typically around 220 °C, while for amorphous cellulose, the Tg is around 160 °C. Cellulose exhibits a melting peak in the DSC thermogram at around 319 °C. This peak depicts the melting point of the crystalline phase in cellulose. The Tm value is sensitive to the molecular weight and the degree of crystallinity of cellulose [41].

NR can crystallize under certain conditions, such as when it is subjected to mechanical stress or cooled slowly from the melt. Certain conditions, such as heating cellulose above its Tg and then slowly cooling it, can cause it to crystallize. DSC can be used to measure the heat of crystallization (ΔH) and the crystallization temperature (Tc). The ΔH value provides information about the degree of crystallinity in NR, while the Tc value indicates the temperature at which crystallization occurs most rapidly [42].

Natural rubber-cellulose microcomposite usually has a lower glass transition temperature (Tg) than pure natural rubber. This is because the cellulose microfibrils in the NR matrix function as physical crosslinks, preventing NR chains from moving freely. The NR content, the cellulose content, and the composite's shape all affect the Tg of NR-cellulose microcomposite. A lower Tg is typically the result of a higher cellulose content because the cellulose microfibrils become more tightly packed and more effectively impede the mobility of NR chains. A larger Tg and more mobility of NR chains are made possible by a more dispersed cellulose phase in the composite, which is another factor. An essential component of the NR-cellulose microcomposite's performance in a range of applications is its Tg. For instance, at low temperatures, a reduced Tg can result in increased flexibility and impact resistance. On the other hand, a lower Tg may also lead to a lower modulus and tensile strength [43, 44].

DSC analysis was carried out on cellulose nanoparticles (CNP) incorporation on the blends of synthetic rubber (SR)/polylactic acid (PLA) to examine the PLA's structure with the addition of SR and CNP [45]. In Fig. 3.3, it is noticed that the Tg of all PR and PNR blends is found at nearly same temperature (70 °C). This implied that the inclusion of SR and CNP had no appreciable impact on PLA's glass transition behavior. Studies delineated similar behavior (smaller change in Tg value with inclusion of SR) and reported average Tg values at 58, 63, and 62 °C, respectively. Higher glass transition values were obtained as a result of the PLA matrix's increased stability brought about by the enhanced interaction between SR and PLA. This study's results were about 10 °C higher than those of the previously stated investigations. PLA exhibited an ageing peak in the 70–90 °C range above Tg, which was associated with the aging of amorphous polymer. The inclusion of SR led to the introduction of crystalline structures (SR and CNP), which made this discovery less sensible for other samples (particularly PNR). PR3's twofold peak melting points, at 151.25 and 153.97 °C, were associated with two distinct melting phases. This mostly affects non-homogeneous biocomposites; at higher temperatures, more perfect crystalline structures melt and produce two peaks. PLA was found to have a melting point (Tm) of 152.43 °C, while PR2 and PNR melted at slightly higher temperatures of 157.18 and 154.49 °C When SR and CNP were added, the crystalline structure of the biocomposite increased along with the flexibility of the polymer chains and the viscosity which was correlated with the little delay in the melting point of PLA [38, 46].

3.2.3 Dynamical Mechanical Analysis

An important technique for characterizing the viscoelastic properties of materials, including natural rubber, is dynamic mechanical analysis or DMA. It calculates a material's loss modulus (E") and storage modulus (E') in relation to frequency or temperature. The material's elastic response is revealed by the storage modulus, but the viscous response is revealed by the loss modulus. A peak in the storage modulus is usually observed in DMA data for natural rubber at the glass transition temperature [47]. The height of this peak is connected to the stiffness of the material. At low temperatures, the loss modulus is often smaller than the storage modulus; but, as temperature rises, it becomes larger. The material's damping capacity is measured by the tan δ that is the ratio of the loss modulus to its storage modulus.

It is commonly acknowledged that the storage modulus is directly correlated with the composite's elastic nature. The storage modulus fluctuation for NR/CNF composites with strain is displayed in Fig. 3.4. In contrast to the gum compound, the modulus rises with increasing CNF loadings at low strain values. The rise in modulus values is connected to both the filler–polymer interaction and the filler-filler, as well

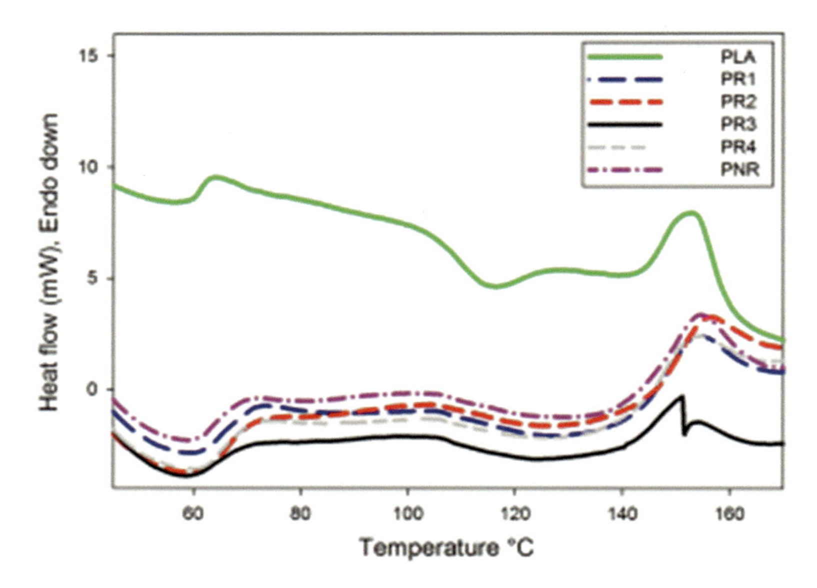

Fig. 3.3 DSC curves of rubber-based blends

as hydrodynamic reinforcement [48]. The modulus values at low strain levels rise only slightly above 3 phr of CNF. All composites' moduli decrease to lower values as the strain intensity rises, indicating a breakdown of the filler network [49]. All the composites have modulus values greater than the rubber signifying reinforcement of NR with the additional CNF [24].

3.2.4 Thermogravimetric Analysis

Thermogravimetric analysis (TGA), is used to examine the degradation temperature of NR, CNR, and their blends. The amount of additives present, the processing circumstances, and the purity of the NR can affect the degradation temperature of NR. Julien Bras et al. revealed the thermal stability of rubber/cellulose whiskers nanocomposites and plain rubber matrix, as illustrated in Fig. 3.5 [50]. Rubber begins to thermally decompose in a nitrogen atmosphere at 380 °C. This is followed by two significant weight losses during the primary processes of volatilization and pyrolysis, which are completed at 550 °C.

NR/cellulose whiskers nanocomposites having 10wt% cellulose whiskers exhibited marginally lower onset degradation temperature ($\sim$265°C) as compared to pure natural rubber [24].

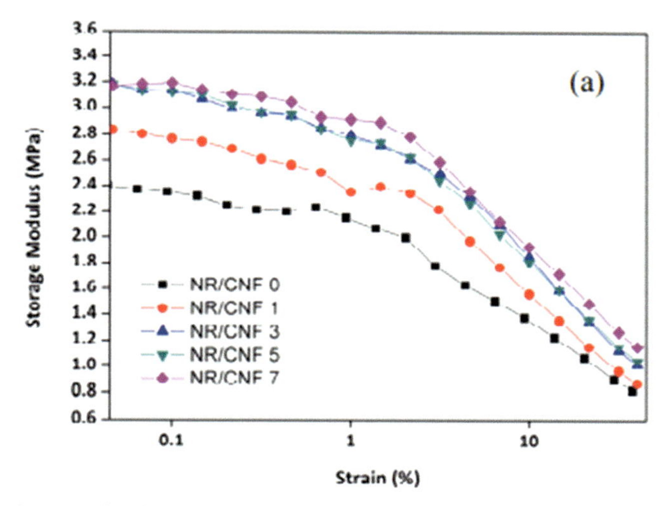

Fig. 3.4 Variation of storage modulus with varying strain [49]

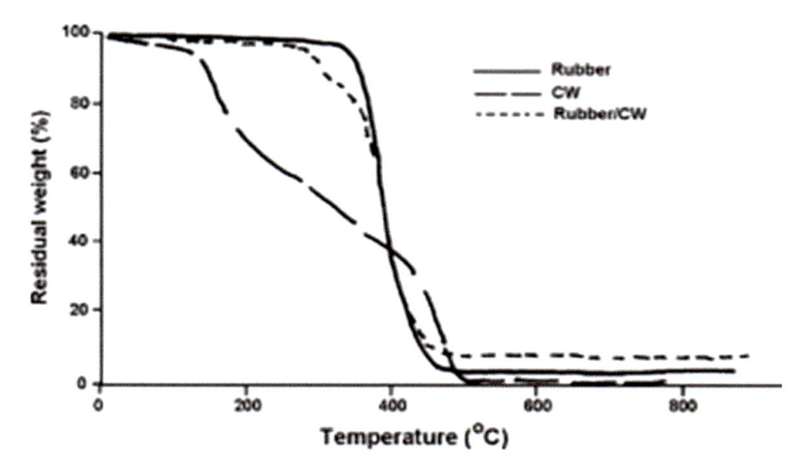

Fig. 3.5 TGA curve of rubber, bagasse cellulose whiskers (CW), and rubber/cellulose whiskers nanocomposites

The lower onset degradation temperature of bagasse whiskers compared to rubber may be the cause of the lower onset degradation seen in the rubber/cellulose whiskers nanocomposites [37]. It was discovered that bagasse whiskers began to degrade at a temperature of roughly 170 °C. Owing to the use of sulfuric acid in the preparation of the whiskers, there may be sulfate groups present on their surface, explaining the comparatively lower onset degradation temperature of bagasse whiskers than that of cellulose [50]. Both rubber and rubber/cellulose nanocrystal nanocomposite exhibited nearly identical thermal breakdown behaviors beginning at around 390 °C.

The primary breakdown of cellulose takes place between 150 and 250 °C as a result of the depolymerization, dehydration, and breakdown of glycosyl units, which is followed by the development of a char. The minor shoulder observed above 300 °C can be attributed to the char's oxidation and decomposition into gaseous molecules with reduced molecular weight [51]. The thermogram of NBR/CNs nanocomposites, as predicted, reveals two stages of degradation: the degradation of NBR is responsible for the second stage, which occurred at approximately 400–480 °C. The first region, at a temperature range of 340–390 °C, is caused by the degradation of CNs, and the weight loss is consistent with the CNs content. It's interesting to note that in NBR/CNs nanocomposites, the degradation temperature related to CNs is significantly greater than that of pure CNs. The establishment of a restricted structure in the NBR/CNs nanocomposites is responsible for the notable improvement of thermal resistance brought about by the existence of NBR matrix. In other words, the NBR matrix protects the CNs, demonstrating the strong bond that exists between the CNs and NBR molecules [52–55].

3.2.5 Scanning Electron Microscopy

The uniform distribution and network behavior of CNCs in the nanocomposite are revealed using SEM. The surfaces of pure NR and ENR exhibit a smooth appearance, whereas the nano-composite of NR and ENR displays typical crumpled morphologies as a result of the presence of CNCs [56]. Even at 1% CNC concentration in NR nanocomposites, there is some agglomeration of CNCs in the NR matrix, and this aggregation gets stronger as the CNC content goes up. This is because the filler and matrix do not interact in the same way as the hydroxyl groups present on the surface of CNCs do through hydrogen bonding. The primary cause for the aggregation of CNCs is the filler-filler network. CNCs are evenly dispersed throughout the matrix of all epoxidized natural rubber (ENR) nanocomposites, showing no signs of aggregation [57]. This is described by the hydrogen bonding that occurs between the epoxy groups of the ENR and the hydroxyl groups present on the surface of CNCs. The low percentage of filler and homogeneous dispersion of CNCs prevent the development of the filler-filler network at a CNC content of 1%. The emergence of the filler-filler network happens when the CNC content hits 3%. The CNCs are more closely bonded to one another at a filler proportion of 10%, creating a more robust filler-filler network [43, 56].

It is found that the widths of individual cellulose fibrils fall between 40 and 50 nm in the nanoscale region. Individual fibers separated as a result of the bleaching, alkali treatment, and acid hydrolysis since the cementing components were eliminated [58]. Figure 3.6 displays SEM images of the tensile fractured surface of the NR/CNF composites incorporated with 3 phr CNF. On the composite's fracture surface,

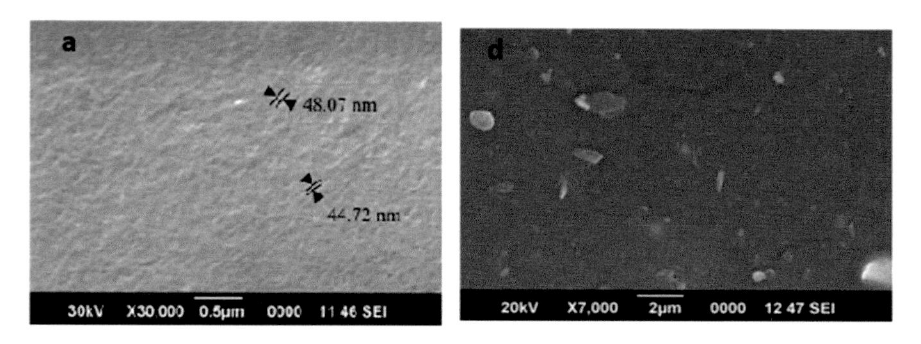

Venugopal and Gopalakrishnan / Materials Today: Proceedings 5 (2018) 16724–16731

Fig. 3.6 **a** SEM image of CNF, **b** SEM image of tensile fractured surface of NR/CNF composite at 3 phr.

the CNF is visible as white spots. Although certain CNFs still persist as aggregates, the NR matrix shows a uniform distribution of CNFs. It is found that the CNFs are incorporated in the matrix, demonstrating strong compatibility between the matrix and the fibers when the bonding agent is present. Additionally, the SEM images revealed that no pull-out cracks or voids, showing strong interfacial adhesion between the matrix and CNFs [24, 59].

3.2.6 Transmission Electron Microscopy

Figure 3.7 displays the TEM micrographs of the CNF nanofibers that were taken out of the coconut spathe. It exhibits CNF fibers with a diameter of 30–60 nm. When assessing the nanofibers' potential to reinforce, one of the most crucial parameters is their aspect ratio, or the ratio of fiber length to diameter. These extracted cellulose nanofibers exhibit a high aspect ratio, as seen by the TEM image [24, 60].

The rubber samples exhibit percolation of CNC (cellulose nanocrystals) as revealed by TEM pictures. NR and SBR samples with low loading (1 phr CNC) may establish very small networks with each other, as shown in Fig. 3.8. The TEM pictures revealed a less uniform dispersion of CNC in SBR, with the CNC agglomerating in tiny places instead of being evenly distributed across the thin sheet. Figure 3.8b shows that fully percolated networks present in NR at high loading (6 phr of CNC). At high CNC loading, percolated networks were seen in SBR; nonetheless, the structure displayed extremely fine webs. Because of the CNC's aggregation in other thin sheet locations, there was a significantly minimum CNC readily available for the construction of a continuous filler network [61].

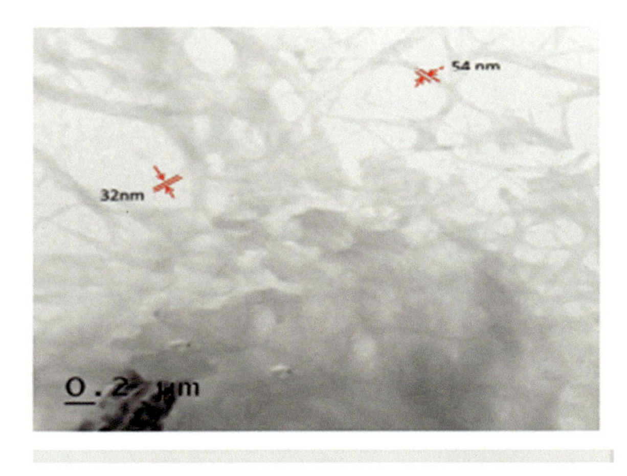

Venugopal and Gopalakrishnan / Materials Today: Proceedings 5 (2018) 16724–16731

Fig. 3.7 SEM image of cellulose nanofiber

3.2.7 Atomic Force Microscopy

AFM is a useful instrument for comprehending the basic characteristics of NR, the surface morphology of cellulose, and the creation of novel NR-based materials with enhanced capabilities for a range of uses. It offers quantitative information and high-resolution images about the adhesion, stiffness, roughness, and topography of NR surfaces [62, 63]. It can also disclose minute characteristics, such as the elasticity and viscoelasticity of NR materials, that are frequently hidden from view by traditional microscopy techniques by probing the material's surface using a fine-tipped cantilever. Figure 3.9 displays an AFM image of the (a) clean matrix, (b) 10% CNF nanocomposite, and (c) cellulose nanofiber.

NR often exhibits a two-phase morphology with amorphous and crystalline domains visible in AFM pictures. In contrast to the crystalline zones, which are hard and inflexible, the amorphous parts are soft and malleable. The overall mechanical characteristics of NR are significantly affected by the size and distribution of these domains, which can also be utilized to map the adhesion forces between NR surfaces and other materials [64]. To ensure enhancement of the mechanical properties of NR composites, it is often beneficial to comprehend the interactions that occur between NR and the reinforcing fillers. The organization of cellulose molecules within the fiber is reflected in the grooves and ridges on the surface of cellulose fibers, which are frequently rough and uneven. When cellulose fibers are imaged using AFM, they usually display long, ribbon-like structures that have a width of one to ten nanometers. The diameter of cellulose fibers and the distance between individual molecules can also be measured with AFM. Cellulose crystals captured by AFM usually exhibit

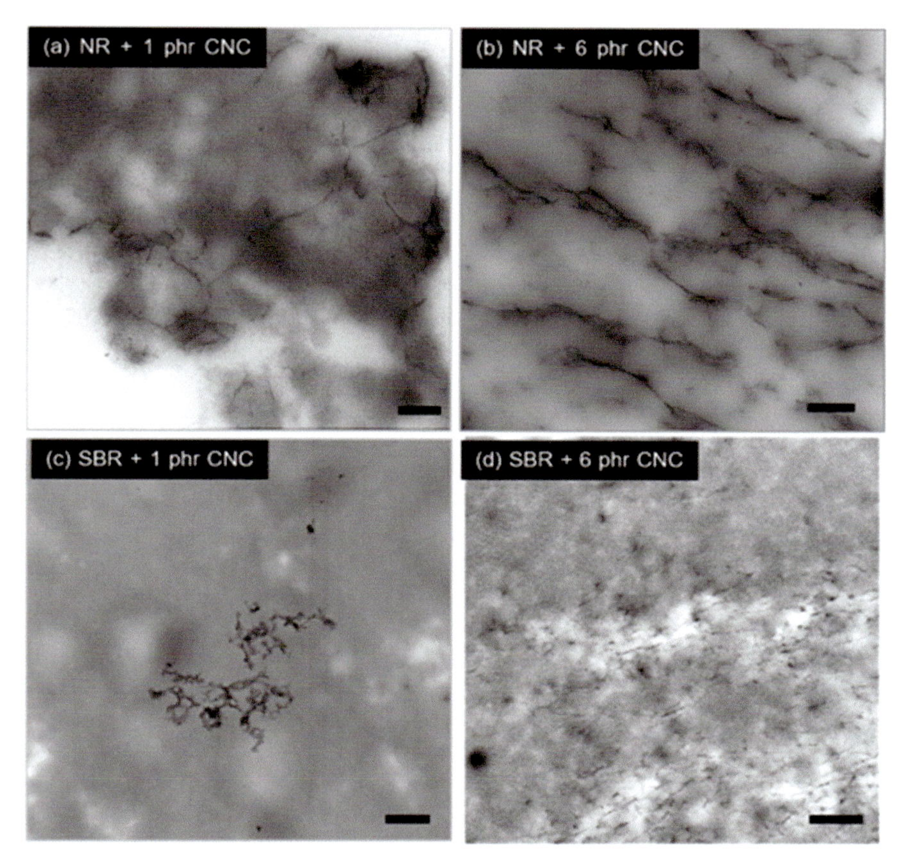

Fig. 3.8 TEM images showing CNC percolation in the thin sheets of rubber. Scale bars represent 1 μm [61]

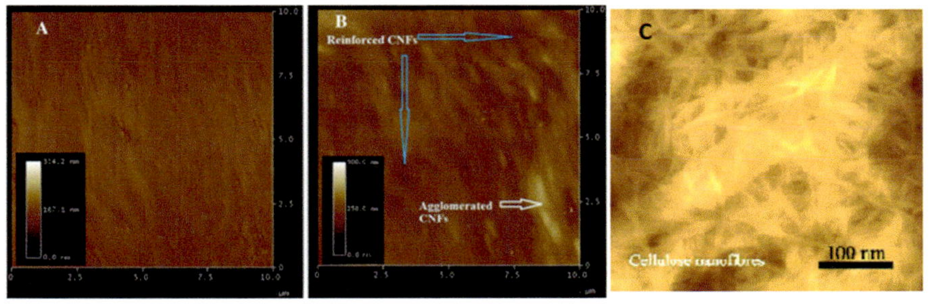

Fig. 3.9 AFM image of the **a** neat matrix, **b** 10% CNF nanocomposite, and **c** cellulose nanofiber

a lattice-like structure with repeating units of about 0.6 nm. Moreover, the size and orientation of cellulose crystals can be determined using AFM [65].

AFM pictures of NR-cellulose microcomposites show that the cellulose microfibers are scattered and embedded in the rubber matrix, forming a heterogeneous structure. The mechanical properties of the composite are greatly affected by the distribution, size, and orientation of these microfibers [66]. The dispersion and alignment of cellulose microfibers inside the NR matrix may be successfully visualized by AFM, yielding vital information for microstructure optimization of the composite. AFM measures the cantilever's deflection by applying a controlled force, which allows it to calculate the composite's elastic modulus at particular points. This method offers important insights into the distribution of stiffness in the composite and the impact of cellulose microfibers on the mechanical behavior as a whole [9, 67].

The adhesion forces between NR-cellulose microcomposites and other materials can also be mapped using AFM. Understanding the interfacial reactions between the cellulose microfibers and the NR matrix, which are essential to the mechanical performance of the composite, is made easier with the help of this knowledge. AFM can distinguish between areas of strong and weak adhesion, revealing information about the efficiency of interfacial bonding as well as possible areas for development. Optimizing the design, performance, and applications of NR-cellulose microcomposites is greatly aided by AFM, which offers high-resolution pictures and quantitative data on the nanoscale morphology, stiffness, and adhesion of these sustainable materials [68, 69].

3.2.8 FT-IR Spectroscopy

Figure 3.10 displays the infrared (IR) spectra of NR, CNC (cellulose nanocrystals), and NR-CNC composite. The infrared spectrum of cellulose shows a number of distinctive absorption bands that reveal important details about the molecular conformation and structure of the material. The hydroxyl groups in cellulose are responsible for the broad and robust absorption band in the 3400–3200 cm^{-1} region [70, 71]. These vibrations are caused by their O–H stretching. The presence of hydroxyl groups in cellulose is shown by this band, which is significant for hydrogen bonding and intermolecular interactions. The C-H stretching vibrations of the alkyl groups in cellulose are represented by two distinct absorption bands in the 2900–2800 cm^{-1} area. The degree of crystallinity and the shape of the alkyl chain in the cellulose sample are revealed by these bands. The weak absorption band observed in the 1640–1620 cm^{-1} area can be attributed to the C=O stretching vibration generated by the carbonyl groups present in cellulose. In samples of oxidized or modified cellulose, this band is more noticeable. The two broad absorption bands in the 1159–1114 cm^{-1} range correspond to the C–O–C stretching vibrations of cellulose [72, 73]. Both the type of glycosidic bond and the cellulose chain's conformation can affect these bands. The hydroxyl groups in cellulose are responsible for the C–O–H bending vibrations that are responsible for the absorption band around 1372–1335 cm^{-1}. The degree of

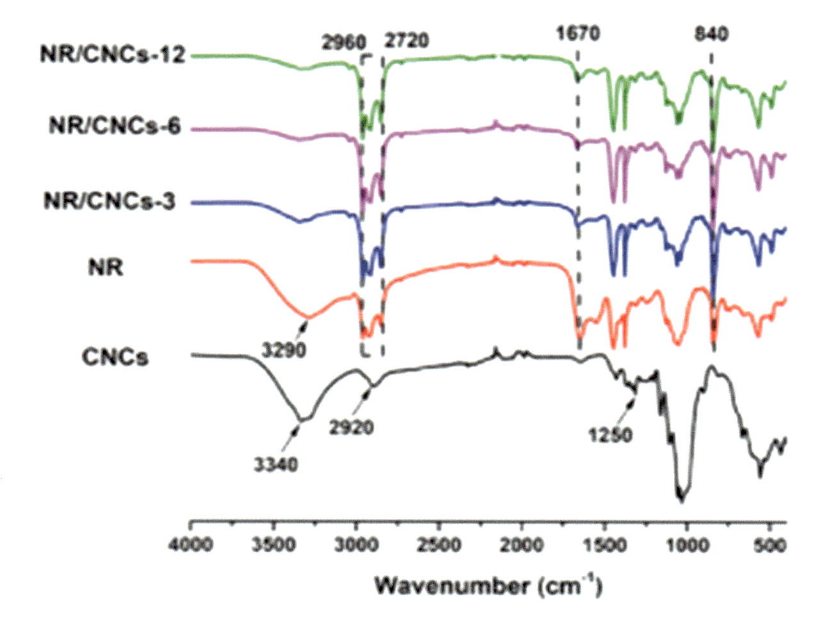

Fig. 3.10 IR Spectrum of CNc, NR, and NR-CNC composite [69]

crystallinity and hydrogen bonding interactions in the cellulose sample are revealed by this band. The weak absorption band noticed at 893 cm^{-1} in cellulose is ascribed to the anomeric C-H bending vibration. The glycosidic linkage type can be utilized to determine the difference between cellulose I and cellulose II polymorphs using this band [74, 75].

The prominent absorption band in the range 1680–1650 cm^{-1} due to its C=C stretching vibration of cis-1,4-polyisoprene backbone of NR. This band signifies the presence of a double bond in the polyisoprene chain, which is responsible for NR's elastomeric properties [76]. A medium-intensity absorption band at 830 cm^{-1} is ascribed to the cis-1,4-polyisoprene backbone of NR's CH wagging vibration. The existence of the double bond in the polyisoprene chain is also indicated by this band. Sharp absorption bands at 3020–2800 cm^{-1} region are associated with the methyl and methylene groups of NR polyisoprene chain's CH stretching vibrations. The degree of crystallinity and aliphatic chain conformation in the NR sample are revealed by these bands. The carbonyl groups in NR have exhibited a weak absorption band in the 1235–1200 cm^{-1} area due to their C–O stretching vibration [77]. This band is more prominent in oxidized or modified NR samples. Several absorption bands in the 1000–900 cm^{-1} region correspond to the C–O–C stretching vibrations in NR. These bands are sensitive to the conformation of the polyisoprene chain and the presence of non-rubber components in NR.

The composite spectrum contains the following stretching vibrations: C=C, CH, O–H, and C–O–C. However, the intensity could change according to the cellulose component concentration [70]. Their location and intensity might shed light on the crystallinity of cellulose and its interactions with NR at the interface. Changes in molecule structure, such as interactions between NR and cellulose through hydrogen bonds, may be indicated by shifts or broadening of absorption bands. The emergence of new absorption bands could indicate the formation of chemical linkages or complexes between NR and cellulose [78].

3.2.9 NMR Spectroscopy

At the molecular level, nuclear magnetic resonance (NMR) spectroscopy sheds light on the composition and characteristics of cellulose. Researchers can ascertain the kinds of links between glucose units, cellulose chain's shape and the amount of crystallinity by examining the chemical shifts and splitting patterns of NMR signals [79]. The various proton types in the isoprene units are represented by a number of distinctive peaks in the ^{1}H NMR spectra of NR. A triplet at around 5.1 ppm represents the proton in the trans configuration of the isoprene unit's double bond A doublet at around 5.6 ppm corresponds to the protons attached to the carbon atom adjacent to the double bond in the isoprene unit. A multiple at around 1.9 ppm corresponds to the protons in the methylene groups of the isoprene unit [80].

The ^{13}C NMR spectrum of cellulose provides more detailed information about the chemical environment of each carbon atom in the glucose unit. A singlet at around 104 ppm, corresponding to the anomeric carbon (C1) on the β-glucose residue. Overlapping triplets at around 73 ppm correspond to the carbons C2, C3, C5 on the β-glucose residue. A singlet at around 83 ppm corresponds to the carbon (C4) on the β-glucose residue. A doublet at around 63 ppm corresponds to the carbon (C6) on the β-glucose residue. It can also be used to find the degree of crystallinity in cellulose. Crystalline cellulose exhibits narrower NMR signals than amorphous cellulose owing to the more ordered arrangement of the polymer chains in the crystalline regions [79, 81].

By exploring the chemical shifts and splitting patterns of NMR signals, researchers can gain insights into the composition, conformation, and interactions between the NR and cellulose phases. The ^{1}H NMR spectrum of an NR-cellulose microcomposite exhibits a combination of peaks from both NR and cellulose. The relative intensities of these peaks can provide information about the relative amounts of NR and cellulose in the composite [82]. The chemical shifts and splitting patterns of the peaks may also be affected by the interactions between the NR and cellulose phases. The ^{13}C NMR spectrum of an NR-cellulose microcomposite also provides information about the composition and interactions in the composite. The chemical shifts of the carbon atoms in NR and cellulose can be influenced by the existence of the other phase.

For example, the chemical shift of the C=C carbon in NR may be shifted due to interactions with cellulose [83, 84].

3.2.10 Raman Spectroscopy

The Raman spectrum of natural rubber is a valuable tool for understanding the structure and properties of this material. The spectrum is characterized by several prominent peaks, each of which corresponds to a specific vibration of the polymer chains. The most prominent peak in the Raman spectrum of natural rubber is at 1650 cm^{-1}. This peak corresponds to the stretching vibration of the carbon–carbon double bond in the cis-1,4-polyisoprene backbone of natural rubber. The peaks noticed at 2918 and 2870 cm^{-1} are related to the stretching vibrations of the methyl groups on the polyisoprene backbone, and the peaks at 2853 and 2930 cm^{-1} show the stretching vibration of –CH_2– groups on the polyisoprene backbone [85]. This peak at 1590 cm^{-1} confirms the stretching vibration of the conjugated diene structure in the cis-1,4-polyisoprene backbone. The Raman spectrum of natural rubber can be used to identify the material and to determine its purity. The impact of processing and degradation on the structure of natural rubber can also be studied using the spectrum [86].

The Raman spectrum of cellulose is characterized by several prominent peaks that correspond to specific molecular vibrations. C–H stretching (376, 2934 cm^{-1}): These peaks arise from the stretching vibrations of the C–H bonds in the cellulose molecule. The peak at 2934 cm^{-1} is particularly strong and is often used to identify cellulose [87]. The broad peaks arise from the stretching vibrations of the O–H bonds observed around 3640–3200 cm^{-1} in the cellulose molecule. The intensity of these peaks can vary depending on the crystallinity of the cellulose. The peak at 1095 cm^{-1} arise from the stretching vibrations of the C–O bonds in the cellulose molecule. The positions and intensities of these peaks are sensitive to the conformation of the cellulose chains. The peaks at 1059 and1121 cm^{-1} arise from the stretching vibrations of the C–C bonds in the cellulose molecule. The peak at 1059 cm^{-1} is particularly strong and is often used to characterize the crystallinity of cellulose [88, 89].

3.2.11 UV–Visible Spectroscopy

Natural rubber (NR) exhibits a UV absorption spectrum with a broad absorption band in the range of 260–350 nm. This absorption is primarily due to the presence of conjugated double bonds in the cis-1,4-polyisoprene backbone of NR. The absorption intensity increases with increasing NR concentration [90]. The UV absorption spectrum of NR is sensitive to the presence of impurities, degradation products, and crosslinking agents. Changes in the absorption intensity and position of the absorption band can give information about the quality and processing of NR. Cellulose exhibits a UV absorption spectrum with a broad absorption band in the range of

250–400 nm, with a maximum peak around 270 nm. This absorption is primarily ascribed to the presence of carbonyl groups in cellulose. The absorption intensity increases with an increase in the concentration of cellulose [91].

3.2.12 EPR or ESR Spectroscopy

The electron spin resonance (ESR) spectrum of natural rubber (NR) reveals valuable insights into the existence and dynamics of free radicals in this elastomeric material. Free radicals are reactive species with unpaired electrons that play a crucial role in various processes, including polymerization, crosslinking, and degradation. The ESR spectrum of NR exhibits several characteristic features that reflect the dynamics and structure of the polymer chains [92]. The primary signal in the NR ESR spectrum is a triplet centered at around g = 2.003. This triplet pattern arises from the interaction of the unpaired electron with the three protons on the carbon adjacent to the double bond in the cis-1,4-polyisoprene backbone of NR.

The triplet pattern in the ESR spectrum of NR is typically broadened due to conformational fluctuations and spin–spin interactions within the polymer chains. The degree of broadening is sensitive to the temperature and mechanical stress applied to the NR sample. In addition to the triplet pattern, weaker signals may be observed in the ESR spectrum of NR, depending on the specific conditions and the presence of impurities or additives [93, 94]. These additional signals can provide further information about the radical species present in the material. The intensity of the triplet signal in the ESR spectrum of NR is proportional to the concentration of free radicals in the material. This intensity is affected by various factors, including the processing conditions, storage time, and exposure to environmental factors like oxygen and ultraviolet radiation. The ESR spectrum of cellulose typically consists of a broad singlet centered at around $g = 2.003$. This singlet arises from the unpaired electrons present in hydroxyl groups along the cellulose chains. The amount of free radicals in the cellulose sample directly correlates with the singlet signal's intensity.

3.2.13 X-Ray Diffraction Analysis

The XRD pattern of natural rubber (NR) is determined by a broad hump centered at around $2\theta = 18°$, which indicates the amorphous nature of the polymer. There are no sharp peaks present, as these are associated with crystalline structures. Its irregular coil conformation, which keeps NR chains from forming regular crystal lattices, is the cause of this lack of crystallinity [95, 96]. The X-ray diffraction (XRD) pattern of cellulose is illustrated by sharp peaks at diffraction angles (2θ) of approximately 14.5°, 16.5°, and 22.5°. These peaks correspond to the interplanar spacings (d-spacings) of 0.789, 0.787, and 0.426 nm, respectively. The existence of these strong peaks implies that cellulose has a high degree of crystallinity, with the majority of the polymer chains structured in regular, ordered patterns. This high crystallinity is responsible for many of the distinctive properties of cellulose, such as its exceptional tensile strength and resistance to biodegradation [1].

The X-ray diffraction (XRD) pattern of a natural rubber-cellulose composite typically exhibits a combination of the features observed for both pure NR and pure cellulose. The broad hump characteristic of amorphous NR is still present, but the sharp peaks associated with crystalline cellulose may also be observed [97]. The relative intensity of these peaks correlated with the amount and nature of cellulose incorporated in the composite. If a small amount of cellulose nanocrystals (CNCs) is dispersed in NR, the XRD pattern may only show a slight broadening of the amorphous hump and a very weak signal from the cellulose peaks. However, if a higher loading of CNCs is used, the cellulose peaks may become more pronounced and the amorphous hump may become less distinct [98].

X-ray diffraction of pattern of natural rubber-cellulose composites with low cellulose loading show moderate improvements and easier processing. In contrast, if a highly crystalline cellulose filler, such as microfibrillated cellulose (MFC), is used, the XRD pattern may show a more dramatic change [99]. The amorphous hump may be almost completely obscured by the sharp cellulose peaks, and the overall diffraction pattern may resemble that of pure cellulose. The presence of crystalline cellulose peaks in the XRD pattern of a natural rubber-cellulose composite indicates that the cellulose has dispersed throughout the NR matrix and has not formed large, separate phases. Ensuring appropriate interfacial interaction between the two components is crucial in improving the composite's characteristics. The extent to which the cellulose peaks are observed in the XRD pattern is also related to the processing conditions used to fabricate the composite. When the composite is prepared using a melt-mixing process, the cellulose may be more dispersed and the cellulose peaks may be more pronounced than if the composite is prepared using a solution-casting process [100]. Hence, the XRD pattern of a natural rubber-cellulose composite renders valuable information about the morphology and crystallinity of the composite and can be used to assess the effectiveness of the cellulose reinforcement.

3.2.14 SAXS and WAXS Analyses

Two complementary techniques that can be used to analyze the structure of natural rubber (NR) are small-angle X-ray scattering (SAXS) and wide-angle X-ray scattering (WAXS). SAXS is used to probe the structure of NR at the nanoscale, typically from 1 to 100 nm. It is useful for determining the distribution and size of filler particles in NR composites, as well as the size and distribution of crosslinks in NR networks. SAXS can also be used to show that the crosslink density of NR increases with increasing irradiation dose. This is because irradiation creates new crosslinks between NR chains [101].

WAXS is used to probe the structure of NR at the atomic scale, typically from 0.1 to 1 nm. It can be used to determine the crystallinity of NR, as well as the type of crystal structure present. WAXS can be used to show that NR is amorphous, meaning that it does not have a regular, ordered crystal structure [102]. However, WAXS can also be used to show that NR can exhibit some degree of crystallinity when it is stretched or subjected to high stress. This is because stretching or stress can cause the NR chains to align more regularly, which can lead to the formation of small crystallites.

The combination of SAXS and WAXS is used to provide an overall view of the structure of NR at different length scales. This information can be used to know the properties of NR and to create new NR-based materials. SAXS and WAXS have been used to show that the addition of fillers can change the distribution and size of filler particles in NR composites, including nature of crosslinks in NR networks [103, 104]. These changes can have a significant impact on the properties of the composites, such as their modulus, tensile strength, and tear resistance. SAXS and WAXS have been used to show that the structure of NR can change over time when it is exposed to heat, light, or oxygen. These changes can reduce the modulus, tensile strength, and tear resistance of the NR. Hence, SAXS and WAXS are the effective tools, which can be used to analyze the characteristics and structure of NR.

SAXS can provide information about the shape, size, and also the arrangement of cellulose microfibrils. Cellulose microfibrils are rod-like structures and it has a diameter of around 2–5 nm and a length up to several micrometers. They are composed of crystalline cellulose chains arranged in a parallel fashion [105]. The arrangement of these microfibrils within the cellulose matrix determines the overall properties of the material. SAXS patterns of cellulose typically exhibit a broad scattering maximum at low scattering angles, which indicates the elongated structure of the cellulose microfibrils. The intensity of the scattering can be used to determine the volume fraction of cellulose microfibrils in the material.

WAXS can provide information about the crystal structure of cellulose. Cellulose has a crystalline structure known as cellulose Iα. This structure is characterized by a repeating unit cell with dimensions of $a = 0.566$ nm, $b = 0.790$ nm, and $c = 0.574$ nm. The distinctive diffraction peaks observed in cellulose WAXS patterns are caused by the unit cell's arrangement of cellulose chains. WAXS patterns of cellulose typically exhibit a series of sharp diffraction peaks at high scattering angles. These peaks are utilized to find out the lattice parameters of the cellulose unit cell and to measure the degree of crystallinity of the material. A crucial element influencing cellulose's tensile strength, modulus, and biodegradability is its degree of crystallinity. SAXS and WAXS can be used together to furnish a wide spectrum of the structure of cellulose at both the nanoscale and the atomic scale. This information can be used to learn the properties of cellulose and to bring in new cellulose-based materials.

SAXS patterns of natural rubber-cellulose composites typically exhibit a broad scattering maximum at low scattering angles, which indicates the elongated structure of the cellulose microfibrils. The intensity of the scattering can also be used to determine the volume fraction of cellulose microfibrils in the composite [103].

Additionally, WAXS can reveal details on the cellulose microfibrils' and the rubber matrix's crystal structures. WAXS patterns of natural rubber-cellulose composites typically exhibit a series of sharp diffraction peaks from the cellulose microfibrils, superimposed on a broad amorphous halo from the rubber matrix. The volume fraction of cellulose in the composite is determined by the intensity of the cellulose peaks relative to the rubber. The degree of crystallinity of the cellulose can also be determined from the width of the cellulose peaks.

SAXS and WAXS can be used together to render an overall view of the structure of natural rubber-cellulose composites at both the nanoscale and the atomic scale. This information can be used to know the properties of the composite and to enable the new composite materials with improved properties for a variety of applications. SAXS and WAXS have been used to show that the size and distribution of cellulose microfibrils in natural rubber-cellulose composites can be controlled by the processing conditions. This control over the microstructure of the composite can be used to tailor its properties, such as its tensile strength, modulus, and biodegradability.

3.2.15 Neutron Scattering

It is a powerful tool for studying the composition of materials, including natural rubber (NR). From the atomic to the macroscopic scales, neutron scattering techniques can be utilized to investigate the structure of NR. It can be used to ascertain the phase behavior of NR, the molecular weight, the architecture of NR networks, the degree of branching of NR chains, and the effect of fillers on the composition and characteristics of rubber composites [104, 106]. The dynamics of NR, which include NR's viscosity and the speed at which the chains pass one another, can also be studied using NSE.

3.2.16 Rheology Measurements

The study of rheology focuses on the flow and deformation of fluids and soft solids under stress. It's a crucial area of research for numerous materials, including natural rubber. Being viscoelastic, natural rubber demonstrates both viscous and elastic properties. Rheological data can be used to forecast how natural rubber will flow during injection molding, as well as to optimize the processing conditions for natural rubber and the performance of rubber goods in use. As a result, it can be applied to describe the composition and characteristics of natural rubber [107, 108].

The rheological behavior of natural rubber is greatly influenced by the addition of cellulose. Because cellulose fibers and rubber chains interact to obstruct the flow of the composite, cellulose addition typically results in an increase in viscosity. Because cellulose fibers and rubber chains form a network structure, cellulose reinforcement enhances the composite's elastic response [109, 110]. Cellulose has the ability to reduce viscous dissipation, which leads to more elastic behavior. This is because the cellulose fibers limit the motion of the rubber chain. The increased viscosity and lower flowability of the cellulose inclusion may cause processing issues that require adjustments to the processing parameters. Hence, rheological data are necessary to completely comprehend and optimize the performance of natural rubber-cellulose composites [111–113]. The development of high-performance materials with specialized features for a range of applications is made possible by these measurements, which offer insightful information about the processing behavior, product performance, and material characterization of these composites.

3.3 Conclusion

Biodegradable materials are becoming more and more popular for a range of applications since they are sustainable, renewable, and safe for the environment. In addition, a few intriguing qualities, like their low density, biocompatibility, biodegradability, and non-toxicity functions served as an accelerant to substitute the synthetic materials. The significance of rubber blends based on cellulose and its microcomposites is outlined in this chapter. The development of stronger NR microcomposites and their blends was made possible in large part by this filler and by combining it with different types of rubber. Moreover, the bio-nanofillers demonstrated greater strength due to their enhanced specific area. This study also examined the ramifications of the cellulose on the mechanical, thermal, biodegradability, surface morphology, and crystalline properties of the resultant rubber composites. Even if the biodegradable materials had advantageous qualities, a few flaws severely restrict how widely they may be used. To overcome this issue, the surfaces of the NR matrix and the fillers must be altered. As a result, many researchers focus on improving the adhesion and interfacial reactions between the NR matrix and natural fillers. They also emphasize the future endeavors of these composite materials in extensive applications. This chapter concludes that the development of efficient bio-based materials would help to stop environmental degradation.

References

1. Luckachan GE, Pillai CKS (2011) Biodegradable polymers—a review on recent trends and emerging perspectives. J Polym Environ 19:637–676
2. Thomas SK, Parameswaranpillai J, Krishnasamy S, Begum PMS, Nandi D, Siengchin S et al (2021) A comprehensive review on cellulose, chitin, and starch as fillers in natural rubber biocomposites. Carbohydr Polym Technol Appl 2:100095
3. Alexandre M, Dubois P, Sun T, Garces JM, Jérôme R (2002) Polyethylene-layered silicate nanocomposites prepared by the polymerization-filling technique: synthesis and mechanical properties. Polymer 43:2123–2132
4. Ghovvati M, Guo L, Bolouri K, Kaneko N (2023) Advances in electroconductive polymers for biomedical sector: structure and properties. Mater Chem Horiz 2:125–137
5. Paul DR (1986) Polymer blends: phase behavior and property relationships. ACS Publications
6. Makhijani K, Kumar R, Sharma SK (2015) Biodegradability of blended polymers: a comparison of various properties. Crit Rev Environ Sci Technol 45:1801–1825

7. Allahvaisi S (2012) Polypropylene in the industry of food packaging. ISSBN
8. Parambath Kanoth B, Claudino M, Johansson M, Berglund LA, Zhou Q (2015) Biocomposites from natural rubber: synergistic effects of functionalized cellulose nanocrystals as both reinforcing and cross-linking agents via free-radical thiol–ene chemistry. ACS Appl Mater Interfaces 7:16303–16310
9. Palanisamy S, Mayandi K, Palaniappan M, Alavudeen A, Rajini N, de Camargo FV et al (2021) Mechanical properties of phormium tenax reinforced natural rubber composites. Fibers 9. https://doi.org/10.3390/fib9020011
10. Sethulekshmi AS, Saritha A, Joseph K (2022) A comprehensive review on the recent advancements in natural rubber nanocomposites. Int J Biol Macromol 194:819–842
11. Eichhorn SJ, Dufresne A, Aranguren M, Marcovich NE, Capadona JR, Rowan SJ et al (2010) Current international research into cellulose nanofibres and nanocomposites. J Mater Sci 45:1–33
12. Nechyporchuk O, Belgacem MN, Bras J (2016) Production of cellulose nanofibrils: a review of recent advances. Ind Crops Prod 93:2–25
13. Dufresne A (2013) Nanocellulose: a new ageless bionanomaterial. Mater Today 16:220–227
14. Gopi S, Balakrishnan P, Chandradhara D, Poovathankandy D, Thomas S (2019) General scenarios of cellulose and its use in the biomedical field. Mater Today Chem 13:59–78
15. Andrew JJ, Dhakal HN (2022) Sustainable biobased composites for advanced applications: recent trends and future opportunities–a critical review. Compos Part C: Open Access 7:100220
16. Bhanderi K, Joshi J, Suthar V, Shah V, Patel GM, Patel J (2023) Eco-friendly polymer nanocomposites based on bio-based fillers: preparation, characterizations and potential applications. Biodegradable and biocompatible polymer nanocomposites. Elsevier, pp 173–203
17. Murugesan TM, Palanisamy S, Santulli C, Palaniappan M (2022) Mechanical characterization of alkali treated Sansevieria cylindrica fibers–Natural rubber composites. Mater Today: Proc 62:5402–5406
18. Kim K, Kim M, Hwang Y, Kim J (2014) Chemically modified boron nitride-epoxy terminated dimethylsiloxane composite for improving the thermal conductivity. Ceram Int 40:2047–2056
19. Shaghaleh H, Xu X, Wang S (2018) Current progress in production of biopolymeric materials based on cellulose, cellulose nanofibers, and cellulose derivatives. RSC Adv 8:825–842
20. Khan F, Ahmad SR (2013) Polysaccharides and their derivatives for versatile tissue engineering application. Macromol Biosci 13:395–421
21. Abdul Khalil HPS, Chong EWN, Owolabi FAT, Asniza M, Tye YY, Rizal S et al (2019) Enhancement of basic properties of polysaccharide-based composites with organic and inorganic fillers: a review. J Appl Polym Sci 136:47251
22. Mokhothu TH, John MJ (2015) Review on hygroscopic aging of cellulose fibres and their biocomposites. Carbohyd Polym 131:337–354
23. Siqueira G, Bras J, Dufresne A (2010) Cellulosic bionanocomposites: a review of preparation, properties and applications. Polymers 2:728–765
24. Bras J, Hassan ML, Bruzesse C, Hassan EA, El-Wakil NA, Dufresne A (2010) Mechanical, barrier, and biodegradability properties of bagasse cellulose whiskers reinforced natural rubber nanocomposites. Ind Crops Prod 32:627–633
25. Khan R, Jolly R, Fatima T, Shakir M (2022) Extraction processes for deriving cellulose: a comprehensive review on green approaches. Polym Adv Technol 33:2069–2090
26. Lu Y, He Q, Fan G, Cheng Q, Song G (2021) Extraction and modification of hemicellulose from lignocellulosic biomass: a review. Green Process Synth 10:779–804
27. Lefatshe K, Muiva CM, Kebaabetswe LP (2017) Extraction of nanocellulose and in-situ casting of ZnO/cellulose nanocomposite with enhanced photocatalytic and antibacterial activity. Carbohyd Polym 164:301–308
28. Zhang C, Zheng X, Wan X, Shao X, Liu Q, Zhang Z et al (2014) The potential use of H102 peptide-loaded dual-functional nanoparticles in the treatment of Alzheimer's disease. J Control Release 192:317–324

29. Liu H, Liu K, Han X, Xie H, Si C, Liu W et al (2020) Cellulose nanofibrils-based hydrogels for biomedical applications: progresses and challenges. Curr Med Chem 27:4622–4646
30. Nayigiziki FX (2016) Physical characterization and antimicrobial properties of PVA-cellulose nanofiber based films
31. Abraham E, Deepa B, Pothan LA, Jacob M, Thomas S, Cvelbar U et al (2011) Extraction of nanocellulose fibrils from lignocellulosic fibres: a novel approach. Carbohyd Polym 86:1468–1475
32. Abol-Fotouh D, Hassan MA, Shokry H, Roig A, Azab MS, Kashyout AE-HB (2020) Bacterial nanocellulose from agro-industrial wastes: low-cost and enhanced production by Komagataeibacter saccharivorans MD1. Sci Rep 10:3491
33. Thatoi H, Behera S (nd) Bacterial cellulose production from renewable feedstock. Bacterial cellulose. CRC Press, pp 22–38
34. Choi SM, Rao KM, Zo SM, Shin EJ, Han SS (2022) Bacterial cellulose and its applications. Polymers 14:1080
35. Joseph S, Appukuttan SP, Kenny JM, Puglia D, Thomas S, Joseph K (2010) Dynamic mechanical properties of oil palm microfibril-reinforced natural rubber composites. J Appl Polym Sci 117:1298–1308
36. Tessanan W, Daniel P, Phinyocheep P (2021) Development of photosensitive natural rubber as a mechanical modifier for ultraviolet-curable resin applied in digital light processing-based three-dimensional printing technology. ACS Omega 6:14838–14847
37. Visakh PM, Thomas S, Oksman K, Mathew AP (2012) Effect of cellulose nanofibers isolated from bamboo pulp residue on vulcanized natural rubber. BioResources 7:2156–2168
38. Ketabchi MR, Ratnam CT, Khalid M, Walvekar R (2018) Mechanical properties of polylactic acid/synthetic rubber blend reinforced with cellulose nanoparticles isolated from kenaf fibres. Polym Bull 75:809–827
39. Burfield DR, Lim KL (1983) Differential scanning calorimetry analysis of natural rubber and related polyisoprenes. Measurement of the glass transition temperature. Macromolecules 16:1170–1175
40. León-Becerra J, Hidalgo-Salazar MÁ, Correa-Aguirre JP, González-Estrada OA, Pertuz AD (2023) Additive manufacturing of short carbon filled fiber nylon: effect of build orientation on surface roughness and viscoelastic behavior. Int J Adv Manuf Technol 2023:1–11
41. Baniasadi H, Trifol J, Lipponen S, Seppälä J (2021) Sustainable composites of surface-modified cellulose with low–melting point polyamide. Mater Today Chem 22:100590
42. Michell RM, Müller AJ (2016) Confined crystallization of polymeric materials. Prog Polym Sci 54:183–213
43. Abraham E, Deepa B, Pothan LA, John M, Narine SS, Thomas S et al (2013) Physicomechanical properties of nanocomposites based on cellulose nanofibre and natural rubber latex. Cellulose 20:417–427
44. Rissanen V (2023) Development of leaf-inspired functional structures from cellulose nanofibers-matrix scaffolds for solid-state photosynthetic cell factories
45. Thomas MS, Koshy RR, Mary SK, Thomas S, Pothan LA (2019) Starch, chitin and chitosan based composites and nanocomposites. Springer
46. Ponnamma D, Sung SH, Hong JS, Ahn KH, Varughese KT, Thomas S (2014) Influence of non-covalent functionalization of carbon nanotubes on the rheological behavior of natural rubber latex nanocomposites. Eur Polym J 53:147–159
47. Chartoff RP, Menczel JD, Dillman SH (2009) Dynamic mechanical analysis (DMA). In: Thermal analysis of polymers: fundamentals and applications, pp 387–495
48. Payne AR (1964) Strainwork dependence of filler-loaded vulcanizates. J Appl Polym Sci 8:2661–2686
49. Venugopal B, Gopalakrishnan J (2018) Reinforcement of natural rubber using cellulose nanofibres isolated from Coconut spathe. Mater Today: Proc 5:16724–16731
50. Wang Y, Tian H, Zhang L (2010) Role of starch nanocrystals and cellulose whiskers in synergistic reinforcement of waterborne polyurethane. Carbohyd Polym 80:665–671

51. Cao X, Habibi Y, Lucia LA (2009) One-pot polymerization, surface grafting, and processing of waterborne polyurethane-cellulose nanocrystal nanocomposites. J Mater Chem 19:7137–7145
52. Cao X, Xu C, Wang Y, Liu Y, Liu Y, Chen Y (2013) New nanocomposite materials reinforced with cellulose nanocrystals in nitrile rubber. Polym Testing 32:819–826
53. Chen Y, Zhang Y, Xu C, Cao X (2015) Cellulose nanocrystals reinforced foamed nitrile rubber nanocomposites. Carbohyd Polym 130:149–154
54. Alexandre M, Dubois P (2000) Polymer-layered silicate nanocomposites: preparation, properties and uses of a new class of materials. Mater Sci Eng R Rep (28)
55. Facio AC, Galindo AS, Cepeda LF, López LL, de León-Gómez RD (2015) Thermal degradation of synthetic rubber nanocomposites. In: Thermal degradation of polymer blends, composites and nanocomposites. Springer, pp 157–191
56. Yihun FA (2022) Nanochitin preparation and its application in polymer nanocomposites: a review. Emergent Mater 5:2031–2060
57. Yang X, Yang L, Teng X, Wang Q, Wang Y, Xu C (2023) Strengthened adipic acid cross-linked epoxidized natural rubber by cellulose nanocrystals. Cellulose 1–18
58. Ferreira SR, de Andrade SF, Lima PRL, Toledo Filho RD (2015) Effect of fiber treatments on the sisal fiber properties and fiber–matrix bond in cement based systems. Constr Build Mater 101:730–740
59. Chang BP, Gupta A, Muthuraj R, Mekonnen TH (2021) Bioresourced fillers for rubber composite sustainability: current development and future opportunities. Green Chem 23:5337–5378
60. Li K, Clarkson CM, Wang L, Liu Y, Lamm M, Pang Z et al (2021) Alignment of cellulose nanofibers: harnessing nanoscale properties to macroscale benefits. ACS Nano 15:3646–3673
61. Jardin JM, Zhang Z, Hu G, Tam KC, Mekonnen TH (2020) Reinforcement of rubber nanocomposite thin sheets by percolation of pristine cellulose nanocrystals. Int J Biol Macromol 152:428–436
62. Murthy NS (2011) Techniques for analyzing biomaterial surface structure, morphology and topography. Surface modification of biomaterials. Elsevier, pp 232–255
63. Gaboriaud F, de Gaudemaris B, Rousseau T, Derclaye S, Dufrêne YF (2012) Unravelling the nanometre-scale stimuli-responsive properties of natural rubber latex particles using atomic force microscopy. Soft Matter 8:2724–2729
64. Badia JD, Strömberg E, Karlsson S, Ribes-Greus A (2012) The role of crystalline, mobile amorphous and rigid amorphous fractions in the performance of recycled poly (ethylene terephthalate) (PET). Polym Degrad Stab 97:98–107
65. Abraham E, Pothan LA, Thomas S (2012) Preparation and characterization of green nanocomposites based on cellulose nanofibre and natural rubber latex. In: 15th European conference on composite materials, Venice, Italy
66. Huang Z-M, Zhang Y-Z, Kotaki M, Ramakrishna S (2003) A review on polymer nanofibers by electrospinning and their applications in nanocomposites. Compos Sci Technol 63:2223–2253
67. Azammi AMN, Ilyas RA, Sapuan SM, Ibrahim R, Atikah MSN, Asrofi M et al (2020) Characterization studies of biopolymeric matrix and cellulose fibres based composites related to functionalized fibre-matrix interface. Interfaces in particle and fibre reinforced composites. Elsevier, pp 29–93
68. Mandriota N, Friedsam C, Jones-Molina JA, Tatem KV, Ingber DE, Sahin O (2019) Cellular nanoscale stiffness patterns governed by intracellular forces. Nat Mater 18:1071–1077
69. Gong X, Liu T, Zhang H, Liu Y, Boluk Y (2021) Release of cellulose nanocrystal particles from natural rubber latex composites into immersed aqueous media. ACS Appl Bio Mater 4:1413–1423
70. Zhbankov RG (2013) Infrared spectra of cellulose and its derivatives. Springer
71. Govindasamy P (2021) Mechanical and water absorption properties of Acacia Arabica bark fiber/polyester composites: effect of alkali treatment and fiber volume fraction. Mater Today: Proc 46:2281–2287
72. Popescu C-M, Popescu M-C, Vasile C (2010) Characterization of fungal degraded lime wood by FT-IR and 2D IR correlation spectroscopy. Microchem J 95:377–387

73. Santhosh N, Selvam S, Reghu R, Sundaran J, Mathew BC, Palanisamy S (2023) Mechanical properties studies on rubber composites reinforced with Acacia Caesia fibre. Mater Today: Proc 72:3172–3176
74. Atalla RH (1976) Raman spectral studies of polymorphy in cellulose. Part I: Celluloses I and II. Appl Polym Symp 28:659–669
75. Palanisamy S, Murugesan TM, Palaniappan M, Santulli C (2023) Characterization of an eco-friendly rubber composite material based on Sansevieria cylindrica fibers. Mater Today: Proc. https://doi.org/10.1016/j.matpr.2023.08.202
76. Parker JR (2002) 2 photoacoustic Fourier transform infrared spectroscopy of rubbers and related materials. In: Spectroscopy of rubbers and rubbery materials, p 49
77. Wiercigroch E, Szafraniec E, Czamara K, Pacia MZ, Majzner K, Kochan K et al (2017) Raman and infrared spectroscopy of carbohydrates: a review. Spectrochim Acta Part A Mol Biomol Spectrosc 185:317–335
78. Cao L, Huang J, Chen Y (2018) Dual cross-linked epoxidized natural rubber reinforced by tunicate cellulose nanocrystals with improved strength and extensibility. ACS Sustain Chem Eng 6:14802–14811
79. Kim SH, Lee CM, Kafle K (2013) Characterization of crystalline cellulose in biomass: basic principles, applications, and limitations of XRD, NMR, IR, Raman, and SFG. Korean J Chem Eng 30:2127–2141
80. Perera MCS (1985) Chemical modification of natural rubber studied by nuclear magnetic resonance spectroscopy. The Australian National University (Australia)
81. Fan TW-M, Lane AN (2008) Structure-based profiling of metabolites and isotopomers by NMR. Progr Nucl Magn Reson Spectrosc 52:69–117
82. Joseph S, PA S, Kenny JM, Puglia D, Thomas S, Joseph K (2010) Oil palm microcomposites: processing and mechanical behavior. Polym Eng Sci 50:1853–1863
83. Ponnamma D, Maria HJ, Chandra AK, Thomas S (2013) Rubber nanocomposites: latest trends and concepts. Adv Elastomers II: Compos Nanocompos 69–107
84. Fliri L, Guizani C, Miranda-Valdez IY, Pitkänen L, Hummel M (2023) Reinvestigating the concurring reactions in early-stage cellulose pyrolysis by solution state NMR spectroscopy. J Anal Appl Pyrol 175:106153
85. Krieg T, Mazzon C, Gómez-Sánchez E (2021) Material analysis and a visual guide of degradation phenomena in historical synthetic polymers as tools to follow ageing processes in industrial heritage collections. Polymers 14:121
86. López-Manchado MA, Biagiotti J, Valentini L, Kenny JM (2004) Dynamic mechanical and Raman spectroscopy studies on interaction between single-walled carbon nanotubes and natural rubber. J Appl Polym Sci 92:3394–3400
87. Punitha S, Uvarani R, Panneerselvam A, Nithiyanantham S (2014) Physico-chemical studies on some saccharides in aqueous cellulose solutions at different temperatures–Acoustical and FTIR analysis. J Saudi Chem Soc 18:657–665
88. Liu H-M, Li H-Y, Li M-F (2017) Cornstalk liquefaction in sub-and super-critical ethanol: characterization of solid residue and the liquefaction mechanism. J Energy Inst 90:734–742
89. Shirazi S, Mafigholami R, Moghimi H, Borghei SM (2023) Feasibility study of microplastic biodegradation in effluents from South Tehran WWTP after quantitative and qualitative measurement of the particles. Appl Water Sci 13:80
90. Sreeja R, Najidha S, Jayan SR, Predeep P, Mazur M, Sharma PD (2006) Electro-optic materials from co-polymeric elastomer–acrylonitrile butadiene rubber (NBR). Polymer 47:617–623
91. Loureiro PEG, Fernandes AJS, Furtado FP, Carvalho MGVS, Evtuguin DV (2011) UV-resonance Raman micro-spectroscopy to assess residual chromophores in cellulosic pulps. J Raman Spectrosc 42:1039–1045
92. Thambiraj S, Shankaran DR (2017) Preparation and physicochemical characterization of cellulose nanocrystals from industrial waste cotton. Appl Surf Sci 412:405–416
93. Yang D, Zhang R, Gai S, Yang P (2023) The fundamental and application of piezoelectric materials for tumor therapy: recent advances and outlook. Mater Horiz 10(4)

94. Reynolds JH (1993) Chemical and spectroscopic properties of 1, 2, 4, 5-tetramethylenebenzene. Yale University
95. Riyajan S (2013) Green natural fibre reinforced natural rubber composites. In: Natural rubber materials, Volume 2: Composites and nanocomposites. The Royal Society of Chemistry, pp 353–400
96. Palanisamy S, Kalimuthu M, Dharmalingam S, Alavudeen A, Nagarajan R, Ismail SO et al (2023) Effects of fiber loadings and lengths on mechanical properties of Sansevieria Cylindrica fiber reinforced natural rubber biocomposites. Mater Res Express 10:85503
97. Thomas MG, Abraham E, Jyotishkumar P, Maria HJ, Pothen LA, Thomas S (2015) Nanocelluloses from jute fibers and their nanocomposites with natural rubber: preparation and characterization. Int J Biol Macromol 81:768–777
98. Chang S, Weng Z, Zhang C, Jiang S, Duan G (2023) Cellulose-based intelligent responsive materials: a review. Polymers 15(19):3905. https://doi.org/10.3390/polym15193905
99. Chuong B, Tung NH, Hung DV, Linh NPD (2017) Invited review. Natural rubber nanocomposites. Vietnam J Chem 55:663
100. Martínez-Sanz M, Lopez-Rubio A, Lagaron JM (2013) Nanocomposites of ethylene vinyl alcohol copolymer with thermally resistant cellulose nanowhiskers by melt compounding (I): Morphology and thermal properties. J Appl Polym Sci 128:2666–2678
101. Shanks RA (2014) Characterization of nanostructured materials. Nanostructured polymer blends. Elsevier, pp 15–31
102. Wongvasana B, Thongnuanchan B, Masa A, Saito H, Sakai T, Lopattananon N (2022) Comparative structure-property relationship between nanoclay and cellulose nanofiber reinforced natural rubber nanocomposites. Polymers 14:3747
103. Fallon JJ, Kolb BQ, Herwig CJ, Foster EJ, Bortner MJ (2018) Mechanically adaptive thermoplastic urethane/cellulose nanocrystal composites: process driven structure property relationships. J Appl Polym Sci 136(4)
104. Varghese J, Jose Chirayil C, Somasekharan L, Thomas S (2013) X-ray, light and neutron scattering studies on natural rubber composites and nanocomposites. Natural Rubber Materials, Volume: 2 Composites and Nanocomposites, Royal Society of Chemistry. https://doi.org/10.1039/9781849737654-00622, https://doi.org/10.1039/9781849737654
105. Hamad WY (2017) Cellulose nanocrystals: properties, production and applications. Wiley
106. Krakovský I (2021) Structure of natural rubber as revealed by X-ray and neutron scattering. In: Chemistry, manufacture, and applications of natural rubber. Elsevier, pp 109–151
107. Lima P, da Silva SPM, Oliveira J, Costa V (2015) Rheological properties of ground tyre rubber based thermoplastic elastomeric blends. Polym Test 45:58–67
108. Polychronopoulos ND, Vlachopoulos J (2018) Polymer processing and rheology. In: Jafar Mazumder M, Sheardown H, Al-Ahmed A (eds) A functional polymers polymers and polymeric composites: a reference series, pp 133–180
109. Ching YC, Ershad Ali M, Abdullah LC, Choo KW, Kuan YC, Julaihi SJ et al (2016) Rheological properties of cellulose nanocrystal-embedded polymer composites: a review. Cellulose 23:1011–1030
110. Mohit H, Arul Mozhi Selvan V (2018) A comprehensive review on surface modification, structure interface and bonding mechanism of plant cellulose fiber reinforced polymer based composites. Compos Interfaces 5:629–667
111. Song K (2017) Interphase characterization in rubber nanocomposites. In: Progress in rubber nanocomposites. Elsevier, pp 115–152
112. Taheri H, Samyn P (2015) Rheological properties and processing of polymer blends with micro-and nanofibrillated cellulose. Agricult Biomass Based Potential Mater 259–291
113. Mavelil-Sam R, Deepa B, Koshy RR, Mary SK, Pothan LA, Thomas S (2017) Rheological properties of nanocomposites based on cellulose nanofibrils and cellulose nanocrystals. Handb Nanocellulose Cellul Nanocomp 2:481–521

Chapter 4
Chitin-Based Rubber Nanocomposites

Anmiya Peter, V Bijina, and K Abhitha

Abstract Bio-fillers and their rubber composites have been topics of interest during the last few decades because of their low cost, eco-friendliness and good mechanical properties. Chitin is a polysaccharide readily available in nature. It is mainly synthesized from anthropods and crustaceans. Due to chitin's solubility restrictions, it has received less attention than starch and cellulose. The current chapter deals with chitin-based rubber nanocomposites where the different modes of chitin extraction and various techniques adopted for incorporating chitin in rubber matrices are disclosed. Essential characterization techniques selected for the composite materials are also discussed.

Keywords Chitin · Rubber · Demineralization · Deproteinization · Tensile testing · Dynamic mechanical analysis · Morphological studies · UV–visible spectroscopy · IR spectroscopy

4.1 Introduction

Biodegradable and environmentally friendly polymers derived from various renewable resources are currently preferred for several uses. Environmental protection is aided by using sustainable materials that decompose quickly in nature [1]. As a result, the number of items made from renewable resources has been rising quickly in recent years. Food packaging, medicine delivery, tissue engineering, medical implants, composite technologies and eco-friendly sorbents are a few examples of applications for eco-friendly products [2–5]. The properties of products can be improved by adding fillers of different sizes and morphology in a suitable ratio.

A. Peter · V. Bijina · K. Abhitha (✉)
Department of Polymer Science and Rubber Technology, Cochin University of Science and Technology, Kerala 682022, India
e-mail: abhithak80@cusat.ac.in

K. Abhitha
Inter University Centre for Nanomaterials and Devices (IUCND), Cochin University of Science and Technology, Kerala 682022, India

Visakh P. M. *Rubber Based Bionanocomposites*, Advanced Structured Materials 210,
https://doi.org/10.1007/978-981-10-2978-3_4

The mechanical properties of the host matrix can be improved by incorporating fibres into it. The reinforcement of rubber with fibres combines the elastic behaviour of the rubber along with the strength and stiffness of the reinforcing phase. Various natural fibres like coir, bamboo and fibres from oil palm can be used as reinforcing materials. Due to their benefits such as biodegradability, low cost and abundance, these bio-fillers have a wide range of uses over synthetic fillers like carbon, glass, aramid, etc. [6]. Even though biocomposites offer many benefits, they also have certain drawbacks, including moisture absorption traits, low heat resistance and poor dimensional stability. It is difficult to combine the filler and rubber matrix uniformly since they have distinct structural qualities. The composite's boundary experiences stress due to the fillers' uneven distribution. It is possible to modify the filler or matrix with any functional groups to lower the stress in composites. This could improve the filler and matrix's compatibility and result in significant changes to the composites' properties. Several materials can be selected for making rubber nanocomposites, including cellulose, starch, chitosan, chitin, etc. Chitin is one of the easily available and low-cost materials.

4.1.1 Chitin

Chitin is the second most important natural polymer in the world. It is poly (β-($1 \rightarrow 4$)-*N*-acetyl-D-glucosamine), a natural polysaccharide of major importance, first identified in 1884. It is the most significant polymer synthesized annually in the world after cellulose [7]. It acts as the vital structural component of the exoskeleton of arthropods and also the cell wall of yeast and fungi. It is also produced by many other living organisms in the lower plant and animal kingdoms, serving many functions where reinforcement and strength are required [8]. The primary commercial sources of chitin extraction are crab and shrimp shells (Fig. 4.1). In industrial processing, calcium carbonate present in the shells is first dissolved by acid treatment to extract chitin from crustaceans, which is then followed by an alkaline extraction to solubilize proteins [9]. Decolourization is frequently included to get rid of any remaining pigments and produce a colourless result. Because of variations in the ultrastructure of the starting materials, these procedures must be customized for each chitin source. The resulting chitin must be evaluated for purity and colour since any remaining protein or pigment may be problematic for further uses, particularly for biomedical items [10]. The most significant chitin derivative in terms of uses is chitosan, which is produced via partial deacetylation under alkaline circumstances [11].

Chitin and its derivatives have been employed as natural flocculants for the treatment of wastewater. The insoluble nature of the chitin caused by its crystalline structure restricts its wide application in different fields. Yet, there is a trend towards creating high-value goods for the pharmaceutical, biotechnology, cosmetic and medical industries [12]. In particular, the benefits of biodegradability, biocompatibility and nontoxicity of chitin are advantageous. The high degree of crystallinity and strong hydrogen bonds between chitin chains render chitin insoluble in water

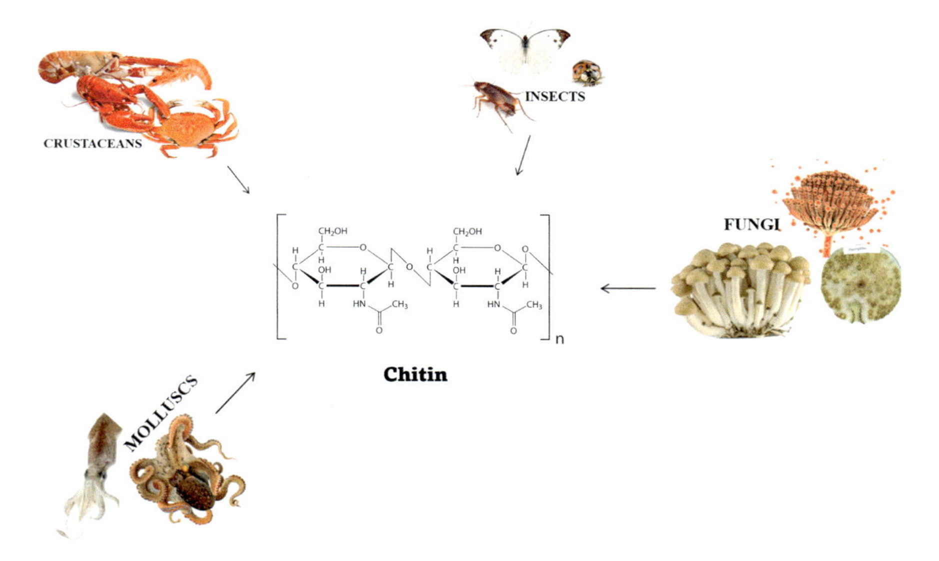

Fig. 4.1 Chitin obtained from different sources

and in many organic solvents; despite this, chitin has found applications in many fields, including textiles, paper making, medicine and wastewater treatment [13].

4.1.1.1 Chemical Structure of Chitin

Due to the presence of bonds between the carbon atoms C-1 and C-4, it has been referred to as a stiff polysaccharide. Chitin's crystal structure was determined by the arrangement of polymer chains, particularly where the 2-N, N′-diacetyl chitobiose units were located in the structure [14]. The chemical structure of chitin is provided in Fig. 4.2. Chitin can be found in different allomorphs depending on its source, such as α and β-chitin forms, can be distinguished using X-ray diffraction, infrared and solid-state NMR spectroscopy [15]. α-chitin refers to when the chains lie antiparallel to one another, β-chitin refers to chitin with parallel chains and γ-chitin is a combination of α- and β-chitin, with some chains parallel and some chains antiparallel to one another[16, 17]. A third allomorph of γ-chitin has also been reported; however, a careful examination suggests that it is simply a different member of the family [14]. The different forms of chitin are provided in Fig. 4.3.

4.1.1.2 Extraction of Chitin

Crustacean shells are mainly made up of three components: chitin, protein and calcium carbonate. The extraction of chitin needs separation of chitin from other

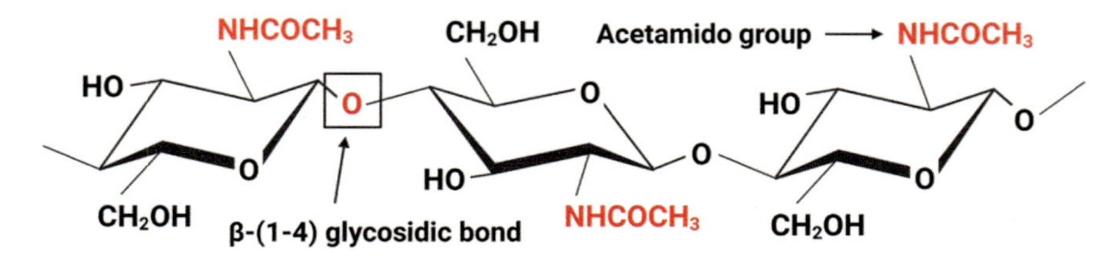

Fig. 4.2 Chemical structure of chitin

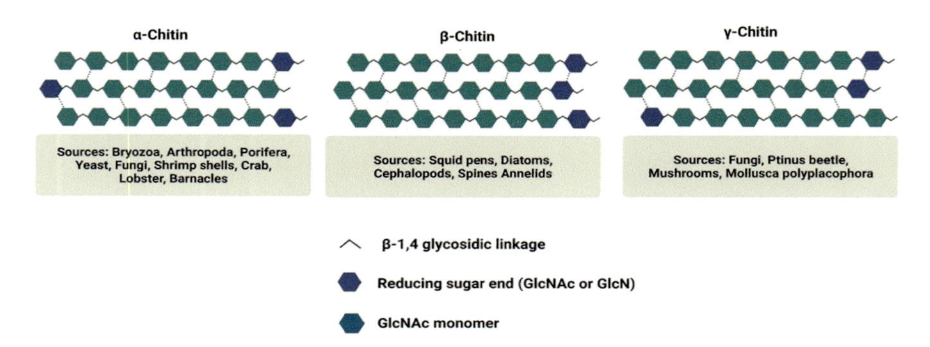

Fig. 4.3 Different allomorphs of chitin [16]

areas of the shell. The extraction methods can be classified mainly as chemical and biological methods [18].

Chemical Extraction of Chitin

Demineralization and deproteinization are typically the initial two processes in the chemical extraction of chitin. Demineralization is frequently done before deproteinization to improve the deproteinization surface area [19]. This is known as the acid–alkali method of chitin extraction because demineralization is often accomplished by subjecting crustacean shells to acidic conditions and deproteinization is accomplished by subjecting the shell to alkaline conditions. Due to the low cost of chemicals and lack of specialized equipment needed for reactions, the acid–alkali method is the most frequently used technique in commercial extraction. The chemical extraction of chitin is provided in Fig. 4.4. Research on alternatives has been sparked by worries about the leakage of hazardous acid and alkaline waste into the environment [20].

a. Demineralization.

Crustacean shells are treated with acid especially hydrochloric acid, to dissolve the minerals in the shells. The entire process was conducted at room temperature for

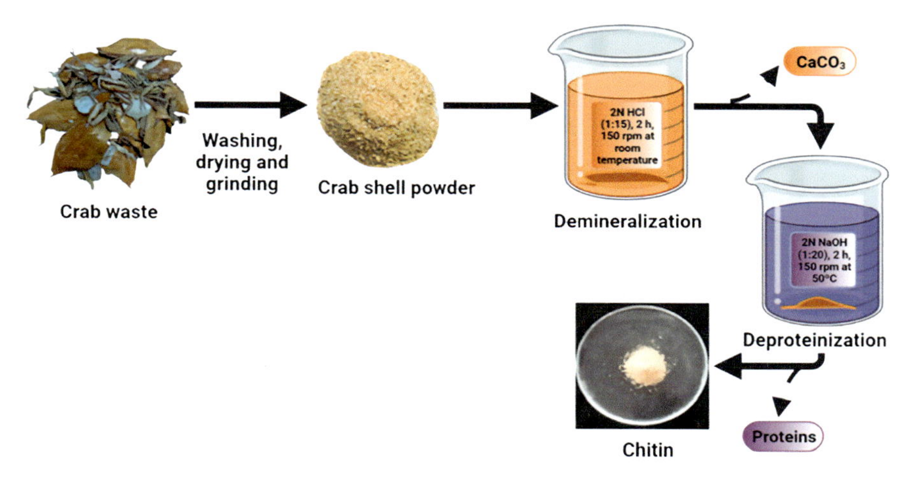

Fig. 4.4 Chemical extraction of chitin by demineralization and deproteinization [16]

a period of 2–3 h. The main shell mineral is calcium carbonate. Adding several acid treatments to the demineralization process has also been modified. Citric acid demineralization was carried out in two steps [21]. Despite calcium citrate's limited solubility, they were able to demineralize the shell to a degree that was comparable to a standard one-step HCl demineralization. By pre-treating whole shrimp shells with diluted hydrochloric acid for 12 and 24 h, divided the demineralization process into two stages as well. This method was not only designed to store shells without an odour but also decreased the ash content of the shells from around 22% to as low as 0.7% before the formal demineralization procedure. The subsequent HCl demineralization further decreased the ash content of the shells [12].

b. Deproteinization.

Chitin is separated from the other organic shell components through a process called chemical deproteinization. In order to make the proteins water-soluble and separate them from the chitin, the most popular technique is to employ strong bases and high temperatures. Due to its economic viability, sodium hydroxide (NaOH) is the most often used base for deproteinization [22]. Although some studies conduct reactions at room temperature and others at higher temperatures of 100°C, reaction temperatures are commonly between 65°C and 100°C. Researchers have turned to other bases with less harmful cations (such as potassium hydroxide) or other deproteinization techniques out of concern that high sodium ion concentrations will pollute the environment. *Yang* et al. substituted reactions with strong bases with hot water in a pressure vessel to achieve the same results as pressure cookers' capacity to soften meat proteins [23]. Although the chitin produced using this method had a lower molecular weight than chitin produced using a standard acid/alkali method, its morphology and purity were comparable.

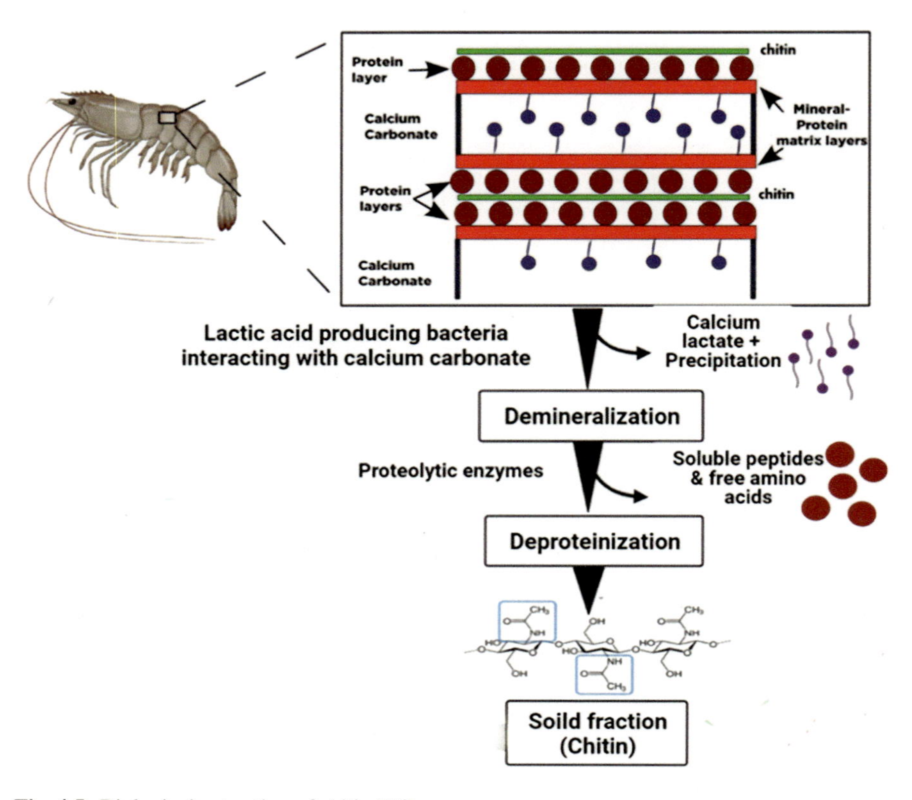

Fig. 4.5 Biological extraction of chitin [16]

Biological Extraction of Chitin

The physicochemical characteristics of chitin could be impacted by the chemical extraction process. Consequently, several research groups have been studying the development of viable biobased ways to overcome this problem and to manufacture value-added products from waste materials. The use of microorganisms in the processing of chitinous waste has attracted considerable interest from the marine industries [24, 25]. It was discovered that using proteolytic fermentation to carry out the deproteinization process was a viable strategy. The methods for chitin biological extraction or microbial fermentation are provided in Fig. 4.5. For instance, *Paenibacillus mucilaginosus* was employed to produce protease from shrimp heads. *Serratia marcescens* and *L. plantarum* were among the species that boosted deproteinization and demineralization activities. Moreover, it has been reported that large yields of chitin, up to 21.35%, were obtained through microbial fermentation. By fermenting crab solid waste with lactic acid bacteria (*L. plantarum*), high-quality chitin was recovered.

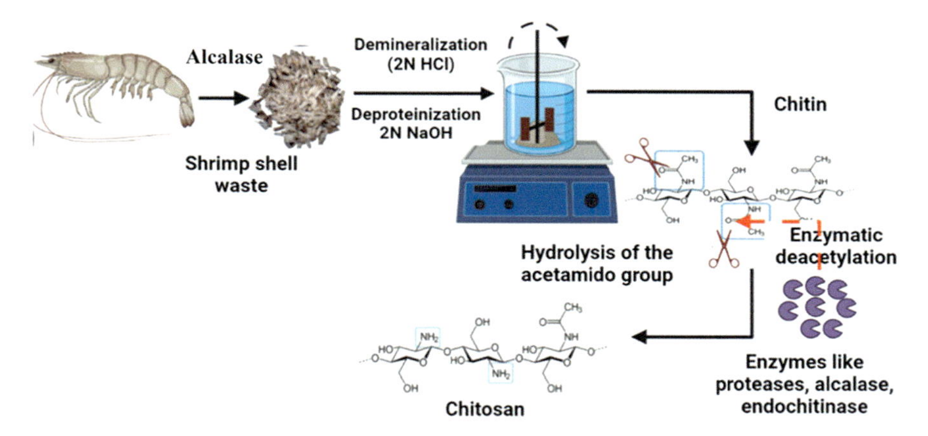

Fig. 4.6 Enzyme-mediated extraction of chitin to produce chitosan [16]

Enzyme-Mediated Extraction

Green extraction methods like enzymatic-mediated extraction are currently gaining much popularity due to their eco-friendly nature [26]. The functionalization of natural polymers using enzymes has been continuously researched as an alternative against harmful and environmental risks. These extractions are more focused, quick and use less water, energy and/or chemicals than conventional methods. It enables the recovery of high-value-added goods, such as chitin, pigments or peptides. Even though chemical treatments produce high-quality chitin, they may result in the depolymerization and deacetylation of polysaccharides. This uneven molecular weight (MW) reduction and deacetylation brought on by the chemical treatment are diminished by the employment of enzymes.

To separate chitin from shell waste, the enzyme-mediated extraction uses a stepwise procedure in conjunction with a mild alkali treatment. Alcalase, a proteolytic enzyme, was used to deproteinize shrimp shell waste, which was then demineralized (10% HCl solution) at 55 °C and a pH of 8.5. At 55 °C, 8.5 pH, protein hydrolysis was carried out on a suspension of the demineralized shells in water (1:2, w/v). Pure chitin was produced as a result of this procedure, and there were fewer amounts of unwanted peptides and amino acids connected to chitin molecules. The degree of hydrolysis (DH) was around 4.4% and a residual protein of 30%. The enzyme-mediated extraction of chitin is provided in Fig. 4.6.

Microwave-Assisted Extraction

Many uses of microwave technology are found in the chemical and food industries. According to numerous research, microwave heating has a greater potential than traditional heating to accelerate chemical reactions, boost reaction yield and

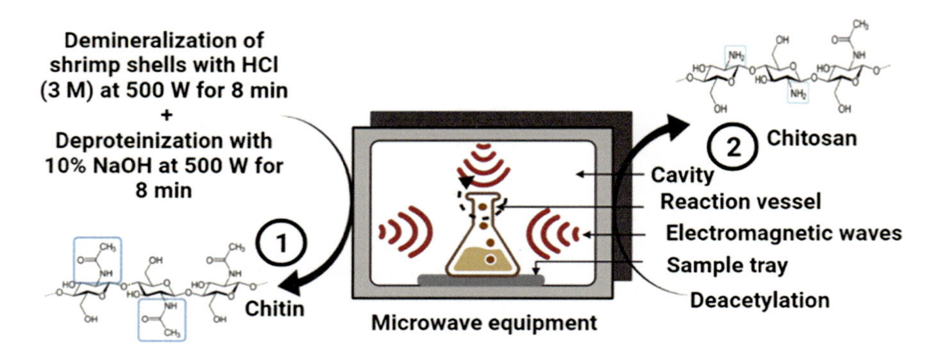

Fig. 4.7 Microwave-assisted preparation of chitin [16]

improve the quality and characteristics of the final product. Here, demineralization of shrimp shells was done by treating them with acid at 500 W for 8 min followed by deproteinization at 500 W for 8 min with 10% NaOH. Microwave-assisted extraction of chitin is given in Fig. 4.7.

4.1.2 Rubber Composites

The rubber can be utilized as a matrix material in composite applications, where it is reinforced with different reinforcing agents. Bio-fillers are the most desirable candidates among several other reinforcing agents which is employed with rubber for enhancing its thermo-mechanical and barrier qualities. Incorporation of bio-fillers over synthetic nanofillers such as carbon black (CB), synthetic silica, graphene, carbon nanotube (CNT) and carbon nanofiber offers many advantages like renewability, abundance, low cost, low density, sustainability and reduced manufacturing, etc. In contrast to synthetic fillers, the major difficulty encountered while using agro-based fillers is their inferior reinforcement characteristics. Adding the bio-fillers results in lower physical and mechanical properties with special emphasis on tensile strength and modulus values. However, the addition of derivatives of chitin, where the positively charged particles in the chitin nanocrystals (ChNCs) are highly crystalline in nature imparting good reinforcing properties to the matrix.

4.1.2.1 Method of Preparation of Chitin/Rubber Nanocomposites

a. Casting method

Aqueous suspension of chitin is combined with natural rubber latex solution or dispersion to create a homogenous dispersion. A chitin-containing nanocomposite is created by evaporating the dispersion after it has been cast into a container. Figure 4.8 represents the images of films obtained after the casting method.

b. Hot pressing and freeze-drying method

This method involves freeze-drying well-distributed aqueous mixes of rubber latex and chitin to produce nanocomposite powders, which are then transformed into specimens by hot pressing. *Gopalan* et al. [27] extracted chitin whiskers from crab shells and reinforced with natural rubber. The limb of the crab (red crab) used in this study is the main source for the production of chitin because it is relatively rich in chitin. The suspensions of chitin whiskers and NR latex were combined in different proportions utilizing freeze-drying, heat pressing and water evaporation techniques. By using a water evaporation process, both unvulcanized and pre-vulcanized NR were employed to create the composites, but when using a hot press, only unvulcanized

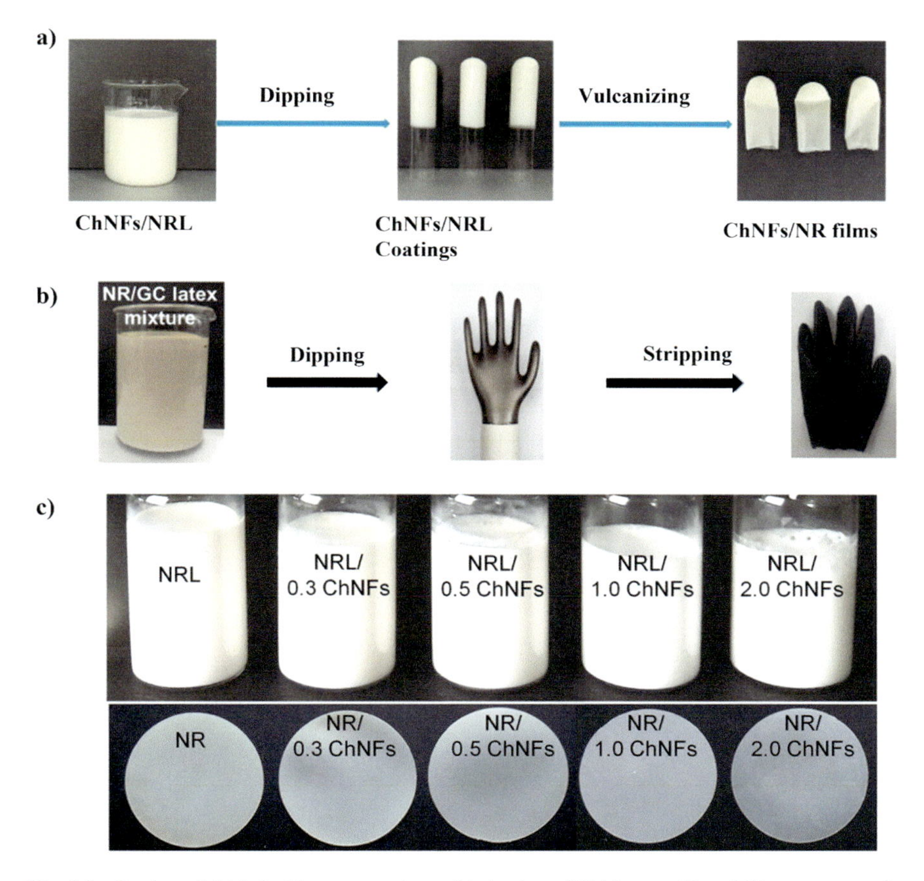

Fig. 4.8 Casting of chitin/ rubber composites **a** fabrication of Chitin nanofibers/NR nanocomposite film, **b** natural rubber latex medical gloves, and **c** NR and NR/Chitin nanofibers (ChNFs) bio-nanocomposites with ChNFs loadings from 0.3 to 2.0 wt% [28–30]

NR was utilized. 2, 5, 10, 15 and 20 weight percent of the chitin was utilized in the colloidal solution. The hot pressing was conducted at 138 bars for 2 min at 100 °C.

c. Dry mixing method

As per the mixing formulation, the natural rubber was ground and combined with the additives using a two-roll mill and the standard procedure outlined in ASTM D 3184. The ingredients are added after the rubber is masticated on the mill, and the process is repeated until the mixture is homogeneous. It is necessary to modify several variables, including the mill roll, nip gap, speed ratio, mixing duration and order of component addition. The chitin is often added towards the conclusion of the mixing process, being careful to keep the compound flowing in the same direction such that the bulk of the chitin mostly follows the flow.

4.2 Characterization Techniques

4.2.1 Tensile Testing

Stress–strain curves mainly explain the mechanical properties of the composites. *Midhun* et al. [31] suggested the acid hydrolysis approach to prepare chitin nanowhiskers (CHNW) from leftover shrimp shells. Shrimp shells were processed using the processes of demineralization, bleaching and acid hydrolysis, then homogenized to create chitin nanowhiskers (CHNW). Using a masterbatch of chitin nanowhisker (CHNW) in NBR latex and dry rubber compounding in two roll mills, chitin nanowhisker (CHNW) reinforced NBR composites were created. Compared to neat NBR, NBR-CHNW2 (2 phr chitin nanowhiskers) composites' tensile strength and tear strength improved by 116% and 54%, respectively. The 'caged' or 'trapped' rubber chains in the three-dimensional rigid hydrogen-bonded network of chitin nanowhiskers are responsible for the improvement in the tensile modulus and cross-link density of NBR-CHNW composites. Surface-modified chitin is introduced (Carbonylated chitin and Zwitterionic chitin) and incorporated in rubber latex further. Based on this, the possible interaction between chitin and rubber (NBR) is depicted below in Fig. 4.9.

The NRL with negatively charged protein and phospholipid layers could only be stable in the basic condition, while chitin nanofibers (ChNFs) with amino groups were only stable in the acid condition. Hence, surface modification of ChNFs with amino groups was necessary to achieve good dispersion of ChNFs in the NRL matrix ensuring enough mechanical reinforcement to the composites [32, 33]. It should be noted that the traditional TEMPO oxidation approach has also been successful in producing carbonylated ChNFs (C-ChNFs), which were water-dispersible in the basic solution. It is anticipated that C-ChNFs with negatively charged COO^- might be diffused steadily in alkaline NRL without generating latex flocculation because

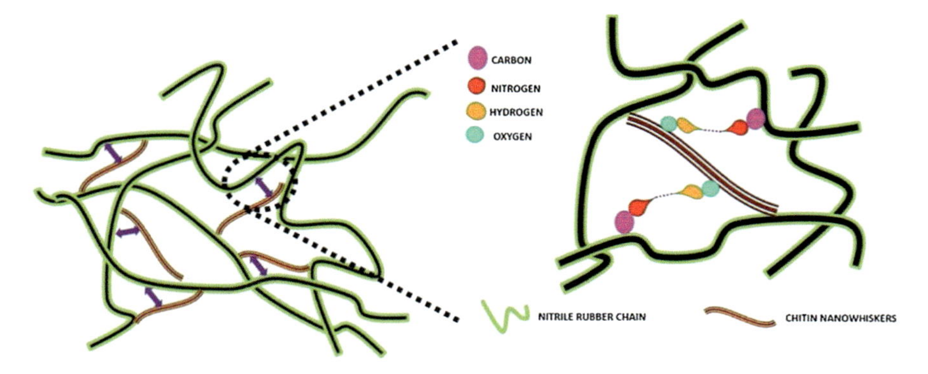

Fig. 4.9 Schematic representation of the interaction between chitin nanowhiskers and nitrile rubber latex [31]

they are stable under basic conditions. The zwitterionic ChNFs (NC-ChNFs) demonstrated stability in a range of pH conditions. By using a dip moulding procedure, NRL was combined with C-ChNFs and NC-ChNFs, respectively. Even though both C-ChNFs and NC-ChNFs are compatible with the NRL suspension in its most fundamental form, the results showed that their distribution in the NRL matrix was completely different. In other words, although NC-ChNFs particles were scattered in the NRL matrix (NC-ChNFs/NR) as aggregated clusters, C-ChNFs nanoparticles were uniformly monodispersed (C- ChNFs/NR). Furthermore, it has been demonstrated that in an aqueous medium with varying pH settings, the mechanical characteristics of C-ChNFs/NR and NC-ChNFs/NR with high-content nanofiller have varied stabilities. With the addition of nanofibers, C- ChNFs/NR nanocomposite film exhibited remarkable increases in stresses at 100% and 300% strain. When the filler content is increased to 3.5 phr, the stress at 100% strain or 300% strain of the C-ChNFs/NR nanocomposites is four times that of NR. The stress at 100% strain or 300% strain of NC-ChNFs/NR nanocomposites, in contrast to those of C-ChNFs/NR nanocomposites, increases slowly with the addition of more nanofibers when the filler content is less than 3.5 phr. In most cases, the formation of a filler network in nanocomposites would result in an improvement in their mechanical characteristics [28]. The stress–strain graphs are given in Fig. 4.10.

The nanocomposites were also prepared using NR latex and regenerated chitin (R-chitin) [34]. The stress–strain curves of NR/ regenerated chitin composites exhibited improved tensile strength and modulus with the addition of regenerated chitin. The rigidity of the composites tends to improve with the addition of R-chitin. For example, when 30 phr R-chitin was added, Young's modulus, stress at 100% strain and tensile strength increased from 0.64, 0.52 and 0.89 for pristine NR to 2.17, 4.94 and 5.85 MPa, respectively. This significant increase in the material's rigidity may be ascribed to.

(i) The addition of rigid chitin to the soft NR matrix. This will undoubtedly increase the composite's overall strength and modulus

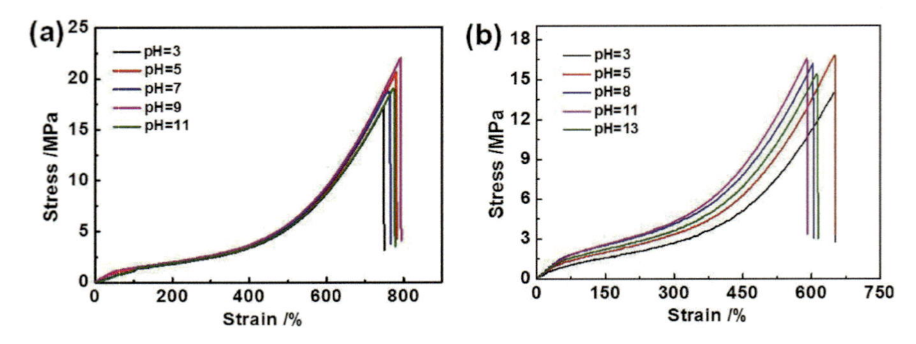

Fig. 4.10 Typical stress–strain curves of (**a**) 3.5 phr NC-ChNFs/NR nanocomposite films and (**b**) 3.5 phr C-ChNFs/NR nanocomposite films treated in aqueous medium, with various pH values [28]

(ii) The trapped rubber would be at least partially 'dead', acting like a filler when the R-chitin content was increased
(iii) The special porous structure of R-chitin could induce strong physical entanglements between both components, with the result that stress would be transferred from the NR to the R-chitin more effectively.

After the tensile testing, the composites' transparency was lost as a result of stress whitening. The transparent nature of the composites corroborated the filler's homogeneous distribution in the NR matrix. Chitin gives the rubber a more compact structure and restricts solvent penetration due to its inclusion. The results of the tensile tests unequivocally demonstrate that R-chitin has a strong reinforcing impact on the elastomer. The observed reinforcing and morphological properties of the composites were caused by the R-chitin's dispersion and distinctive porous structure.

Kawano et al. prepared porous and self-assembling chitin nanofiber (CNF) and further reinforced in natural rubber (CNF-NR) to form composite sheets. The nanofiller formed a network structure over the natural rubber (NR) particles as a result of self-assembling them. The addition of the chitin enhanced the tensile strength, modulus and tear strength. At 4 weight percent filler content, the highest tensile strength, elongation at break and tear strength were noted. The reinforcing effect of CNFs in the sheets was verified by tensile testing [35]. *Hu* et al. investigated the interactions that existed between composites made with deproteinized natural rubber (DNR) and chitin nanocrystals (ChNCs). ChNC concentrations ranged from 1 to 6 weight percent. ChNCs' stiff 3D network skeleton may self-assemble to strengthen DNR latex films. The composites' DNR-4 (deproteinized natural rubber-4 phr chitin nanocrystals) ideal tensile and tear strengths were 25.00 MPa and 42.42 KN/m, respectively, which were 211.3% and 145.9% and greater than those of pure DNR film, respectively. In comparison to pure DNR film, the elongation at break of DNR/ChNCs films (DNR-4) was increased by 15% [36].

Yin et al. prepared high-content ChNFs/NR nanocomposites for reversible plasticity shape memory polymers [37]. The mechanical characteristics of NR could be significantly impacted by the addition of nanofillers. ChNFs can be added up to a maximum filler loading of 30 weight percent, giving ChNFs/NR composite films a special mechanical performance. Tensile experiments are used to examine the mechanical characteristics of ChNFs/NR composite films with varying ChNFs contents. The usual stress–strain curves are depicted in Fig. 4.11 below. As the ChNFs content increases from 2 wt% to 5 wt%, the modulus and stress at 100% strain of ChNFs/NR composite films exhibited marginal improvement than those of neat NR. This phenomenon is brought on by the nanofiller's reinforcing role in NR cross-linking via ChNFs. The rod-shaped chitin nanocrystals (ChNCs) were extracted [38] and added to the rubber matrix at concentrations of 1, 2.5, 5, 7.5 and 10%. The incorporation of ChNCs greatly boosts the tensile strength of NR/ChNCs composites. For instance, the composite containing 10 wt% ChNCs has a tensile strength of 5.75 MPa, which is 6.25 times more than pure NR. Both the tensile strength and the elongation at break are increased simultaneously when the amount of ChNCs in the composites is less than 1 wt%. The tensile strength of the NR/ChNCs composite keeps growing while the elongation at break starts to decline with increasing ChNCs loading over 2.5 wt%.

Steam explosion and mild oxalic acid hydrolysis were used to create the chitin nanofibers CHNFs (12–30 nm) from chitin powder [39]. To create NR/CB/CHNF

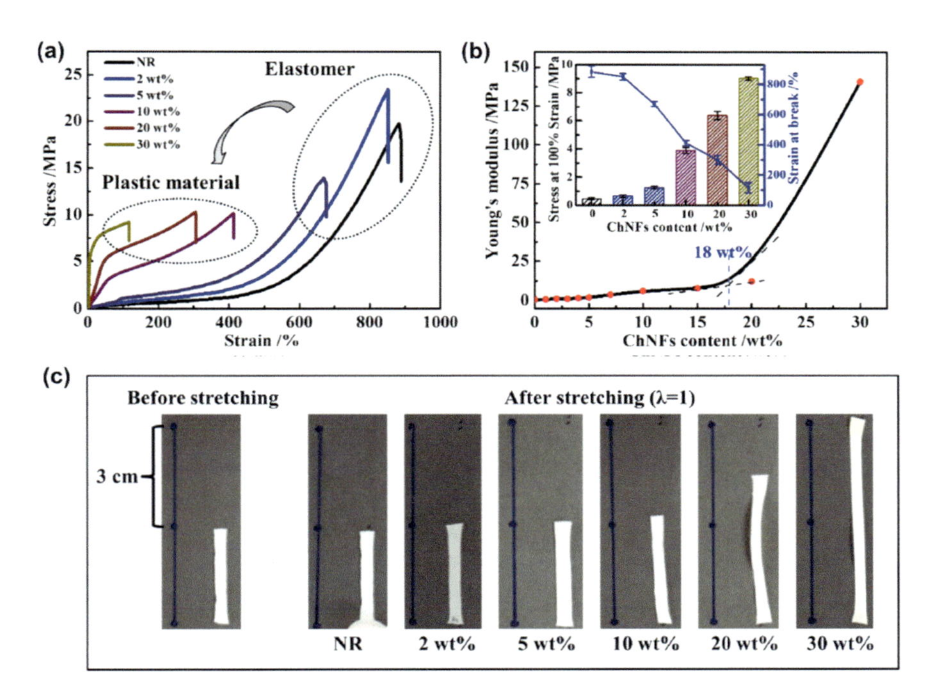

Fig. 4.11 **a** Stress–strain curves of ChNFs/NR composites with different content of ChNFs. **b** The effect of ChNFs percentage on the tensile properties **c** Digital pictures showing the unstretched and stretched ChNFs/NR composites with a stretching ratio (λ) of 1 [37]

composites, the CHNFs were uniformly dissolved in natural rubber (NR) latex, dried and combined with carbon black (CB) in a two-roll mill. The addition of CHNFs to CB resulted in a notable improvement in the tensile strength. The creation of 3D network structures as a result of the interaction of CHNFs and CB with the NR matrix may be the cause of the enhanced tensile strength of NR/ CB/CHNF composites. With the –NH and –OH functional groups of CHNFs, it can create intermolecular hydrogen bonds. Strong 3D network architectures may result from these interactions inside the polymer.

The high-performance elastomers in the natural rubber (NR) sector without sulphur vulcanization were done by *Liu* et al. [29]. To strengthen unvulcanized NR, graphene oxide (GO)/zwitterionic chitin nanocrystals (NC) hybrids (GC) are first added to NR latex. The NR/GC nanocomposite has a huge breaking elongation of 825.9% and a high tensile strength of 19.2 MPa. Due to its strong attraction to both GO sheets and NR particles, NC with amino and carboxyl groups can serve as a macromolecular bridge and improve the interfacial interaction between them. In addition, the hybrid-synergetic effect after adding NC greatly enhances the dispersion of GO in NR. Consequently, the produced uniform hybrid nanofiller networks actively absorbed on NR macromolecules can greatly increase the mechanical characteristics of nanocomposites. Hence, natural latex medical gloves with high water vapour permeability, strong biocompatibility and good reprocessing ability are made using NR/GC films.

4.2.2 Differential Scanning Calorimetry (DSC)

The glass transition temperature (Tg) of a chitin-reinforced carboxylated styrene-butadiene rubber matrix increases with increasing the content of chitin. A smaller increase is also seen as a result of annealing, which was explained by the matrix stiffening. The calorimetric data indicate that even in the presence of chitin nanoparticles, the mobility of the polymer chains is marginally modified by the annealing at 100 °C maintained for 90 min [40]. The glass transition temperature of epoxidized natural rubber rises with the addition of chitin nanocrystals (CNCs), showing that the CNC particles restrict the mobility of the ENR chains in the composites. The T_g of the pristine ENR is −31.2 °C, but as the CNCs filler concentration reaches 10 wt% and 20 wt% in the composite, an increase in Tg was observed with values—-30.9 and −30.7 °C. The creation of hydrogen bonds improved the connections between polymeric chains in supramolecular networks, which is primarily responsible for the increased thermal stability of ENR/CNC composites. Additionally, a portion of the CNCs were grafted onto the ENR molecule, creating a potent interfacial connection between the two molecules improving the thermal stability [41]. The DSC of chitin-incorporated ENR is given in Fig. 4.12.

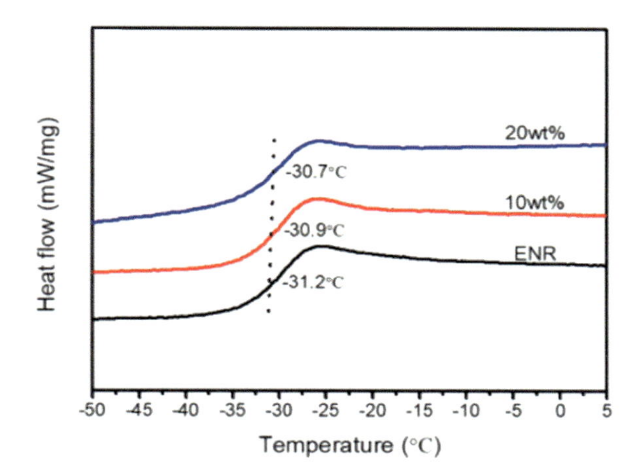

Fig. 4.12 DSC heating curves of ENR/CNCs composites with various CNC contents [41]

4.2.3 Dynamic Mechanical Analysis (DMA)

The viscoelastic response of the polymer composites is evaluated using dynamic mechanical analysis under various conditions like temperature sweep, strain sweep, frequency sweep, etc. Several important parameters like glass transition temperature (T_g) and loss factor which is the ratio of loss modulus to storage modulus commonly termed tan δ are measured using this technique by applying cyclic stress to the polymer material at different conditions. The incorporation of chitin nanofibers along with the carbon black fillers in the natural rubber enhances the crystallinity of the composites resulting in increased storage moduli due to the presence of physical crosslinks and percolation networks formed by these chitin nanofibers. The T_g of the chitin rubber composites exhibited a value of -44.5 ˚C compared to the neat NR elucidating a temperature rise of 7.5 ˚C. This is ascribed to the restricted mobility of the composites which facilitates improved filler–polymer interaction.

The storage modulus at 40 °C for (Acrylonitrile-butadiene rubber) NBR-CHNW2 (2phr chitin nanowhiskers) composite was found to be 1473 MPa which is greater than NBR gum (1189 MPa), in the glassy zone. Even below the glass transition temperature, the generated NBR nanocomposites' improved storage modulus demonstrated how well CHNW reinforced NBR. The storage modulus was found to be practically constant and a plateau was established at high temperatures (T > glass transition temperature) (i.e. the rubbery region). The addition of 2 phr CHNW to the NBR matrix resulted in a 68% improvement in the rubbery modulus. The material's degree of crystallinity has a significant impact on the rubbery modulus. The NBR chains are physically connected by the crystalline domains of chitin nanowhiskers [31]. The addition of modest amounts of chitin micro-sized particles, generated from shellfish waste, to a carboxylated styrene-butadiene rubber matrix has been examined [40]. The polymeric matrix's increased filler content has the effect of raising the G' modulus (storage modulus) throughout the entire temperature range. The tan delta

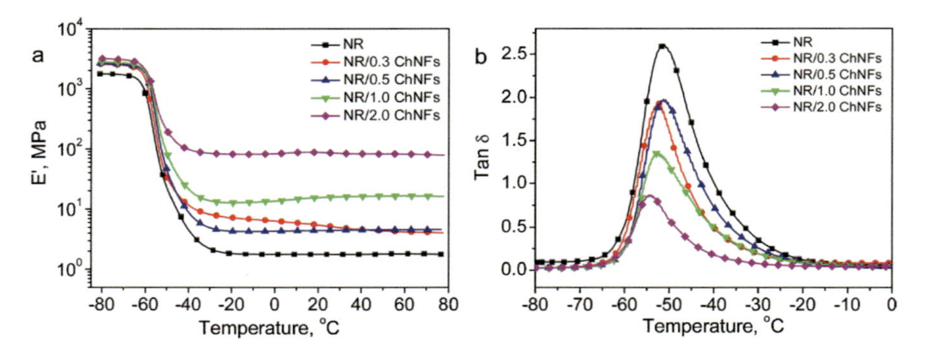

Fig. 4.13 DMTA temperature sweeps that show (**a**) tensile storage moduli E' and (**b**) loss tangents tanδ of NR and NR/ChNFs bio-nanocomposites with ChNFs loadings from 0.3 to 2.0 wt% [30]

values are nearly identical for the various systems. Even with the introduction of the smallest amount of chitin, the amplitude of the tan delta peak in terms of ΔT at half of the maximum value, denoted as $T_{0.5}$, somewhat increases. In as much as the stiffening that results from the annealing procedure only slightly increases the G' values across the entire temperature range, it has a minimal impact on the loss modulus profile.

In the case of XSBR-CNC, the peak value of the tan delta shows a diminishing characteristic on increasing the chitin nanocrystals (CNCs) due to the restriction imparted by the crystals on the flexibility of the polymer chain leading to lower mobility. This is due to the formation of the rigid network in the XSBR matrix. In the rubbery region, more enhanced values of storage modulus were observed with CNC loading with effective stress transfer to the matrix media [43].

Ding et.al studied the dynamic mechanical analysis of the NR and chitin nanofiber incorporated (ChNF) natural rubber samples at different ChNFs loadings. A strong reinforcing effect of ChNFs with the polymer matrix was observed with the tensile modulus of the samples indicating strong filler–polymer interactions. Further, the Tg of the composites showed a decreasing trend on increasing the concentration of ChNF from 0.3 to 2% due to the mechanical coupling effect, it is given in Fig. 4.13 [30].

4.2.4 Thermogravimetric Analysis (TGA)

The thermal degradation behaviour pointing to the thermal stability of the composites is evaluated using the TGA. The incorporation of regenerated chitin (R-chitin) in the latex medium decreases the thermal stability of NR [34]. As the R-chitin content increased, a decreasing trend towards thermal stability was observed. There is only one main mass loss step in the thermal degradation behaviour of all NR/R-chitin composites, and no distinct degradation stage is readily apparent in any

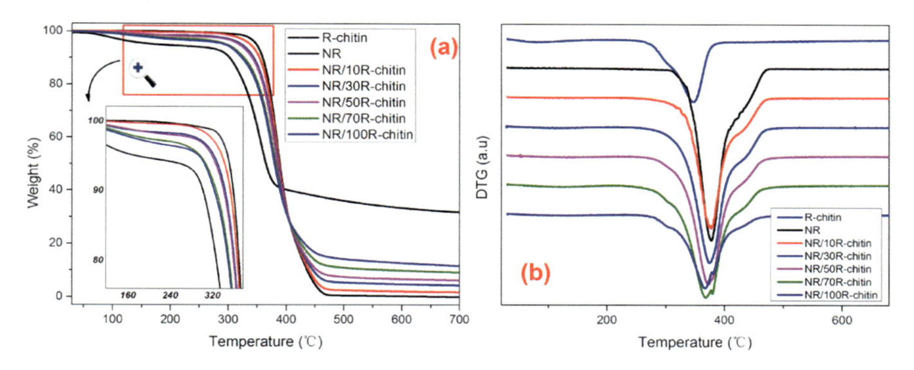

Fig. 4.14 **a** Thermogravimetric analysis and **b** Differential thermal analysis of NR, R-chitin and NR/R-chitin composites [34]

DTG curve. When the R-chitin content is increased, a consistent shift in the degradation temperature to lower values is shown in NR/R-chitin composites which is provided in Fig. 4.14. Similarly, chitin nanowhiskers were added by *Dominic* et al. to acrylonitrile-butadiene rubber. Using thermogravimetric analysis, it was discovered that all of the manufactured composites degraded in a single stage. Chain scission and cross-link breakage were the primary causes of degradation in all of the samples between 250 and 400 °C. NBR gum, NBR-CHNW2 (2 phr chitin nanowhiskers), NBR-CHNW4 and NBR-CHNW6 all had onset degradation temperatures (T_{on}) of 300 °C, 318 °C, 354 °C and 374 °C, respectively. T_{on} rose to 74 °C when 6 phr CHNW was added in comparison to neat NBR. This demonstrates how well CHNW works to delay the heat breakdown of NBR. Char and volatiles are produced during the breakdown of CHNW. These leftovers serve as a barrier that prevents mass and energy from the burning surface from reaching the connected polymeric chains. The diffusion of breakdown products from the material is slowed down by the NBR phase's reduced mobility near CHNW [31].

In another study, it is found that TGA and DTG analysis show that the inclusion of chitin does not significantly alter the thermal stability of the styrene-butadiene rubber (XSBR) matrix [40]; in other words, the maximum weight loss rate is not greatly impacted by the annealing process. While the peak caused by matrix degradation essentially stays the same, the existence of the chitin degradation peak is seen in the DTG profile as a shoulder of the main peak at 360–375°C. The system's thermal stability is maintained, according to these data, even with varying amounts of naturally occurring filler that exhibits thermally unstable behaviour. This can be viewed as a successful result because it elucidates that the addition of the naturally occurring filler preserves the thermal stability of the XSBR rubber at these quantities.

The addition of chitin nanofibers (CHNFs) in NR caused the onset degradation temperature (T_{on}) values to slightly increase. This was ascribed to the well-distributed dispersion of the CHNFs in the NR matrix, which allowed them to resist heat degradation. By acting as a barrier, the distributed CHNFs prevented the transit of volatile materials, and the matrix's effective heat-absorbing ability allowed it to postpone thermal degradation. Higher onset degradation values for the elaborated natural rubber/ carbon black/ chitin nanofiber (NR/CB/CHNF) composites result from the CHNFs and CB immobilizing the NR macromolecular chains. T_{50} is the temperature at which half of the sample begins to deteriorate. The prepared composites' T_{50} values (thermal stability) are unaffected by the addition of CHNFs with CB [39].

In chitin nanocrystals reinforced epoxidized natural rubber (ENR), the expanded supramolecular network improved the thermal stability of ENR/CNC composites. Due to the lack of a crosslink network, neat ENR exhibits a maximum weight loss temperature at 389 °C, but the addition of CNCs causes the maximum weight loss temperature of ENR/CNC composites to shift upwards. For instance, the insertion of 20 weight percent CNCs raises the highest temperature at which weight may be lost to 408 °C, indicating that the addition of CNCs may significantly boost the thermal stability of ENR [41].

4.2.5 Scanning Electron Microscopy (SEM)

The surface morphology of the composite films can be easily identified by SEM images. The addition of chitin whiskers from crab shells in natural rubber can be easily identified by SEM images. The prediction of surface morphology is clear from SEM images in Fig. 4.15. Chitin whisker/NR composites' cryo cracked surface was also studied by *Nair* et al. [27]. Unvulcanized NR latex evaporated film surfaces have a homogeneous phase morphology, in contrast to freeze-dried and then hot-pressed NRL surfaces, which exhibit a non-uniform phase morphology as given in Fig. 4.15a and b. In NRL, the quenching stage before freeze-drying prevents rubber chains from positioning themselves uniformly. As a result, the molecular weight distribution becomes uneven, leading to uneven surface phase morphology in freeze-dried sample NRL as given in Fig. 4.15b. It is discovered that the surface of pre-vulcanized NR evaporated (PNR_{ev}) film in Fig. 4.15c is not as uniform as that of NR_{ev}, because it contains some spherical granular domains. This might be because the surface molecules of the individual particles present in the pre-vulcanized latex crosslink more quickly, retaining some of the original particulate structure that was created. Figure 4.15d-f represents fractured surfaces of composites with 20 wt% chitin whiskers added. In Fig. 4.15d and e (unvulcanized evaporated NCH20ev and freeze-dried & hot pressed NCH20L) the chitin whiskers appear as white dots, which are distributed evenly throughout the unvulcanized evaporated matrix. However in pre-vulcanized rubber composite (PCH20ev), the cross-linking between rubber particles will prevent the uniform distribution of chitin nanowhiskers throughout the matrix (Fig. 4.15f).

Kawano et al. [35] studied the preparation of porous materials and self-assembling chitin nanofiber (CNF)/natural rubber composite (CNF-NR) sheets. Redispersion of the self-assembled CNFs was achieved by combining the CNF with ammonia (aq.). To create CNF-NR composite sheets, the CNF dispersion with ammonia (aq.) was combined with NR latex stabilized with ammonia. This was followed by drying under

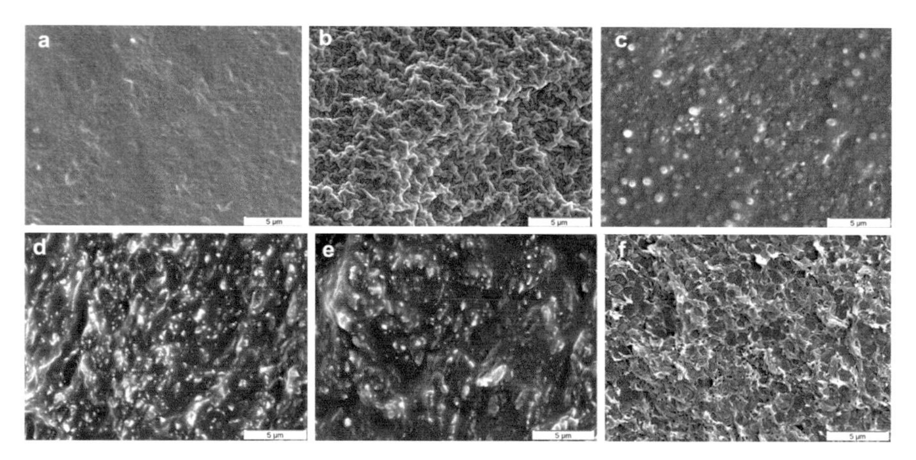

Fig. 4.15 Scanning electron micrographs of the cryo-fractured surfaces of (**a**) NR latex evaporated film (NRev), (**b**) freeze-dried and hot-pressed film (NRL), (**c**) Prevulcanized NR evaporated film (PNRev), (**d**) Composites with 20 wt% chitin (NCH20ev), (**e**) NCH20L and (**f**) PCH20ev films [27]

decreased pressure. The relatively independent nanofiber morphologies in areas of NR solids are seen in SEM pictures of the resultant sheets with various weight ratios. These findings imply that CNFs were evenly distributed throughout the NR latex. Also, the SEM pictures of the composite sheets show the nanofiber morphologies more clearly when the weight ratio of CNFs rises.

In the SEM images of deproteinized natural rubber (DNR) and chitin nanocrystals (ChNCs) composites, the pre-vulcanized DNR latex film control sample displays distinct microscale latex globules, which can be attributed to the molecules on the surface of DNR latex particles crosslinking more quickly. The cryo-fractured images also underwent some extremely intriguing alterations when ChNC variations were added. The DNR/ChNCs composite cryo-fractured images showed equally distributed wavy-like steps in comparison to the DNR latex film, and it then notably becomes dense with the further addition of ChNCs, showing that ChNCs are more evenly spread in the DNR latex matrix throughout the film process [36]. In another study, after toluene absorption of chitin-reinforced carboxylated styrene-butadiene rubber (XSBR/CW), the weak interface between the XSBR and aggregates of chitin nanowhiskers was observed in the SEM images by an exfoliation of the rubber matrix. The annealing process does not greatly exacerbate this impact, indicating once more that brief contact with temperatures of 100 °C does not dramatically alter the material's characteristics [40].

Environmentally safe and antibacterial natural rubber latex foam (NRLF) was created using chitin, a natural antibacterial ingredient, as a loading filler. Comparing NRLF composites to pure NRLF may increase antibacterial activity by up to 181.3%. The microstructure and mechanical characteristics of the chitin-NRLF composite alter somewhat when loaded with 3 phr of chitin, and the antibacterial activity more than doubled that of pure NRLF. The performance of the NRLF composite is influenced by the loading of chitin as compared to that of pure NRLF. As the chitin content rises, the cells enlarge and change shape along with it. The bubbles could burst and the rubber walls could crack as a result of overexpansion [42].

4.2.6 Transmission Electron Microscopy (TEM)

The rubber particles in the bio-nanocomposites were close to one another in the TEM pictures of chitin nanofibers (ChNF) in NR. NR phase contacts did not exhibit any obvious ChNF aggregates because of the size differences between the two phases. The NR/ChNFs bio-nanocomposites did not include any obvious chitin aggregates, but the addition of ChNFs caused the NR particles' spherical morphologies to shift to fusiform ones. The large size disparity between ChNFs and NR particles may be a result of this phenomenon. As a result, the ChNFs phase has strong interfacial interaction with the NR phase and is nanoscale in size [30]. The TEM images of the chitin nanofibres incorporated composite material is provided in Fig. 4.16.

Chitin nanocrystals (CNC) combined with epoxidized natural rubber (ENR) are prepared for self-healing purposes. In the TEM images of ENR/CNCs-10 and ENR/CNCs-20, it is obvious that the CNC, which has an average diameter of around 200 nm and is evenly dispersed throughout the ENR matrix, is a type of heteromorphic nanoscale particle. No clustering is seen and the majority of CNC particles have a single dispersion [41]. Based on the TEM images of carbonylated chitin nanofibers (C-ChNFs/NR) and zwitterionic chitin nanofibers (NC-ChNFs/NR) modified forms, *Yin* et al. made a possible mode of interaction between the chitin with natural rubber latex (NRL) as given in Fig. 4.17 [28]. The C-ChNFs were dispersed uniformly into

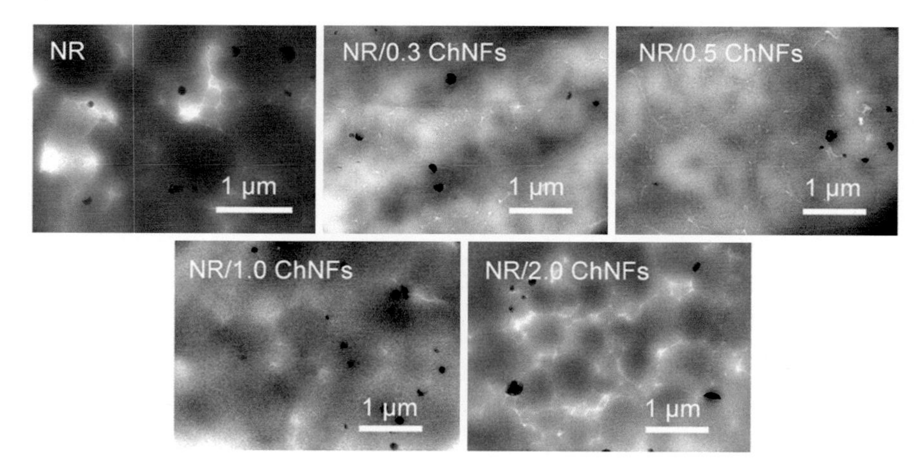

Fig. 4.16 TEM images of NR and ChNFs bio-nanocomposites with 0.3 to 2.0 wt% filler [30]

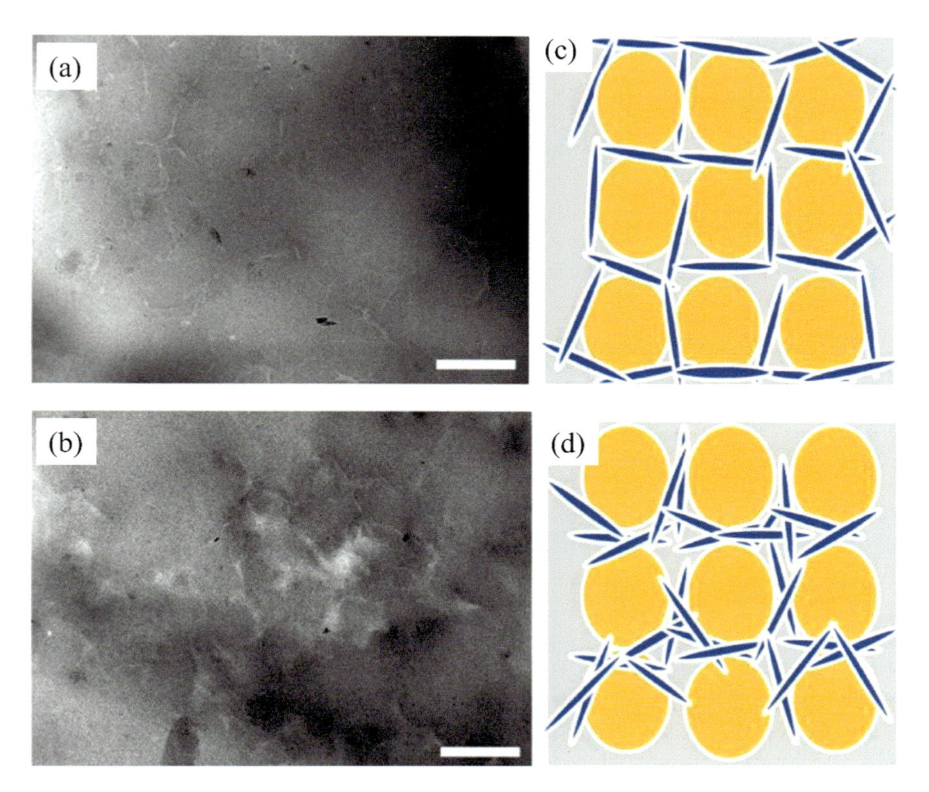

Fig. 4.17 TEM images of the ultrathin cryosection of (**a**) 3.5 phr C-ChNFs/NR and (**b**) 3.5 phr NC-ChNFs/NR (**c**) & (**d**) Proposed structural model for describing the dispersion and distribution of C-ChNFs and NC-ChNFs in their NRL nanocomposites [28]

NRL, while NC-ChNFs were dispersed into NRL in the form of small aggregated clusters, which is evident from the TEM images.

4.2.7 Atomic Force Microscopy (AFM)

AFM works well for analysing the surface morphologies of films. The self-assembled rigid chitin nanofibers' (ChNFs') reinforcing impact on NR latex was studied by *Ding* et al. The composites were prepared by adding 0.3, 0.5, 1 and 2 weight percent chitin. The inclusion of ChNFs decreased the size of the rubber particles, as evidenced by AFM [30]. The NR/ChNFs bio-nanocomposite contains spherical NR particles that are encircled by ChNFs. It is important to note that after the addition of ChNFs, the size of NR particles becomes smaller and more uniform, indicating that the growth of NR particles was somewhat restricted. Moreover, it is conceivable that ChNFs in the NR/ChNFs bio-nanocomposite will weave a percolating network.

In chitin rubber composite, various amounts of chitin nanocrystals (ChNCs) were added to pure pre-vulcanized deproteinized natural rubber latex film. The crosslinked DNR (deproteinized natural rubber) latex particles and porosity are primarily responsible for the uneven surface of the pure pre-vulcanized DNR latex film. The crosslinked rubber particles create saddle-like pits amongst themselves in the pure pre-vulcanized DNR latex film, which results from an accumulation of vulcanized rubber particles. When 2 weight percent of the ChNCs were introduced to the DNR latex, the surface morphology demonstrated a more uniform distribution [36].

4.2.8 FT-IR Spectroscopy

The FTIR experiments revealed that the chemical interactions between the polymer and filler may be a factor in increasing the elaborate composites' characteristics. The FT-IR spectra of ChNFs and NR/ChNFs bio-nanocomposites show distinctive bands for -chitin at 3355 cm^{-1} and 3270 cm^{-1}, respectively, as well as a large peak for the amide I band at 1645 cm^{-1} and amide II bands at 1556 cm^{-1}. The C–C stretching peak is at 1662 cm^{-1}, the C–H_2 deformation peaks are observed at 1447 cm^{-1} and 1375 cm^{-1} and the C–H bending peak is at 834 cm^{-1} in the spectrum of NR. The spectra of the NR/ChNFs bio-nanocomposites showed the typical peaks of both ChNFs and NR with no derivative peaks, indicating a physical and homogenous blending process for the ChNFs suspension and natural rubber latex (NRL) [30]. The ATR spectra of composite films are provided in Fig. 4.18.

Nearly no bands in the FTIR spectrum of chitin nanocrystal rubber composites are observed when the ChNCs load is very low (0.1 wt%), but as the ChNCs loading increases, the peaks of the amide I and amide II bands become more noticeable[36].

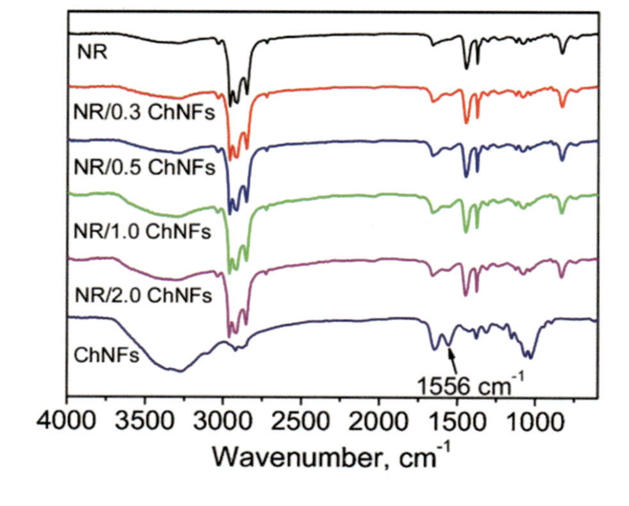

Fig. 4.18 ATR spectra of NR, ChNFs and NR/ChNFs bio-nanocomposites with ChNFs loading from 0.3 to 2.0 wt% [30]

Mathew et al. combined chitin with carbon black and used this material for green tyres. When chitin nanofibers (CHNFs) are added to rubber, all bands significantly shift upwards, indicating the development of a new bond between the (CHNFs) and the NR. The O–H stretching in the NRCBCHNF0.5 composite, which was not evident in the NRCB50, is indicated by the absorption peak at 3336 cm^{-1}. The inclusion of hydrophilic filler increased the strength and sharpness of the –OH peaks, resulting in larger intensities of surface –OH groups in the composite [39]. The presence of CHNFs in the NRCBCHNF0.5 composite is guaranteed by the absorption bands at 1582, 1424 and 886 cm^{-1} which are suggestive of amide I, amide II and amide III. In chitin nanofibers, the peaks corresponding to amide I, amide II and amide III were found at 1660, 1559 and 929 cm^{-1}. Due to interactions with carbon black and rubber, these amide peaks for the CHNF composite, however, displayed a red shift.

The self-healing capability of epoxidized natural rubber (ENR)/CNC composites was examined [41]. The distinctive peak at 870 cm^{-1} in the clean ENR is related to the oxygenous groups' stretching vibrations (C–O–C). The presence of hydroxyl groups is responsible for the large peak in the 3200–3500 cm^{-1} range. Several epoxy ring-opened products, such as –OH, C–O and others, have reportedly been generated at low pH and high temperatures during the synthesis of ENR. The peak at 870 cm^{-1} for the prepared ENR/CNC composites steadily diminishes as the CNC content rises. The continuous growth of the 1115 cm^{-1} ether bond stretching vibration peak confirms the chemical reaction between the epoxy groups on the ENR and the hydroxyl groups of the CNCs. The FTIR spectra of chitin-reinforced epoxidized natural rubber composite and possible mode of interaction are given in Fig. 4.19.

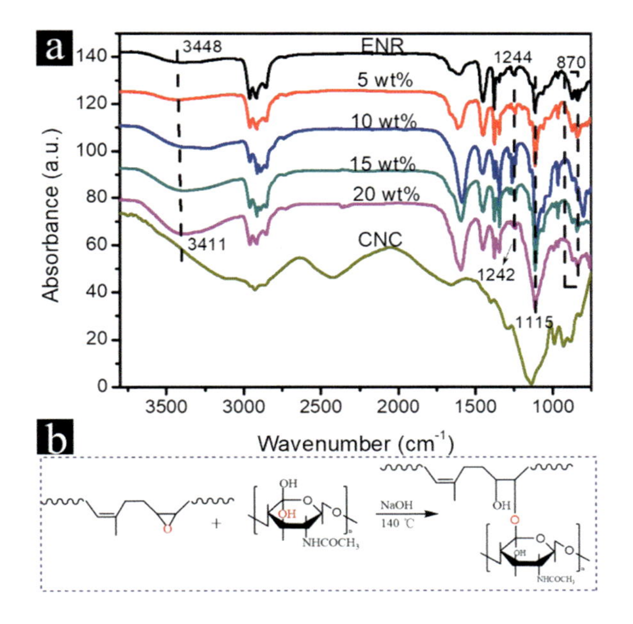

Fig. 4.19 (**a**) FTIR spectra of neat ENR, CNC and ENR/CNCs composites with various CNCs ratios; (**b**) the chemical reaction of hydroxyl groups on CNCs and epoxy groups on ENR [41]

By using the solution-casting process, carboxylated styrene-butadiene rubber (XSBR) composites with CNCs were prepared [43]. The amide I band moves to a lower wavenumber in the FTIR of XSBR/CNC nanocomposites, showing the presence of extra hydrogen bonds between the C–O of chitin and the hydroxyl groups of XSBR. The shift of the amide II band (1558 cm^{-1}) in the composite provides another hint of the chitin-XSBR hydrogen bonding. The –NH bending and stretching modes together with the -CN stretching mode are represented by the chitin's amide II band. Overall, the FTIR data support the composite's strong hydrogen bonding interactions between CNCs and XSBR.

4.2.9 UV–Visible Spectra

The transmittance spectra of carboxylated styrene-butadiene rubber (XSBR) composites [43] containing chitin nanocrystals (CNCs) were analysed. The installation of CNCs causes a minor reduction in the light transmittances. The transmittances of the nanocomposite had a maximum reduction value of 6.5% when compared to pure XSBR film in the visible light spectrum (400–760 nm). The UV–visible spectra of the rubber film are provided in Fig. 4.20.

The nanocomposite films do not contain any aggregated CNCs that can be seen with the naked eye. This is explained by the fact that the CNC suspension underwent ultrasonic treatment before being dispersed completely in the XSBR matrix. The mechanical and thermal performance of rubber nanocomposites is enhanced by the homogeneous dispersion of nanofillers.

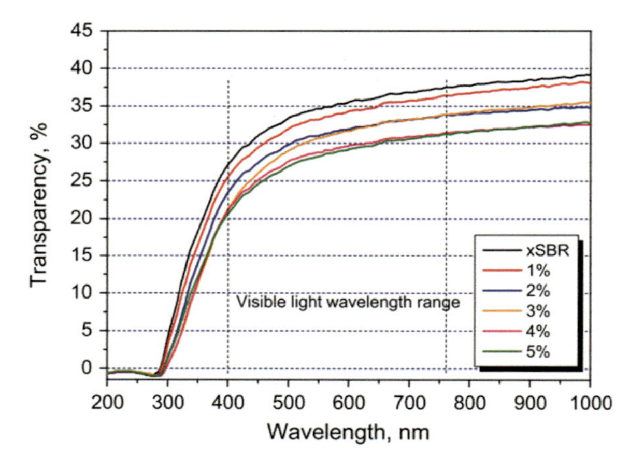

Fig. 4.20 UV–vis transmittance spectra of XSBR/CNC nanocomposites with different CNC contents [43]

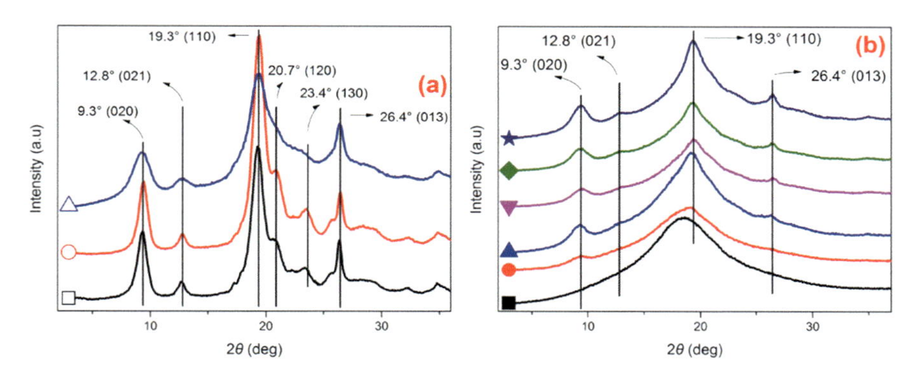

Fig. 4.21 XRD patterns of (**a**) chitin (regenerated chitin (Δ), pure chitin (O) raw chitin (□) and (**b**) NR and regenerated chitin rubber composites (NR (·), NR/ 10R-chitin (•), NR/30R-chitin (▲), NR/50R-chitin (▼), NR/70R-chitin (◆), NR/100R-chitin ()) [34]

4.2.9.1 X-ray Diffraction Analysis (XRD)

The X-ray diffraction patterns for chitin and the regenerated chitin (R-chitin) elastomer composites are depicted in Fig. 4.21. NR doesn't show any diffraction peaks and behaves as a completely amorphous polymer material. It is distinguished by a wide hump at about 18°–19°. When R-chitin is added, the hybrids show the characteristic peaks for chitin at 9.3° (020), 12.8° (021), 19.3° (110) and 26.4° (013). As would be expected, as the sample's R-chitin content rises, these peaks get stronger. The XRD results showed that R-chitin was successfully absorbed into the NR matrix and that its crystalline chitin structure was preserved. This indicates that the rigid chitin will likely reinforce the NR material [34].

Chitin nanocrystals (CNCs) exhibit strong scattering peaks at 2θ of 9.6° (020 plane) and 19.5° (110, 040 planes) in carboxylated styrene-butadiene rubber (XSBR) composites with CNCs, as well as three other weak peaks at 21° (101 plane), 23° (130 plane) and 26° (013 plane). There is no peak attributed to CNCs when the loading of CNCs is less than 3 wt%. The composite then progressively shows a minor peak of about 9.6° assigned to the CNCs' 020 plane. Since the XRD pattern is a statistical outcome, the more CNCs are loaded, the greater the peak intensity. The XRD data indicate that the XSBR has successfully incorporated CNCs [43].

4.2.9.2 Rheology

The basic rheological characterizations of the rubber composites generally comprise the properties like scorch time, cure time and cure rate index of the vulcanizates elucidating the processing features of the rubber composites. After compounding, the moulding conditions of the rubber specimens are evaluated using the rheological properties like cure time where the time required to form the crosslinked network is evaluated using the rubber process analyser. Usually, the time to achieve 90%

cure is taken as the cure time, whereas the time to achieve 10% of the crosslinks is termed as scorch time. The processing safety of the rubber compound is determined using this scorch safety period or induction period where premature vulcanization occurs. Rheological properties can be regulated by using processing additives like retarders, accelerators and vulcanizing agents [44]. *Midhun* et al. observed that the cure time of the NBR-CHNW (chitin nanowhiskers) is higher than that of the NBR gum due to the surface absorption of accelerator molecules by the reactive functional groups of CHNW, retarding the vulcanization effect. Similarly, the induction period also increases on increasing the filler content offering enough scorch safety ensuring the processing feasibility to the composites. The cure investigation reveals that nanocomposites with CHNW loading have a faster optimal cure time (t_{90}) than NBR gum. As a result of composites' excellent scorch time, it was discovered that as CHNW loading increases, the processing safety of composite was ensured significantly. The NBR-CHNW composites' cure rate index value decreased indicating that CHNW was unable to exert an activating influence on the NBR curing reaction [31]. Chitin nanofibers that are carbonylated and zwitterionic are incorporated into rubber molecules. It was found that zwitterionic chitin nanofibers (NC-ChNFs) and carbonylated chitin nanofibers (C-ChNFs) affected the vulcanization behaviour of NR. With increasing ChNFs content, C-ChNFs/NR nanocomposites exhibit a more pronounced rise in crosslinking density than NC-ChNFs/NR nanocomposites resulting in a remarkable improvement in the mechanical characteristics of C-ChNFs/NR due to enhanced filler dispersion and improved cross-link density [28].

4.3 Conclusion

The application of bio-fillers in different fields is significantly emerging due to their easy availability, low cost, biodegradability, biocompatibility, sustainability, etc. This chapter discussed the importance of chitin, a natural polysaccharide, as a bio-filler in rubber composites. Various methods can be adopted to extract chitin from natural sources. The specific surface area of chitin along with their reinforcing characteristics make them viable for various rubber composite applications like green tyres, gloves, etc. The mechanism behind the interactions between rubber matrix and chitin filler was explained. Characterization techniques helped to evaluate the uniform dispersion of the filler–matrix system and reinforcement. The thermal stability and surface morphology of the composites were also evaluated.

References

1. Blanchard R, Ogunsona EO, Hojabr S, Berry R, Mekonnen TH (2020) Synergistic Cross-linking and reinforcing enhancement of rubber latex with cellulose nanocrystals for glove applications. ACS Appl Polym Mater 2:887–898. https://doi.org/10.1021/acsapm.9b01117
2. Tang B, Chen X, He Y, Zhou J, Zhao H, Chen W, Wang J, Wang X (2021) Fabrication of kapok fibers and natural rubber composites for pressure sensor applications. Cellulose 28:2287–2301. https://doi.org/10.1007/s10570-020-03647-z
3. Ma X, Lv M, Anderson DP, Chang PR (2017) Natural polysaccharide composites based on modified cellulose spheres and plasticized chitosan matrix. Food Hydrocoll 66:276–285. https://doi.org/10.1016/j.foodhyd.2016.11.038
4. Żółtowska-Aksamitowska S, Bartczak P, Zembrzuska J, Jesionowski T (2018) Removal of hazardous non-steroidal anti-inflammatory drugs from aqueous solutions by biosorbent based on chitin and lignin. Sci Total Environ 612:1223–1233. https://doi.org/10.1016/j.scitotenv.2017.09.037
5. Freier T, Montenegro R, Koh HS, Shoichet MS (2005) Chitin-based tubes for tissue engineering in the nervous system. Biomater 26:4624–4632. https://doi.org/10.1016/j.biomaterials.2004.11.040
6. Salaberria AM, Labidi J, Fernandes SCM (2014) Chitin nanocrystals and nanofibers as nano-sized fillers into thermoplastic starch-based biocomposites processed by melt-mixing. Chem Eng J 256:356–364. https://doi.org/10.1016/j.cej.2014.07.009
7. Rudall KM, Kenchington W (1973) The chitin system. Biol Rev Camb Philos Soc 48:597–633. https://doi.org/10.1111/j.1469-185x.1973.tb01570.x
8. Pakizeh M, Moradi A, Ghassemi T (2021) Chemical extraction and modification of chitin and chitosan from shrimp shells. Eur Polym J 159:110709. https://doi.org/10.1016/j.eurpolymj.2021.110709
9. Rudall KM (1969) Chitosan Its Association with Other Molecules. J Polym Sci 102:83–102
10. Soetemans L, Uyttebroek M, Bastiaens L (2020) Characteristics of chitin extracted from black soldier fly in different life stages. Int J Biol Macromol 165:3206–3214. https://doi.org/10.1016/j.ijbiomac.2020.11.041
11. Atkins E (1985) Conformations in polysaccharides and complex carbohydrates. J Biosci 8:375–387. https://doi.org/10.1007/BF02703990
12. Tao F, Cheng Y, Shi X, Zheng H, Du Y, Xiang W, Deng H (2020) Applications of chitin and chitosan nanofibers in bone regenerative engineering. Carbohydr Polym 230:115658. https://doi.org/10.1016/j.carbpol.2019.115658
13. Kozma M, Acharya B, Bissessur R (2022) Chitin, Chitosan, and Nanochitin: Extraction, Synthesis, and Applications. Polymers (Basel). 14:1–28
14. Khajavian M, Vatanpour V, Castro-Muñoz R, Boczkaj G (2022) Chitin and derivative chitosan-based structures—Preparation strategies aided by deep eutectic solvents: A review, Carbohydr. Polym. 275. https://doi.org/10.1016/j.carbpol.2021.118702
15. Jang MK, Kong BG, Il Jeong Y, Lee CH, Nah JW (2004) Physicochemical characterization of α-chitin, β-chitin, and γ-chitin separated from natural resources, J Polym Sci Part A Polym Chem 42 3423–3432. https://doi.org/10.1002/pola.20176
16. Mohan K, Ganesan AR, Ezhilarasi PN, Kondamareddy KK, Rajan DK, Sathishkumar P, Rajarajeswaran J, Conterno L (2022) Green and eco-friendly approaches for the extraction of chitin and chitosan: A review. Carbohydr Polym 287:119349. https://doi.org/10.1016/j.carbpol.2022.119349
17. (Gabriel) Kou S, Peters LM, Mucalo MR (2021) Chitosan: A review of sources and preparation methods, Int J Biol Macromol 169 85–94. https://doi.org/10.1016/j.ijbiomac.2020.12.005
18. Thomas SK, Parameswaranpillai J, Krishnasamy S, Begum PMS, Nandi D, Siengchin S, George JJ, Hameed N, Salim NV, Sienkiewicz N (2021) A comprehensive review on cellulose, chitin, and starch as fillers in natural rubber biocomposites. Carbohydr Polym Technol Appl 2:100095. https://doi.org/10.1016/j.carpta.2021.100095

19. Moussian B (2019) Chitin: structure, chemistry and biology a cuticle a body shape a evolution Á Barrier. Springer Singapore. https://doi.org/10.1007/978-981-13-7318-3
20. Mohan K, Ganesan AR, Muralisankar T, Jayakumar R, Sathishkumar P, Uthayakumar V, Chandirasekar R, Revathi N (2020) Recent insights into the extraction, characterization, and bioactivities of chitin and chitosan from insects. Trends Food Sci Technol 105:17–42. https://doi.org/10.1016/j.tifs.2020.08.016
21. Santos VP, Marques NSS, Maia PCSV, de Lima MAB, Franco L de O, Campos-Takaki GM de (2020) Seafood waste as attractive source of chitin and chitosan production and their applications, Int J Mol Sci 21 1–17. https://doi.org/10.3390/ijms21124290
22. Casadidio C, Peregrina DV, Gigliobianco MR, Deng S, Censi R, Di Martino P (2019) Chitin and chitosans: Characteristics, eco-friendly processes, and applications in cosmetic science, Mar Drugs 17. https://doi.org/10.3390/md17060369
23. Yadav M, Goswami P, Paritosh K, Kumar M, Pareek N, Vivekanand V (2019) Seafood waste: a source for preparation of commercially employable chitin/chitosan materials, Bioresour. Bioprocess. 6. https://doi.org/10.1186/s40643-019-0243-y
24. Wang J, Tavakoli J, Tang Y (2019) Bacterial cellulose production, properties and applications with different culture methods – A review. Carbohydr Polym 219:63–76. https://doi.org/10.1016/j.carbpol.2019.05.008
25. Razzaq A, Shamsi S, Ali A, Ali Q, Sajjad M, Malik A, Ashraf M (2019) Microbial proteases applications, Front. Bioeng. Biotechnol 7:1–20. https://doi.org/10.3389/fbioe.2019.00110
26. Chen W, Cao P, Liu Y, Yu A, Wang D, Chen L, Sundarraj R, Yuchi Z, Gong Y, Merzendorfer H, Yang Q (2022) Structural basis for directional chitin biosynthesis. Nature 610:402–408. https://doi.org/10.1038/s41586-022-05244-5
27. Gopalan Nair K, Dufresne A (2003) Crab shell chitin whisker reinforced natural rubber nanocomposites. 1. Processing and swelling behavior, Biomacromolecules. 4 657–665. https://doi.org/10.1021/bm020127b
28. Yin J, Hou J, Huang S, Li N, Zhong M, Zhang Z, Geng Y, Ding B, Chen Y, Duan Y, Zhang J (2019) Effect of surface chemistry on the dispersion and pH-responsiveness of chitin nanofibers/natural rubber latex nanocomposites. Carbohydr Polym 207:555–562. https://doi.org/10.1016/j.carbpol.2018.12.025
29. Liu C, Huang S, Hou J, Zhang W, Wang J, Yang H, Zhang J (2021) Natural rubber latex reinforced by graphene oxide/zwitterionic chitin nanocrystal hybrids for high-performance elastomers without sulfur vulcanization. ACS Sustain Chem Eng 9:6470–6478. https://doi.org/10.1021/acssuschemeng.1c01461
30. Ding B, Huang S, Shen K, Hou J, Gao H, Duan Y, Zhang J (2019) Natural rubber bio-nanocomposites reinforced with self-assembled chitin nanofibers from aqueous KOH/urea solution. Carbohydr Polym 225:115230. https://doi.org/10.1016/j.carbpol.2019.115230
31. M. Dominic C.D., Joseph R, Sabura Begum PM, Raghunandanan A, Vackkachan NT, Padmanabhan D, Formela K (2020) Chitin nanowhiskers from shrimp shell waste as green filler in acrylonitrile-butadiene rubber: Processing and performance properties, Carbohydr. Polym. 245 116505. https://doi.org/10.1016/j.carbpol.2020.116505
32. Nawamawat K, Sakdapipanich JT, Ho CC, Ma Y, Song J, Vancso JG (2011) Surface nanostructure of Hevea brasiliensis natural rubber latex particles. Colloids Surfaces A Physicochem Eng Asp 390:157–166. https://doi.org/10.1016/j.colsurfa.2011.09.021
33. Rochette CN, Crassous JJ, Drechsler M, Gaboriaud F, Eloy M, De Gaudemaris B, Duval JFL (2013) Shell structure of natural rubber particles: Evidence of chemical stratification by electrokinetics and cryo-TEM. Langmuir 29:14655–14665. https://doi.org/10.1021/la4036858
34. Yu P, He H, Luo Y, Jia D, Dufresne A (2017) Elastomer Reinforced with Regenerated Chitin from Alkaline/Urea Aqueous System. ACS Appl Mater Interfaces 9:26460–26467. https://doi.org/10.1021/acsami.7b08294
35. Kawano A, Yamamoto K, Kadokawa JI (2017) Preparation of self-assembled chitin nanofiber-natural rubber composite sheets and porous materials. Biomol 7:18–21. https://doi.org/10.3390/biom7030047

36. Jiarui Hu YY, Xiaohui Tian, Jinyu Sun, Jianyong Yuan (2019) J of Applied Polymer Sci—2020—Hu—Chitin nanocrystals reticulated self-assembled architecture reinforces deproteinized.pdf, 1–11
37. Yin J, Hu J, Han Y, Chen Y, Hu J, Zhang Z, Huang S, Duan Y, Wu H, Zhang J (2022) Facile fabrication of high nanofiller-content natural rubber nanocomposites for reversible plasticity shape memory polymers. Compos Sci Technol 221:109349. https://doi.org/10.1016/j.compscitech.2022.109349
38. Liu Y, Liu M, Yang S, Luo B, Zhou C (2018) Liquid crystalline behaviors of chitin nanocrystals and their reinforcing effect on natural rubber. ACS Sustain. Chem. Eng. 6:325–336. https://doi.org/10.1021/acssuschemeng.7b02586
39. Mathew M, Midhun Dominic CD, Neenu KV, Begum PMS, Dileep P, Kumar TGA, Sabu AA, Nagane D, Parameswaranpillai J, Badawi M (2023) Carbon black and chitin nanofibers for green tyres: Preparation and property evaluation, Carbohydr. Polym. 310 120700. https://doi.org/10.1016/j.carbpol.2023.120700
40. Visakh PM, Monti M, Puglia D, Rallini M, Santulli C, Sarasini F, Thomas S, Kenny JM (2012) Mechanical and thermal properties of crab chitin reinforced carboxylated SBR composites. Express Polym Lett 6:396–409. https://doi.org/10.3144/expresspolymlett.2012.42
41. Nie J, Mou W, Ding J, Chen Y (2019) Bio-based epoxidized natural rubber/chitin nanocrystals composites: Self-healing and enhanced mechanical properties. Compos Part B Eng 172:152–160. https://doi.org/10.1016/j.compositesb.2019.04.035
42. Zhang N, Cao H (2020) Enhancement of the antibacterial activity of natural rubber latex foam by blending it with chitin. Materials (Basel). 13:1–15. https://doi.org/10.3390/ma13051039
43. Liu M, Peng Q, Luo B, Zhou C (2015) The improvement of mechanical performance and water-response of carboxylated SBR by chitin nanocrystals. Eur Polym J 68:190–206. https://doi.org/10.1016/j.eurpolymj.2015.04.035

Chapter 5
Hemicellulose Rubber Composites and Rubber Bionanocomposites

Malu Kottayil Madhavan, Vaishak Nambaithodi, Anand Krishnamoorthy, and Sivasubramanian Palanisamy

Abstract In this chapter, different types of hemicellulose-based rubber composites and rubber bio-nanocomposites are discussed and this extends to elucidate more topics such as applications of hemicellulose in natural rubber composites, natural rubber nanocomposites and synthetic rubber composites. It encompasses topics including applications in biomedical, membrane technology, packaging, structural, aerospace and defense and also in tyre and coating industries.

Keywords Hemicellulose · Rubber composites · Rubber bionanocomposites · Natural rubber · Nanocomposites · Biomedical · Packaging etc.

5.1 Introduction

Hemicellulose, such as cellulose and lignin, are complex carbohydrate polymer found in cell walls of plant structure. Plant cell walls consist of sugar units resulting in a heterogenous polysaccharide. Hemicellulose gives the cell wall structural stability by acting as a cementing agent in between cellulose fibers [1]. In rubber composites,

M. K. Madhavan
Technical Consultancy Division, Rubber Research Institute of India, Rubber Board, P.O, Kottayam Kerala-686009, India

V. Nambaithodi
International and Inter University Centre for Nanoscience and Nanotechnology, Mahatma Gandhi University, Priyadarsini Hills, Kottayam, Kerala 686560, India

A. Krishnamoorthy (✉)
Department of Basic Sciences & Humanities, Adi Shankara Institute of Engineering and Technology, Kerala 683574, India
e-mail: anandkrishnamoorthy20@gmail.com

S. Palanisamy (✉)
Department of Mechanical Engineering, P T R College of Engineering & Technology, Thanapandiyan Nagar, Madurai-Tirumangalam Road, Madurai 62500, Tamilnadu, India

Visakh P. M. *Rubber Based Bionanocomposites*, Advanced Structured Materials 210,
https://doi.org/10.1007/978-981-10-2978-3_5

hemicellulose can function as a reinforcing filler when it is appropriately treated and distributed. It can enhance the rubber's mechanical qualities, including its modulus, tensile strength, and resistance to tearing. Because hemicellulose is fibrous and works well with rubber matrices, it can improve the composite material's structural integrity. In rubber composites with additional natural or synthetic fillers, hemicellulose can also act as a compatibilizer. The interaction among the filler and rubber matrix can be enhanced by them. Hemicellulose can lessen the chance of delamination or failure at the interface, avoid filler agglomeration, and enhance dispersion by fostering improved interaction between the rubber and other components [2]. Various mechanical properties of natural rubber composites such as force at break, elongation at break, the modulus, and its abrasion properties have been shown to be improved by hemicellulose. Hemicellulose's fibrous character and its compatibility with natural rubber contribute to the reinforcement effect [3]. Sufficient processing methods are necessary to guarantee that the hemicellulose is evenly distributed throughout the rubber matrix. The mechanical properties of the composite can be enhanced by uniformly dispersing the hemicellulose filler. Additionally, the bonding of filler with rubber can be increased by incorporating hemicellulose in the natural rubber composite [4]. Hemicellulose provides a sustainable and biodegradable filler for natural rubber composites. It is made from renewable biomass sources and is more environmentally friendly than synthetic fillers [5].

Plants, animals, or minerals are made up of natural fibers which is used to derive hemicellulose. Cellulose is derived from plants whereas protein is derived from animal body parts such as from their hair, silk, wool, etc. Natural fiber has a load bearing property in rubber composite which has increased its utilization in future due to its cost-effectiveness, recyclability and mainly due to its strength to weight ratio. The mechanical property of cellulose fibrils increases when it is oriented in the direction of fiber length. Depending upon the crystalline characteristic and properties of cellulose the reinforcing potential in natural fibers changes [6]. Primarily hemicellulose is classified into four categories depending upon the structure [7].

Modified lignin particles have remained discovered as reinforcements in polymer compounds for potential use in the rubber product manufacturing area. Xiao and colleagues synthesized a complex of lignin incorporated double hydroxide and employed it in the production of styrene-butadiene rubber lignin composites through a process known as melt compounding. By incorporating the novel complex into the rubber, the researchers were able to enhance the mechanical properties of the composites. This breakthrough allowed the composites to outperform traditional styrene-butadiene lignin composites. The increased mechanical properties of the new composites not only demonstrated the effectiveness of the novel complex but also pointed towards potential future applications in various industries. The successful integration of the novel complex provides a promising outlook for the development of advanced materials with superior performance characteristics. The improvement in mechanical properties was attributed to several factors. First, the presence of the lignin–double hydroxide complex generated shear zones under tension, which likely contributed to the material's enhanced stress resistance. Additionally, the composite benefited from improved stress transfer at the interface, thanks to the high aspect ratio of the lignin-LDH complex, which facilitated better load distribution throughout the material. Furthermore, the LDH particles were superiorly dispersed within the rubber matrix, ensuring a more uniform distribution of the reinforcing particles and, consequently, a more consistent material strength throughout the composite. The development of rubber composites with superior mechanical properties was led to by Xiao and their team through a meticulous combination of material engineering and process innovation, marking a significant advancement in the field of materials science.

In contrast, rubber reinforced with hexamethylenetetramine (HMT)-modified lignin did not match the performance of carbon black/rubber compound in another study. This was attributed to particle size differences and weak interfacial adhesion between lignin's polar groups and the hydrophobic matrix [8].

5.2 Applications of Hemicellulose in Natural Rubber Composites

Hemicellulose, a complex polysaccharide found in plant cell walls, has sparked interest in its possible applications in a variety of sectors, including rubber composites. When integrated into natural rubber composites, hemicellulose can provide reinforcement, enhanced mechanical qualities, and environmental sustainability. Here are several uses of hemicellulose in natural rubber composites.

I. Reinforcement and Improved Mechanical Properties:

   In natural rubber composites, hemicellulose can be added as a filler or reinforcement to enhance physical properties. Liu and colleagues deliberate the effect of hemicellulose derived from corn stalks on the properties of natural rubber composites, finding significant increases in tensile strength and modulus [9].

II. Biodegradability and Environmental Sustainability:

   Hemicellulose, as a natural and biodegradable substance, improves the environmental sustainability of natural rubber composites. The usage of hemicellulose as a sustainable filler in natural rubber composites was examined by Saba and his colleague in 2020, with emphasis placed on its biodegradability and eco-friendliness [10].

III. Thermal Stability and Flame Retardancy:

   Hemicelluloses can be combined with natural rubber compounds to enhance their thermal activity and flame resistance, making them ideal for applications that require resistance to high temperatures and fire. For example, the flame retardancy of natural rubber composite filled with hemicellulose was studied in a study published in the Journal of Flammability by Yuniarto, 2019. The study showed improved flame retardancy as well as improved thermal stability [11].

IV. Cost-Effectiveness and Sustainability:

Hemicellulose filler in natural rubber composites is an affordable, environmentally friendly substitute for traditional fillers that lowers production costs and improves resource efficiency. In their evaluation of hemicellulose's viability as a filler in natural rubber composites, Rizal et al. (2021) emphasized the material's potential for low-cost, environmentally friendly rubber products [12].

5.3 Applications of Hemicellulose in Natural Rubber Nanocomposites

The potential practice of hemicellulose in natural rubber nanocomposites has also been investigated. It offers special benefits like better mechanical qualities, better nanoparticle dispersion, and environmental sustainability.

I. Enhanced Mechanical Properties.

The utilization of hemicellulose as a compatibilizer or dispersing agent can enhance the spreading and interfacial bond among the rubber matrix and nanofillers, such as nanoparticles, in natural rubber nanocomposites. Thakur et al. (2019) found that hemicellulose significantly improved the tensile strength, modulus, and tear resistance of natural rubber nanocomposites when used as a green dispersing agent for nanoclay [13].

II. Improved Dispersion of Nanoparticles.

Natural rubber matrix with hemicellulose can facilitate the good dispersion of nanoparticles, improving mechanical strength, barrier qualities, and thermal stability, among other attributes. Sarkawi et al.'s (2020) study examined the function of hemicellulose in natural rubber nanocomposites as a dispersing agent for nanoclay, emphasizing how well it improved dispersion and mechanical properties [14].

III. Enhanced Barrier Properties.

Hemicellulose can be added to natural rubber nanocomposites to improve their barrier qualities, which qualifies them for uses that call for impermeability or gas barrier qualities. Numan et al.'s research from 2021 examined the hemicellulose-filled natural rubber nanocomposites' gas barrier characteristics and found that they performed better than unfilled rubber [15].

5.4 Applications of Hemicellulose in Synthetic Rubber Composites

Lignin, recognized as a biopolymer, is known for imparting relatively higher tensile strength characteristics to SBR copolymer in comparison to carbon filler [16]. Setua et al. conducted an investigation where lignin was utilized as a reinforcing filler in NBR. The vulcanizates were evaluated for thermal, physical, mechanical, and fuel resistance properties, with a comparison made between vulcanizates with basic lignin and those containing carbon black (CB), modified lignin, and phenolic resins. Modification of lignin was carried out using benzoyl peroxide. Vulcanizates with modified lignin showed a significant increase in tensile strength compared to those with unmodified lignin, with a measurement of 1.53 MPa versus 1.51 MPa. Additionally, the elongation at break was improved from 175 to 250% with modified lignin in nitrile rubber. The vulcanizates containing modified lignin also demonstrated the highest thermal stability. Properties such as hardness and compression set were found to be superior in vulcanizates containing modified lignin compared to those with other fillers [17]. Wang and colleagues explored replacing half of the CB content with lignin to develop advanced NBR/poly(vinyl chloride) (NBR/PVC) composites. The Zn^{2+} was given between lignin and the rubber matrix, by employing interfacial Zn^{2+}-based coordination bonds, with lignin serving as a natural ligand. Additionally, lignin improved the composites' resistance to thermo-oxidative aging, with control achievable through the amounts of ZnCl2 and sulfur. The composites also showed better resistance to high-temperature oils due to half replacement of carbon black with lignin and the inclusion of metal coordination bonds. This study introduces a straightforward method for incorporating eco-friendly lignin into rubber–plastic composites using traditional rubber compounding techniques, highlighting its potential for environmentally friendly applications [18]. Synthetic rubber composites based on hemicellulose are used to produce vibration-dampening components with improved damping qualities for industrial and automotive applications [19]. Hemicellulose-reinforced synthetic rubber composites are employed in anti-vibration mounts for machinery and equipment, offering improved resilience and damping characteristics [20]. Hemicellulose-modified synthetic rubber composites are used in the manufacturing of gaskets and seals for automotive and aerospace applications, providing enhanced sealing performance [21]. Chen, L., et al. (2015) state that hemicellulose-incorporated synthetic rubber composites are utilized to produce flexible packaging films for food and consumer items, offering mechanical strength and barrier qualities [22]. Synthetic rubber composites treated with hemicellulose are used to create adhesives that provide better cohesive strength and adhesion when connecting different surfaces [23]. To increase air retention and decrease permeability, hemicellulose-reinforced synthetic rubber composites are used in tyre inner liners, improving tyre performance and lifetime [24].

5.5 Biomedical Applications

Hemicelluloses have garnered attention for biological applications due to their biocompatibility, biodegradability, and capacity for functional modification. Below are several biomedical applications of hemicelluloses. Since their biocompatibility property and capability to form hydrogels, hemicelluloses offer potential for constructing drug delivery systems. These hydrogels can encapsulate medications and deliver them in a regulated manner [25]. Hemicellulose-based materials can be used to make wound dressings with features including good moisture retention, gas permeability, and biocompatibility. These dressings promote tissue regeneration by fostering an environment that accelerates wound healing. Numerous polymers—natural, synthetic, and semi-synthetic—have been researched for their possible application as dressing and wound-healing materials. Considerable interest has been garnered by natural polymers due to their low toxicity and excellent biocompatibility. However, inferior mechanical strength and processability are often exhibited by them compared to synthetic alternatives. Fascinating properties are possessed by cellulose, a natural polysaccharide, and it finds applications in various scientific fields and industries due to its excellent biocompatibility [26]. The surface modification of cellulose fibers was explored in a review conducted by Kalia and colleagues, using a range of techniques. Hemicelluloses can be treated to provide porous scaffolds with appropriate mechanical characteristics for tissue engineering applications. These scaffolds create a 3D environment for cell adhesion, proliferation, and differentiation [27]. Hemicellulose derivatives have been studied for hemostatic characteristics. They can help to produce blood clots and stop bleeding in surgical and traumatic wounds [28].

The bonding between deproteinized natural rubber (DNR) and chitin nanocrystals (ChNCs) in composite materials was explored in their 2020 study by Hu and colleagues. DNR particles were surrounded by self-assembled ChNCs, forming a network structure embedded within the DNR matrix. Enhancements in tensile strength, modulus, and tear strength of the composites were observed as a result of this interaction. Optimal mechanical properties, including maximum force at break, elongation at break, and tear strength, were exhibited by the composite materials at a 4 wt.% filler content. However, beyond this concentration, a decline in these properties was noted. The improvements in properties were attributed to the formation of a network structure, which facilitated efficient stress transfer from the matrix to the nanofillers. These composites, which promise potential for biomedical applications due to their low protein content and improved mechanical properties, are being shown. [29]. The enhancements achieved by incorporating self-assembled chitin nanofibers into NR latex were further investigated in their study. The proliferation of mouse bone mesenchymal stem cells within the composites was promoted by the addition of chitin nanofibers. As a result, these composites exhibit potential for biomedical applications, including the development of blood vessels and human diaphragms. [30]. These surgical adhesives are made from hemicellulose and are used in surgical procedures. They are used to attach and seal tissue and replace the use of traditional sutures or staples [31].

5.6 Packaging Applications

Hemicelluloses' high hydrophobicity and readily digestible nature position them as excellent candidates for the production of edible films and food coatings. Moreover, research demonstrates their capability to bind directly with cholesterol in the gut, thereby enhancing bowel movements. This evidence further supports their widespread application in the food industry, indicating their significant role not only in food preservation but also in promoting digestive health [32]. Advantageous properties for packaging applications are offered by hemicelluloses due to their biodegradability, renewability, and barrier capabilities. They can be utilized in the production of biodegradable films and coatings for food packaging, providing moisture and gas barriers. Remarkably low levels of oxygen and vapor permeability have been shown by nanomaterial-bound hemicellulose. Additionally, the potential for long-term moisture management is demonstrated by lignin, with an absorption capacity of up to 189%. [33]. Hemicellulose is introduced into edible packing material which increases the shelf life of the package and reduces the plastic waste. New possibilities are presented by the nanoporous structure, combined with its specific surface area. Microfibrillated cellulose (MFC) holds promise as a solution to society's demand for improved packaging. Its renewable nature generates significant interest as an alternative to mitigate the ongoing oil crisis. Additionally, its biodegradability aligns well with environmental regulations and waste management challenges. This innovative material, derived from plant fibers, offers an eco-friendly option that is gaining attention for its potential to replace traditional petroleum-based packaging materials. Its unique properties, such as high strength and barrier qualities, make it an attractive choice for industries looking to enhance product sustainability while minimizing environmental impact. Additionally, its ability to reinforce barrier properties provides superior product protection and preservation, potentially extending food shelf life. Despite its promising attributes, MFC faces several hurdles. Industrial-scale production remains a challenge, requiring advancements before practical industrial applications can be realized. While various strategies exist for creating effective barrier materials, fully harnessing MFC's potential with paper for industrial coatings remains unrefined. There is also room for enhancing the integration of MFC's barrier properties. [34]. The potential of hardwood xylan hemicellulose in creating uniform films and serving as a biopolymer for paper coatings is examined in this study. Xylan-coated paper and film samples were tested for various properties, including water, air, and water vapor permeability, water solubility, mechanical strength, and antimicrobial activity against pathogenic bacteria. A high abundance of hydroxyl groups in xylan hemicelluloses was revealed in structural analysis, contributing to their strong water affinity. Despite being a natural polymer, xylan's potential in packaging has been relatively underexplored and inadequately researched. Xylan hemicelluloses, the second most abundant polysaccharides after cellulose, possess intrinsic barrier properties essential for food packaging papers and films. The functional properties of xylan-coated papers are influenced by this characteristic, which also provides opportunities for chemical modifications to enhance hydrophobicity and broaden its applications. It is suggested by the findings that there is a promising potential for using this material in food packaging as a sustainable and competitive alternative to petroleum-based polymers. Additionally, hemicellulose-based films can be enhanced with antimicrobial agents to inhibit microbial growth, thus extending the shelf life of packaged foods [35]. Hemicellulose-based films can be engineered to release active compounds such as antioxidants or antimicrobials, providing additional functionality to packaging materials. Lignin boasts remarkable attributes such as its abundance, lightweight nature, environmental friendliness, and properties like antioxidant, antimicrobial, and biodegradability. Its CO2 neutrality and reinforcing capabilities further enhance its appeal as an excellent candidate for crafting innovative polymer composite materials. Diverse polymer composites reinforced with lignin are examined, encompassing both synthetic and biodegradable polymer matrices, with a focus on recent progress in lignin's multifaceted applications. The structural attributes and capabilities of lignin/polymer composite systems are explored in each segment. Additionally, current research directions in lignin-based materials for engineering applications are discussed, encompassing areas such as lignin modification techniques, the production of thermoset, thermoplastic, biodegradable, rubber, and foam composites, as well as lignin's function as a compatibilizer [36]. Hemicellulose-based films can be designed to be biocompatible, making them suitable for packaging medical and pharmaceutical products [37]. Hemicellulose can be applied as a protective coating to cardboard packages to improve moisture and grease resistance, which improves the performance of the packaging material [38]. Biodegradable blister packs, crafted from hemicellulose-based materials, serve as protective measures against counterfeiting in pharmaceuticals and consumer goods. Films derived from native cellulose nanofibrils, comprising both crystalline and non-crystalline cellulose components, were examined to comprehend their enzymatic breakdown. Throughout the process, alterations in frequency and energy dissipation were monitored using a Quartz Crystal Microbalance (QCM). Under the specified experimental conditions, the enzymatic degradation of these nanofibril films occurred rapidly. Atomic Force Microscopy (AFM) images of the cellulose substrates, both pre- and post-enzymatic treatment, illustrated the complete removal of the nanofibril film within a brief degradation period. Comparison with three other cellulose model types—LS and SC regenerated cellulose, and cellulose NC cast film—revealed distinctive enzyme interactions and substrate degradation in the nanofibril films. The enzymatic degradation process was significantly influenced by the characteristics of the cellulose substrate, including crystallinity and morphology [39]. The morphological, crystallographic, and mechanical properties of the resulting nanocomposites produced by casting were investigated by Abraham et al. Nanofibrillated cellulose (NFC), obtained from banana fibers via steam explosion, was employed as reinforcement in natural rubber latex. The NFC filler was found to exhibit uniform dispersion within the nanocomposites. It was observed that higher NFC content correlated with increased Young's modulus and tensile strength, but decreased the rubber-like flexibility and crystallinity of the material. Additionally, the biodegradability and solvent transport properties of the films were assessed. It was found that the nanocomposites were fully biodegradable, with nanocellulose promoting degradation within the composite's inner portion. The diffusion coefficient decreased with increasing nanofiber concentration, indicating reduced solvent permeability. On the contrary, solvent resistance rose as a result of the limitations posed by the hydrophilic hydroxyl groups against hydrophobic organic solvents [40]. Natural fibers are gaining traction as potential reinforcements in polymer composites owing to their favorable mechanical properties, processing benefits, and environmental advantages. Nonetheless, the hydrophilic characteristics of these fibers may diminish their compatibility with the matrix, resulting in less-than-ideal mechanical properties in the composites.

The vital role played by chemical treatment in diminishing the hydrophilicity of fibers, thereby improving their adhesion to the matrix, is emphasized. The structure and surface morphology of the fibers are altered by these treatments, with hydrophilic hydroxyl groups being eliminated through diverse chemical reactions. As a result, more reactive hydroxyl groups are formed, leading to better bonding with the matrix. The application of various chemical treatment methods to the reinforcing fibers has been shown to significantly enhance the mechanical properties of the composites [41].

5.7 Structural Applications

Recent advancements in renewable and bio-based fillers for rubber composite applications are explored in this review, with a focus on incorporating hemicellulose into rubber tyre treads to enhance performance and durability. Biofillers can be used to effectively reinforce rubber composites through appropriate processing, purification, and occasional surface modifications, boosting grip and wear resistance. Significant potential is presented to replace or complement petroleum-derived carbon black and high-density mineral fillers. Particularly promising as alternative reinforcing agents for both natural and synthetic rubbers are micro- and nanoscale cellulose. However, the industrial implementation of these fillers is hindered by the challenge of balancing cost and performance [42]. The market price of CNCs, for instance, can be soared up to USD 50/kg (from a Canadian supplier, CelluForce Inc.), while CNF can be exceeded USD 200/kg, despite over a decade of intensive research and development. This elevated cost primarily arises from the intricate extraction and purification processes, along with scalability limitations. Consequently, achieving large-scale production and cost reduction is imperative for the widespread integration of nanocellulose in rubber composite applications. Additionally, the added advantage of expediting biodegradation is offered by CNCs, chitin, chitosan, NF, and other polysaccharides when employed as fillers with NR, owing to their faster degradation rate compared to NR [43].

The potential of lignin in high-performance rubber composite products is crucial to be unlocked by understanding its structural attributes. Encouraging composite properties, such as weight reduction advantages and efficient reinforcement, are demonstrated by both lignin and NFs through appropriate modifications and surface treatments. Enhancing the performance of these composites can be achieved by blending them with other fillers like CB. Progress in the field is greatly reduced by challenges in the advancement of practical high-performance rubber biocomposites, stemming from issues such as the thermal stability of bio-fillers and compatibility concerns between hydrophilic bio-fillers and hydrophobic polymers [44].

The surge in interest in the development of hydrophilic filler/rubber composites has indeed been seen, especially with the emergence of biochar-filled rubber composites as competitive alternatives to carbon black-based ones. An environmentally friendly and cost-efficient solution is offered by biochar, derived from renewable biomass and waste materials. Promise has been shown in its incorporation into rubber composites, particularly in tyre and rubber technology, where it replaces carbon black in certain proportions. One key advantage is found in the versatility achieved by adjusting the pyrolysis process parameters during biochar synthesis. The creation of biochars with tailored physical properties, graphitization levels, surface functionalities, and carbon structures is permitted. However, in order to fully exploit the potential of biochar as a filler in rubber, understanding how its functionality interacts with different types of rubber—both natural and synthetic—is crucial. Improving compatibility and dispersion of biochar within rubber matrices are considered essential for enhancing composite performance. This necessitates a deeper understanding of the surface interactions between biochar and rubber. Bridging these knowledge gaps could lead to the acceleration of the transition from non-renewable carbon black to biochar-based fillers. Additionally, advancements in synthesizing nanoscale biochar in various carbon structure forms are held to have transformative potential. Unique properties and performance benefits could be offered by nanoscale biochars, further elevating the capabilities of bio-based carbon materials in rubber composites and other applications [45].

While promising performance in tyres and other rubber products is shown by elastomeric compound incorporated with bio-based fillers, fundamental material properties, like tensile strength & modulus, are primarily focused on in many studies. A demand exists for research that addresses critical performance parameters specific to rubber goods, such as the tyre magic triangle properties. Therefore, further research is deemed necessary to customize bio-fillers for specific applications, such as green tyre technology. Additionally, current research on rubber recycling processes predominantly revolves around carbon black-filled rubber products. More studies on bio-filler-based rubber composites and hybrid filler systems are seen as needed to integrate bio-fillers into commercial rubber products effectively.

Despite continuous research and innovation in bio-filler reinforced rubber composites aimed at advancing sustainable development there is an urgent requirement for a streamlined, cost-effective green processing technology that can provide superior product performance. This necessity is essential for guiding the rubber industry toward a more sustainable future [46].

Rubber composites packed with hemicellulose can dampen vibration, which makes them useful for structural components that reduce noise and vibration in machinery and automotive applications [46]. Rubber composites reinforced with hemicellulose can be used as shock-absorbing materials in a variety of structural applications, including industrial machinery and automobile suspension systems [47]. Rubber compounds containing hemicellulose can be utilized in structural bushings and mounts to provide support, cushioning, and isolation for mechanical systems and car suspensions. N. Saba et al.'s review paper offers valuable insights for future research and the detailed application of Dynamic Mechanical Analysis (DMA) to assess the properties of fibers of natural resource-reinforced polymeric compounds/hybrid composites compared to those based on synthetic fibers. Subsequent endeavors could focus on producing fully sustainable composite and nanocomposite materials from natural fibers as fillers with biodegradable resin polymeric matrices, aiming to enhance dynamic thermal properties [48]. Rubber composites containing hemicellulose can be employed in piping and ducting systems for their flexibility, durability, and resistance to corrosion and abrasion in various industrial applications. Biopolymers have increasingly become the preferred matrix for these composites. Review papers literature from 2000 to 2010 on natural fibers and biopolymers commonly used in biocomposites. It covers the properties of reinforcing fibers, their source, nature, structure, fiber-to-polymer ratio, and mechanical properties.

Several process for modifying fibers are explored in the paper, encompassing both physical techniques like corona and plasma treatment, as well as chemical approaches such as silane, alkaline, acetylation, maleated coupling, and enzyme treatment. The most prevalent matrices derived from both petrochemical and renewable sources are discussed. Processing techniques for biocomposites and the factors influencing them, including moisture content, fiber type, and content. The significance of semi-finished product manufacturing in the biocomposite process is emphasized, alongside the examination of processing technologies for thermoplastic matrices like compression molding, extrusion, injection molding, LFT-D-method, and thermoforming. Processes like resin transfer molding and sheet molding compound are examined for thermosets. The impact of these processes on mechanical performance, including tensile, flexural, and impact properties, is assessed [49]. Hemicellulose-stuffed rubber compounds may be used as curler coverings in commercial equipment for it's resilience, abrasion resistance, and capacity to offer traction and grip [50].

5.8 Military Applications

1. Ballistic Protection: Hemicellulose-enhanced rubber composites find utility in body armor and vehicle armor applications, offering a combination of lightweight construction and formidable ballistic defense. The integration of hemicellulose serves to bolster the impact resistance and energy absorption properties of these composite materials [51].

2. Tactical Gear and Equipment: Hemicellulose-reinforced rubber composites offer versatility in crafting a range of tactical gear and equipment like helmets, knee pads, and backpacks, ensuring military personnel benefit from enhanced durability, flexibility, and impact resistance [52].
3. Seals and Gaskets for Military Vehicles: Hemicellulose-derived rubber seals and gaskets can be utilized in military vehicles to ensure a secure seal, protecting against dust, water, and various contaminants. This enhances the durability and lifespan of vehicle parts, especially in challenging environmental conditions. [53].
4. Fuel and Oil-Resistant Components: Hemicellulose-reinforced rubber composites offer the potential to fabricate fuel and oil-resistant components like hoses, seals, and O-rings for military vehicles and equipment. This ensures consistent performance even in challenging operational environments [54].
5. Shock-Absorbing Materials: Rubber composites containing hemicellulose can be utilized in military scenarios as shock-absorbing materials, offering cushioning and vibration dampening for delicate electronic devices, ammunition, and transportation infrastructure. The mechanical and physical characteristics of sisal fiber are dependent on factors such as its origin, location, age, and experimental conditions, including fiber diameter, gauge length, strain rate, and test temperature. This makes sisal fiber a reliable reinforcement material for a variety of matrices, including polymers, rubber, gypsum, and cement, expanding its application beyond conventional uses like ropes, carpets, and mats.

   Various surface treatments can improve the bond between sisal fiber and the matrix while also reducing water absorption. These treatments include the use of coupling agents like silane to modify fiber hydrophilicity, peroxide to stimulate grafting reactions, permanganate, and alkali to enhance fiber surface roughness, and thermal treatment.

   The performance of composites reinforced with sisal fiber is significantly affected by issues such as processing techniques, fiber length, fiber orientation, and fiber-volume ratio. Additionally, combining sisal and glass fibers can result in hybrid composites that capitalize on the superior qualities of both materials [55].
6. Tactical Footwear: Hemicellulose-infused rubber composites offer a viable option for crafting tactical footwear tailored to military needs, ensuring soldiers experience comfort, stability, and resilience across diverse landscapes and climates. [56].
7. Aircraft Components: Rubber composites strengthened with hemicellulose can be utilized in aircraft parts like seals, gaskets, and vibration dampeners, enabling them to endure the rigorous conditions of flight operations, including elevated temperatures, pressures, and dynamic forces. [57].
8. Underwater Equipment: Hemicellulose-infused rubber composites find application in crafting underwater equipment and diving gear for military divers, providing durability against saltwater corrosion, abrasion, and hydrostatic pressure. [58].

9. Tent and Shelter Materials: Rubber composites incorporating hemicellulose can be utilized to construct military tents and shelters, offering resistance to weather, and UV radiation, and ensuring durability for temporary accommodation during field operations. [59].
10. Ammunition Packaging: Hemicellulose-reinforced rubber composites have the potential to be utilized in military logistics operations for packaging ammunition and explosives. By enhancing cushioning and shock absorption properties, they contribute to the safe transportation and storage of these materials [60].

5.9 Tyre Industry

1. Tyre Tread Compounds: Incorporating hemicellulose into rubber compounds for tyre treads can enhance traction, wear resistance, and overall performance [61].
2. Tyre Sidewalls: Utilizing hemicellulose-reinforced rubber composites in tyre sidewalls can bolster stiffness, durability, and resilience against environmental factors like heat and UV radiation. The results of the investigation showed that lignin surface properties affect curing kinetics as well as the compatibility and interfacial adhesion of fillers with rubber. The compatibility between the filler and the polymer was improved by using a surface modification technique with different coupling agents, resulting in the enhancement of rubber properties [62].
3. Inner Liners: Additives derived from hemicellulose show promise in improving the impermeability and air retention properties of inner liners in tyres, thus reducing pressure loss over time. Three methods were employed by the author to characterize fibers, which were considered suitable for assessing structural differences among fiber components. Through FTIR spectroscopy analysis, higher extractive contents in Mezilaurus itauba and buriti fibers were found to be correlated with prominent bands around 2920 and 2850 cm − 1, cautioning against the use of these bands, among others, for comparing crystallinity between lignocellulosic fibers. Results from X-ray diffractometry indicated that Dipteryx odorata and curaua fibers possess more organized cellulose chains with larger crystal sizes, potentially resulting in increased crystallinity. The collective findings suggest that lower extractive and bound water levels, linked to higher crystallinity and larger crystallite size, contribute to slowing down the degradation process and enhancing the thermal stability of lignocellulosic fibers [63].
4. Tyre Compounding: Hemicellulose offers utility as a filler or reinforcing agent in tyre rubber compounds, elevating mechanical properties such as tensile strength and tear resistance [64].
5. Tyre Retreading Materials: Rubber compounds containing hemicellulose can extend tyre lifespan through tyre retreading processes, diminishing waste and promoting sustainability within the tyre industry [65].

6. Winter Tyres: Winter tyre performance can be boosted by incorporating hemicellulose-reinforced rubber compounds, improving traction and grip on snowy and icy surfaces [66].
7. Tyre Vulcanization: Additives derived from hemicellulose can aid in the vulcanization process during tyre manufacturing, enhancing cross-linking density and overall mechanical properties of the rubber [67].
8. Run-Flat Tyres: Materials based on hemicellulose can be integrated into run-flat tyre systems to provide support and stability in case of a puncture, allowing for continued driving over a limited distance [68].
9. Green Tyre Technologies: Hemicellulose-based additives align with the growing trend of green tyre technologies, offering sustainable alternatives to traditional tyre materials and contributing to reduced environmental impact [69].
10. Off-Road Tyres: Enhancing the durability and puncture resistance of off-road tyres is achievable through hemicellulose-reinforced rubber composites, improving performance in challenging terrains and harsh conditions [70].

5.10 Coating Industry

1. Water-Based Paints: Hemicellulose rubber composites serve as binders in water-based paints, offering enhanced adhesion, film formation, and moisture resistance [71].
2. Corrosion Protection Coatings: Coatings formulated with hemicellulose-based rubber composites provide effective corrosion protection for metal surfaces, acting as a barrier against moisture and corrosive agents [72].
3. Wood Coatings: As coatings for wood surfaces, hemicellulose rubber composites enhance durability, water resistance, and UV protection, prolonging the lifespan of wooden structures [73].
4. Paper Coatings: Hemicellulose-based rubber composites improve surface smoothness, printability, and grease resistance when used as coatings for paper products, elevating the quality of printed materials [74].
5. Textile Coatings: Applied as coatings on textiles, hemicellulose rubber composites impart water repellence, flame resistance, and antimicrobial properties, enhancing the functionality of textile materials [73].
6. Metal Coatings: Coatings containing hemicellulose-based rubber composites offer corrosion protection, lubrication, and wear resistance when applied on metal surfaces, extending the longevity of metal components [75].
7. Concrete Coatings: Hemicellulose rubber composites, when used as coatings on concrete surfaces, enhance durability, water resistance, and resistance to chemical attack, improving the performance and lifespan of concrete structures [76].

8. Roof Coatings: Roof coatings formulated with hemicellulose-based rubber composites provide waterproofing, UV protection, and thermal insulation, enhancing energy efficiency and extending the lifespan of roofing materials [77].
9. Automotive Coatings: Applied as coatings on automotive surfaces, hemicellulose rubber composites enhance scratch resistance, weather resistance, and aesthetic appearance, providing durable protection and gloss [78].
10. Food Packaging Coatings: Hemicellulose-based rubber composites serve as coatings for food packaging materials, providing moisture barrier properties, grease resistance, and biodegradability to enhance the protection and freshness of packaged edible items [79].
11. Anti-fouling Coatings: Coatings containing hemicellulose rubber composites prevent fouling by marine organisms when applied on marine surfaces, reducing drag and fuel consumption for ships and marine structures [80].
12. Aerospace Coatings: In aerospace applications, coatings formulated with hemicellulose-based rubber composites provide thermal insulation, corrosion resistance, and protection against extreme environmental conditions, ensuring the integrity and performance of aerospace components [81].
13. Medical Coatings: Hemicellulose rubber composites, as coatings on medical devices and implants, improve biocompatibility, tissue adhesion, and antimicrobial properties, enhancing the safety and efficacy of medical treatments [82].
14. Furniture Coatings: Used as coatings for furniture surfaces, hemicellulose-based rubber composites offer scratch resistance, moisture resistance, and aesthetic appeal, enhancing the durability and appearance of furniture products [83].
15. Electronic Coatings: Coatings containing hemicellulose rubber composites provide insulation, moisture resistance, and protection against electrical and thermal stress when applied on electronic components and circuits, ensuring the reliability and longevity of electronic devices [84].

5.11 Membrane Technology Applications

1. Water Filtration Membranes: Utilizing hemicellulose rubber composites enables the production of water filtration membranes, facilitating the elimination of contaminants and impurities from aquatic bases [85].
2. Ultrafiltration Membranes: Hemicellulose-based rubber composites serve as viable materials for ultrafiltration membranes, allowing the separation of macromolecules and particles from liquids [86].
3. Reverse Osmosis Membranes: Hemicellulose rubber composites have potential applications in reverse osmosis membranes for desalination and water purification by eliminating salts and other dissolved substances [87].

4. Gas Separation Membranes: Hemicellulose-based rubber composites are applicable in gas separation membrane fabrication, beneficial for carbon dioxide capture and natural gas purification [88].
5. Nanofiltration Membranes: Hemicellulose rubber composites are employed in the creation of nanofiltration membranes, enabling the separation of solutes based on size and charge, beneficial in water treatment and industrial processes [89].
6. Pervaporation Membranes: Hemicellulose-based rubber composites find potential in pervaporation membranes for separating liquid mixtures via membrane permeation [90].
7. Selective Separation Membranes: Tailoring hemicellulose rubber composites enables the creation of selective separation membranes for specific applications, such as separating ions or molecules from complex mixtures [91].
8. Antibacterial Membranes: Functionalizing hemicellulose-based rubber composites with antibacterial agents allows the creation of membranes with antimicrobial properties, suitable for water treatment and biomedical applications [92].
9. Ionic Separation Membranes: Engineering hemicellulose rubber composites facilitate the creation of membranes capable of selectively separating ions from solutions, beneficial in electrodialysis and ion exchange applications [93].
10. Hemodialysis Membranes: Hemicellulose-based rubber composites may be employed to fabricate hemodialysis membranes for blood purification, offering biocompatibility and selective filtration properties [94]
11. Vapor Permeation Membranes: Hemicellulose rubber composites are suitable for vapor permeation membranes used in dehydration and solvent recovery processes [95].
12. Membrane Distillation Membranes: Hemicellulose rubber composites are employed in membrane distillation membranes for desalination and concentration processes, offering thermal stability and resistance to fouling [96].
13. Selective Permeation Membranes: Hemicellulose rubber composites are engineered to create membranes with selective permeation properties, allowing specific molecules or solvents to pass while rejecting others, beneficial in separation and purification processes [97].
14. Environmental Remediation Membranes: Hemicellulose-based rubber composites are utilized in membranes for environmental remediation applications, such as removing pollutants and contaminants from air and water streams [98].

5.12 Aerospace Applications

1. Aircraft Seals and Gaskets: Employing hemicellulose rubber composites in the production of seals and gaskets for aircraft components ensures dependable sealing against fluid leaks and preserves pressure integrity [99].

2. Aircraft Interior Components: Hemicellulose rubber composites find application in crafting interior components such as panels, trim, and insulation for aircraft, providing lightweight, fire-resistant, and eco-friendly alternatives [100].
3. Aircraft Flooring: Utilizing hemicellulose rubber composites as materials for aircraft flooring offers durability, impact resistance, and reduced weight compared to conventional materials [101].
4. Aircraft Interior Soundproofing: Hemicellulose rubber composites can be utilized in soundproofing materials for aircraft interiors, diminishing noise transmission and enhancing passenger comfort during flight [102].
5. Aircraft Cargo Liners: Incorporating hemicellulose rubber composites into cargo liners for aircraft cargo compartments ensures protection for transported goods while reducing overall weight and environmental impact [103].
6. Aircraft Ducting and Ventilation Components: Hemicellulose rubber composites are utilized in fabricating ducting and ventilation components for aircraft air conditioning and environmental control systems, offering lightweight and corrosion-resistant alternatives [104].
7. Aircraft Radomes: Hemicellulose rubber composites serve as materials for radomes, safeguarding radar systems on aircraft from environmental factors while maintaining electromagnetic transparency [105].
8. Aircraft Engine Mounts: Utilizing hemicellulose rubber composites in the fabrication of engine mounts for aircraft engines provides vibration isolation and shock absorption, reducing structural fatigue [106].
9. Aircraft Wing and Fuselage Components: Hemicellulose rubber composites are employed in manufacturing wing and fuselage components, presenting lightweight, high-strength alternatives to traditional materials [107].
10. Aircraft Fairings: Hemicellulose rubber composites serve as materials for aircraft fairings, delivering aerodynamic shaping and protection for aircraft components while reducing weight and environmental impact [108].
11. Aircraft Winglets: Incorporating hemicellulose rubber composites into winglets for aircraft wings enhances aerodynamic efficiency and fuel economy by minimizing drag and vortex formation [109].
12. Aircraft Landing Gear Components: Hemicellulose rubber composites are utilized in fabricating landing gear components for aircraft, providing lightweight, corrosion-resistant alternatives to traditional materials [110].
13. Aircraft Antenna Housings: Hemicellulose rubber composites serve as materials for antenna housings on aircraft, delivering electromagnetic shielding and protection for sensitive communication equipment [111].
14. Aircraft Insulation Materials: Utilizing hemicellulose rubber composites in the fabrication of insulation materials for aircraft enhances thermal insulation and fire resistance, improving passenger comfort and safety [56].
15. Aircraft Composite Structures: Hemicellulose rubber composites are employed in constructing composite structures for aircraft, providing lightweight, high-strength alternatives to traditional metal alloys [114].

5.13 Conclusion

The integration of hemicellulose in both natural and synthetic rubber nanocomposites presents a multifaceted and promising avenue across a myriad of industries. Through a comprehensive review of current literature, this paper has highlighted the diverse applications of hemicellulose-infused rubber nanocomposites, spanning from biomedical to aerospace sectors.

In biomedical applications, the biocompatibility and biodegradability of hemicellulose offer unique advantages for tissue engineering, drug delivery systems, and wound dressings. Moreover, the mechanical properties and chemical versatility of hemicellulose contribute to its utilization in packaging, structural, and military applications, enhancing performance and sustainability simultaneously.

The tyre industry benefits from the improved mechanical properties and reduced environmental impact achieved through hemicellulose incorporation, paving the way for more resilient and eco-friendly tyre products. Furthermore, the coating industry finds value in the enhanced adhesion, barrier properties, and corrosion resistance offered by hemicellulose-based rubber nanocomposites.

Innovations in membrane technology leverage hemicellulose to enhance filtration efficiency, water purification, and gas separation processes, addressing critical challenges in environmental sustainability and resource management. Lastly, the aerospace sector capitalizes on the lightweight nature, thermal stability, and fire-retardant properties of hemicellulose-infused rubber nanocomposites to advance aerospace applications, from structural components to insulation materials.

Overall, this review underscores the versatility and potential of hemicellulose in shaping the future of rubber nanocomposites across various sectors, driving innovation, sustainability, and performance excellence. As research continues to progress, further exploration and optimization of hemicellulose-based formulations will undoubtedly unlock new opportunities and propel advancements in materials science and engineering.

References

1. Scheller HV, Ulvskov P (2010) Hemicelluloses. Annu Rev Plant Biol 61:263–289. https://doi.org/10.1146/annurev-arplant-042809-112315
2. Abdul Khalil HPS, Bhat AH, Ireana Yusra AF (2012) Green composites from sustainable cellulose nanofibrils: A review. Carbohyd Polym 87(2):963–979. https://doi.org/10.1016/j.carbpol.2011.08.078
3. Jana S, Samanta S, Halder SK (2017) Natural rubber/hemicellulose composite: Mechanical, thermal, and swelling behavior. Int J Biol Macromol 105(Pt 1):364–372. https://doi.org/10.1016/j.ijbiomac.2017.07.062
4. Lee SY, Chun SJ (2019) Study on the mechanical properties of natural rubber composites reinforced with hemicellulose microfibers. Compos Interfaces 26(5):451–461. https://doi.org/10.1080/09276440.2018.1520242

5. Sreekala MS, Kumaran MG, Joseph S, Jacob M, Thomas S, Thomas S (2002) Oil palm fibers: Morphology, chemical composition, surface modification, and mechanical properties. J Appl Polym Sci 83(3):1799–1809. https://doi.org/10.1002/app.2279
6. John MJ, Anandjiwala RD, Thomas S (2009) Lignocellulosic fiber reinforced rubber composites. np: Old City Publishing
7. Rennie EA, Scheller HV (2014) Xylan biosynthesis. Curr Opin Biotechnol 26:100–107
8. Ten E, Vermerris W (2015) Recent developments in polymers derived from industrial lignin. J Appl Polym Sci, 132(24)
9. Liu Y, Ma J, Yu J, Zhou Z, Ma J (2018) Effect of hemicellulose extracted from corn stalk on the mechanical properties of natural rubber composites. Ind Crops Prod 125:25–31
10. Saba N, Tawakkal ISMA, Jawaid M, Alothman OY (2020) Hemicelluloses as sustainable fillers in natural rubber composites. In Handbook of Sustainable Polymers (pp. 379–398). Springer, Cham
11. Yuniarto K, Muchtar A, Aziz AA, Sapuan SM, Wahyuni S (2019) Flame retardancy of hemicellulose-filled natural rubber composites. Polymers 11(10):1664
12. Rizal S, Muchtar A, Yuniarto K (2021) Hemicellulose as filler in natural rubber composites: A review. Materials Today: Proceedings
13. Thakur VK, Thakur MK, Raghavan P, Kessler MR, Kaur I (2019) Hemicelluloses-based materials: Advances in green chemistry and technology. Springer Nature
14. Sarkawi SS, Hamid ZAA, Sulaiman SA, Sam ST, Umemura K (2020) The effect of hemicellulose on properties of natural rubber nanocomposites reinforced with nanoclay. Polym Testing 86:106478
15. Numan A, Jawaid M, Paridah MT, Nasir M (2021) Hemicellulose-filled natural rubber nanocomposites: Processing, characterization and gas barrier properties. Compos B Eng 221:109035
16. Rahman MZ, Mace BR, Jayaraman K (2016) Vibration damping of natural fibre-reinforced composite materials. In Proceedings of the 17th European Conference on Composite Material, Munich, Germany (pp. 26–30). Faruk O, et al. (2014) Bio-composites for industrial applications. CRC Press
17. Park CW et al (2020) Hemicellulose-based elastomeric composites for high-performance sealants. ACS Appl Polym Mater 2(6):2510–2519
18. Li S et al (2018) Hemicellulose-modified rubber composites with enhanced damping properties. Compos Sci Technol 165:223–230
19. Zhu H et al (2016) Hemicellulose-based latex for anti-corrosive coating applications. Ind Crops Prod 84:367–375
20. Yang W, Liu J-J, Wang L-L, Wang W, Yuen ACY, Peng S, Yu B, Lu H-D, Yeoh GH, Wang C-H (2020) MultifunctionalMXene/NaturalRubberCompositeFilmswithExceptionalFlexibilityandDurability. Compos. Part. B-Eng. 188:107875
21. Peng X, Du F, Zhong L (2019) Synthesis, characterization, and applications of hemicelluloses based eco-friendly polymer composites. Sustain Polym Compos Nanocomposites, 1267–1322
22. Liu G, Shi K, Sun H (2023) Research progress in hemicellulose-based nanocomposite film as food packaging. Polymers 15(4):979
23. Wu X, Qi Z, Li X, Wang H, Yang K, Cai H, Han X (2024) Polymerizable deep eutectic solvent treated lignocellulose: Green strategy for synergetic production of tough strain sensing elastomers and nanocellulose. Int J Biol Macromol 264:130670
24. Pabasara WGA, Saputhanthree A, Weragoda VSC (2022) Utilization of paddy husk carbon black as an alternative filler material in rubber industry. In 2022 Moratuwa Engineering Research Conference (MERCon) (pp 1–6). IEEE
25. Klemm D, Heublein B, Fink HP, Bohn A (2005) Cellulose: Fascinating biopolymer and sustainable raw material. Angew Chem Int Ed 44(22):3358–3393
26. Abazari MF, Gholizadeh S, Karizi SZ, Birgani NH, Abazari D, Paknia S, Delattre C (2021) Recent advances in cellulose-based structures as the wound-healing biomaterials: a clinically oriented review. Appl Sci 11(17):7769

27. Kalia S, Dufresne A, Cherian BM, Kaith BS, Avérous L, Njuguna J, Nassiopoulos E (2011) Cellulose-based bio-and nanocomposites: a review. Int J Polym Sci
28. Tudoroiu EE, Dinu-Pîrvu CE, Albu Kaya MG, Popa L, Anuța V, Prisada RM, Ghica MV (2021) An overview of cellulose derivatives-based dressings for wound-healing management. Pharmaceuticals 14(12):1215
29. Hu J, Tian X, Sun J, Yuan J, Yuan Y (2020) Chitin nanocrystals reticulated self-assembled architecture reinforces deproteinized natural rubber latex film. J Appl Polym Sci 137(39):49173
30. Ding B, Huang S, Shen K, Hou J, Gao H, Duan Y, Zhang J (2019) Natural rubber bio-nanocomposites reinforced with self-assembled chitin nanofibers from aqueous KOH/urea solution. Carbohyd Polym 225:115230
31. Jia Z, Zeng H, Ye X, Dai M, Tang C, Liu L (2023) Hydrogel-based treatments for spinal cord injuries. Heliyon
32. Mudgil D, Barak S (2013) Composition, properties and health benefits of indigestible carbohydrate polymers as dietary fiber: A review. Int J Biol Macromol 61:1–6
33. Dey N, Vickram S, Thanigaivel S, Subbaiya R, Kim W, Karmegam N, Govarthanan M (2022) Nanomaterials for transforming barrier properties of lignocellulosic biomass towards potential applications–A review. Fuel 316:123444
34. Lavoine N, Desloges I, Dufresne A, Bras J (2012) Microfibrillated cellulose—Its barrier properties and applications in cellulosic materials: A review. Carbohydr Polym 90:735–764
35. Nechita P, Mirela R, Ciolacu F (2021) Xylan Hemicellulose: A renewable material with potential properties for food packaging applications. Sustainability 13(24):13504
36. Thakur VK, Thakur MK, Raghavan P, Kessler MR (2014) Progress in green polymer composites from lignin for multifunctional applications: a review. ACS Sustain Chem & Eng 2(5):1072–1092
37. Aulin C, Gällstedt M, Lindström T (2010) Oxygen and oil barrier properties of microfibrillated cellulose films and coatings. Cellulose 17:559–574
38. Najafi A, Khademi-Eslam H (2011) Lignocellulosic filler/recycled HDPE composites: Effect of filler type on physical and flexural properties. BioResources 6(3):2411–2424
39. Ahola S, Turon X, Osterberg M, Laine J, Rojas OJ (2008) Enzymatic hydrolysis of native cellulose nanofibrils and other cellulose model films: effect of surface structure. Langmuir 24(20):11592–11599
40. Vilela C, Pinto RJ, Figueiredo AR, Neto CP, Silvestre AJ, Freire CS 1 Development and applications of cellulose nanofibres based polymer nanocomposites
41. Kabir MM, Wang H, Lau KT, Cardona F (2012) Chemical treatments on plant-based natural fibre reinforced polymer composites: An overview. Compos B Eng 43(7):2883–2892
42. Chang BP, Gupta A, Muthuraj R, Mekonnen TH (2021) Bioresourced fillers for rubber composite sustainability: current development and future opportunities. Green Chem 23(15):5337–5378
43. Barrera CS, Cornish K (2017) Ind Crops Prod 107:217–231
44. Barana D (2017) Lignin-based elastomeric composites for sustainable tyre technology
45. Asonye NN (2021) Experimental characterization of sustainable jute fiber hybrid composites in tyre tread (doctoral dissertation, southern university and agricultural and mechanical college).
46. Mago J, Negi A, Pant KK, Fatima S (2022) Development of natural rubber-bamboo biochar composites for vibration and noise control applications. J Clean Prod 373:133760
47. Zhao X, Bhagia S, Gomez-Maldonado D, Tang X, Wasti S, Lu S, Ozcan S (2023) Bioinspired design toward nanocellulose-based materials. Materials Today. properties of natural fibre reinforced polymer composites. Constr Build Mater 106:149–159
48. Faruk O, Bledzki AK, Fink HP, Sain M (2012) Biocomposites reinforced with natural fibers: 2000–2010. Prog Polym Sci 37(11):1552–1596
49. Ngo TD, Kashani A, Imbalzano G, Nguyen KTQ, Hui D (2018) Additive manufacturing (3D printing): A review of materials, methods, applications and challenges. Compos Part B Eng 143:172–196

50. Cheung HY, Ho MP, Lau KT, Cardona F, Hui D (2009) Natural fibre-reinforced composites for bioengineering and environmental engineering applications. Compos B Eng 40(7):655–663
51. Sonawane P, Utikar VA, Deshmukh D, Jadhav SS, Jain SS, Bhadekar PP, Karle A, Application of nanocellulose in military sector: a review
52. Mikkonen KS (2012) Recent studies on hemicellulose-based blends, composites and nanocomposites. In Advances in Natural Polymers: Composites and Nanocomposites (pp. 313–336). Berlin, Heidelberg: Springer Berlin Heidelberg
53. Bouafif H, Koubaa A, Perré P, Cloutier A (2007) Analysis of viscoelastic properties of natural fiber reinforced thermoplastic composites using the time-temperature superposition principle. Mater Sci Eng A 459:155–161
54. Li Y, Mai Y-W, Ye L (2000) Sisal fibre and its composites: A review of recent developments. Compos Sci Technol 60:2037–2055
55. Satyanarayana KG, Arizaga GG, Wypych F (2009) Biodegradable composites based on lignocellulosic fibers—An overview. Prog Polym Sci 34(9):982–1021
56. Asim M, Saba N, Jawaid M, Nasir M (2018) Potential of natural fiber/biomass filler-reinforced polymer composites in aerospace applications. In Sustain Compos Aerosp Appl (pp. 253–268). Woodhead Publishing
57. Shalwan A, Yousif BF (2013) In state of art: mechanical and tribological behaviour of polymeric composites based on natural fibres. Mater Des 48:14–24
58. Dittenber DB, Gangarao HVS (2012) Critical review of recent publications on use of natural composites in infrastructure. Compos Part A Appl Sci Manuf 43:1419–1429
59. Jincy EM, Femina KS (2023) Heteropolymer in biomass: hemicellulose extraction and modifications. In Handbook of Biomass (pp. 1–32). Singapore: Springer Nature Singapore
60. Kabir MM, Wang H, Lau KT, Cardona F (2013) Effects of chemical treatments on hemp fibre structure. Appl Surf Sci 276:13–23
61. Sekar, P. (2020). Design of bio-based filler system for tyre tread application.
62. Poletto M, Ornaghi Junior HL, Zattera AJ (2014) Native cellulose: structure, characterization and thermal properties. Materials 7(9):6105–6119
63. Nair AB, Joseph R (2014) Eco-friendly bio-composites using natural rubber (NR) matrices and natural fiber reinforcements. In Chemistry, manufacture and applications of natural rubber (pp. 249–283). Woodhead Publishing
64. Alahapperuma KG (2021, July) Optimize technical properties with master batch blends in RSS/Scrap rubber tyre retread compounds. In 2021 Moratuwa Engineering Research Conference (MERCon) (pp. 549–554). IEEE
65. Das S, Satpathi H, Roopa S, Gupta SD (2021) Sustainability of the tyre industry: Through a material approach. In Applied biopolymer technology and bioplastics (pp. 53–98). Apple Academic Press
66. Borazan AA, Alkan A (2019) The effect of plantbased oils and cellulose fillers on the rheological and physico mechanical properties of tyre tread. Feb-Fresenius Environ Bull, 7227
67. Ganster J, Fink HP (2011) Man-made cellulose short fiber reinforced oil and bio-based thermoplastics. Cellulose Fibers: Bio-and Nano-Polymer Composites: Green Chemistry and Technology, 479–506
68. Dominic M, Joseph R, Begum PS, Kanoth BP, Chandra J, Thomas S (2020) Green tyre technology: Effect of rice husk derived nanocellulose (RHNC) in replacing carbon black (CB) in natural rubber (NR) compounding. Carbohyd Polym 230:115620
69. Formela K, Kurańska M, Barczewski M (2022) Recent advances in development of waste-based polymer materials: A review. Polymers 14(5):1050
70. Huang Y, Feng Q, Ye C, Nair SS, Yan N (2020) Incorporation of ligno-cellulose nanofibrils and bark extractives in water-based coatings for improved wood protection. Prog Org Coat 138:105210
71. Azani NFSM, Haafiz MM, Zahari A, Poinsignon S, Brosse N, Hussin MH (2020) Preparation and characterizations of oil palm fronds cellulose nanocrystal (OPF-CNC) as reinforcing filler in epoxy-Zn rich coating for mild steel corrosion protection. Int J Biol Macromol 153:385–398

72. Nypelö, T., Laine, C., Aoki, M., Tammelin, T., & Henniges, U. (2016) Etherification of wood-based hemicelluloses for interfacial activity. Biomacromol 17(5):1894–1901
73. Farhat W, Venditti R, Quick A, Taha M, Mignard N, Becquart F, Ayoub A (2017) Hemicellulose extraction and characterization for applications in paper coatings and adhesives. Ind Crops Prod 107:370–377
74. da Silva AE, Marcelino HR, Gomes MCS, Oliveira EE, Nagashima T Jr, Egito EST (2012) Xylan, a promising hemicellulose for pharmaceutical use. Prod Appl Biopolym 7:61–84
75. Manigandan S, Praveenkumar TR, Al-Mohaimeed AM, Brindhadevi K, Pugazhendhi A (2021) Characterization of polyurethane coating on high performance concrete reinforced with chemically treated Ananas erectifolius fiber. Prog Org Coat 150:105977
76. Hess D, Baar J, Tippner J (2020) Different types of coatings used for wooden shingles exposed to natural and artificial aging. Table of content Monday, July 13th, 40
77. Prasanth SM, Kumar PS, Harish S, Rishikesh M, Nanda S, Vo DVN (2021) Application of biomass derived products in mid-size automotive industries: A review. Chemosphere 280:130723
78. Zhao Y, Sun H, Yang B, Weng Y (2020) Hemicellulose-based film: potential green films for food packaging. Polymers 12(8):1775
79. Morsy A, Mahmoud AS, Soliman A, Ibrahim H, Fadl E (2022) Improved anti-biofouling resistances using novel nanocelluloses/cellulose acetate extracted from rice straw-based membranes for water desalination. Sci Rep 12(1):4386
80. Gonçalves GC (2022) Design and elaboration of anti-corrosion hybrid bio-sourced coatings for aircraft industries via EISA methodology (Doctoral dissertation, Université de Pau et des Pays de l'Adour)
81. Liu H, Chen T, Dong C, Pan X (2020) Biomedical applications of hemicellulose-based hydrogels. Curr Med Chem 27(28):4647–4659
82. Ren Y, Yang Y, Zhang J, Ge S, Ye H, Shi Y, Zhang Z (2022) Innovative conversion of pretreated Buxus sinica into high-performance biocomposites for potential use as furniture material. ACS Appl Mater Interfaces 14(41):47176–47187
83. Altay BN, Aksoy B, Banerjee D, Maddipatla D, Fleming PD, Bolduc M, Demir M (2021) Lignin-derived carbon-coated functional paper for printed electronics. ACS Appl Electron Mater 3(9):3904–3914
84. Egues I, Sanchez C, Mondragon I, Labidi J (2012) Separation and purification of hemicellulose by ultrafiltration. Ind Eng Chem Res 51(1):523–530
85. Krawczyk H, Arkell A, Jönsson AS (2011) Membrane performance during ultrafiltration of a high-viscosity solution containing hemicelluloses from wheat bran. Sep Purif Technol 83:144–150
86. Roy Y, Top RW, de Vos WM, Schuur B (2023) Organic solvent reverse osmosis (OSRO) for the recovery of hemicellulosic derivatives after wood-pulping with a deep eutectic solvent. Chem Eng Sci 267:118367
87. Amusa AA, Ahmad AL, Jimoh AK (2021) Enhanced gas separation prowess using functionalized lignin-free lignocellulosic biomass/polysulfone composite membranes. Membranes 11(3):202
88. Ajao O, Rahni M, Marinova M, Chadjaa H, Savadogo O (2015) Retention and flux characteristics of nanofiltration membranes during hemicellulose prehydrolysate concentration. Chem Eng J 260:605–615
89. Sagehashi M, Nomura T, Shishido H, Sakoda A (2007) Separation of phenols and furfural by pervaporation and reverse osmosis membranes from biomass–superheated steam pyrolysis-derived aqueous solution. Biores Technol 98(10):2018–2026
90. Lee SC (2014) Purification of xylose in simulated hemicellulosic hydrolysates using a two-step emulsion liquid membrane process. Biores Technol 169:692–699
91. Ahmad N, Ahmad MM, Alruwaili NK, Alrowaili ZA, Alomar FA, Akhtar S, Elkomy MH (2021) Antibiotic-loaded psyllium husk hemicellulose and gelatin-based polymeric films for wound dressing application. Pharmaceutics 13(2):236

92. Abejón R, Rabadán J, Lanza S, Abejón A, Garea A, Irabien A (2018) Supported ionic liquid membranes for separation of lignin aqueous solutions. Processes 6(9):143
93. Duolikun T, Ghazali N, Leo BF, Lee HV, Lai CW, Johan MRB (2020) Asymmetric cellulosic membranes: current and future aspects. Symmetry 12(7):1160
94. Hartman J, Albertsson AC, Sjöberg J (2006) Surface-and bulk-modified galactoglucomannan hemicellulose films and film laminates for versatile oxygen barriers. Biomacromol 7(6):1983–1989
95. Hou D, Li T, Chen X, He S, Dai J, Mofid SA, Ren ZJ (2019) Hydrophobic nanostructured wood membrane for thermally efficient distillation. Science Advances, 5(8), eaaw3203
96. Parsin S, Kaltschmitt M (2024) Processing of hemicellulose in wheat straw by steaming and ultrafiltration–A novel approach. Biores Technol 393:130071
97. Ibrahim H, Sazali N, Salleh WNW, Abidin MNZ (2021) A short review on recent utilization of nanocellulose for wastewater remediation and gas separation. Materials Today: Proceedings 42:45–49
98. Merkys B, Nugaras J, Karpenko M, Dubovas A (2023) Manufacturing and implementation of cork-based composites in aviation
99. Arockiam NJ, Jawaid M, Saba N (2018) Sustainable bio composites for aircraft components. In Sustainable composites for aerospace applications (pp. 109–123)
100. Publishing W, Scarponi C, Santulli C, Sarasini F, Tirillò J (2017) Green composites for aircraft interior panels. Int J Sustain Aviat 3(3):252–270
101. Zhang J, Shen Y, Jiang B, Li Y (2018) Sound absorption characterization of natural materials and sandwich structure composites. Aerospace 5(3):75
102. Jamir MR, Majid MS, Khasri A (2018) Natural lightweight hybrid composites for aircraft structural applications. In Sustain Compos Aerosp Appl (pp. 155–170). Woodhead Publishing
103. Chai M (2014) Flammability performance of bio-derived composite materials for aircraft interiors (Doctoral dissertation, ResearchSpace@ Auckland)
104. Haris MY, Laila D, Zainudin ES, Mustapha F, Zahari R, Halim Z (2011) Preliminary review of biocomposites materials for aircraft radome application. Key engineering materials, 471, 563–567
105. Lazo N, Vodenitcharova T, Hoffman M (2015) Optimized bio-inspired stiffening design for an engine nacelle. Bioinspir Biomim 10(6):066008
106. Arockiam NJ, Jawaid M, Saba N (2018) Sustainable bio composites for aircraft components. In Sustainable composites for aerospace applications (pp 109–123). Woodhead Publishing
107. Suhr J (2022) Lightweight, energy ansorbing sandwich composite structures with Bamboo Fiber Reinforced Plastic (BFRP) core by using additive manufacturing. Sungkyunkwan Univ Res Bus
108. Gül Ç, Kocak ED (2021) Composite material design for aircrafts from sustainable lignocellulosic Fibers—A Review. Sustain Des TextEs Fash 23–37
109. Leao AL, Cesarino I, Chanes M, Botelho EC, Dias OAT, Jawaid M (2023) Ecologically enhanced natural/synthetic polymer hybrid composites for Aviation-Interior and Secondary Structures. Green Hybrid Compos Eng Non-Eng Appl, 43–59
110. Nayak NV (2014) Composite materials in aerospace applications. Int J Sci Res Publ 4(9):1–10
111. Huang QY, Xiong F, Fan K, He YC (2013) An experimental study on thermal insulation performance of straw wire aircraft sandwich panel. Adv Mater Res 639:1307–1312

Chapter 6
Starch-Based Rubber Nanocomposites

Aparna Jayan, V. Bijina, and K. Abhitha

Abstract There has been an increasing need to replace the conventional fillers used in rubber with the bio-based materials, which are easily available, renewable and biocompatible. Nanobiofillers are gaining even more importance due to their better reinforcing capability with much less loading. In this context, we aim to include the basic structure, properties, and synthesis methodology of starch nanoparticles. We have also incorporated the preparation methods for fabricating rubber-based composites of starch and the various techniques for characterising these composites like XRD, FTIR, DSC, DMA, SEM, TEM and so on. This chapter paves a way for getting a better understanding regarding the structure and properties of starch nanocrystals and its rubber-based composites.

Keywords Starch · Rubber · Nanocomposites · Biocomposite · Biofiller · Structure · Properties · Synthesis

6.1 Introduction

A considerable amount of work is happening worldwide to develop alternatives for conventionally used fillers in elastomers. Biofillers of plant origin like cellulose and starch have always attracted immense research interest due to their biodegradability, renewability and low material cost. The incorporation of biofillers into elastomers enables the development of 'green elastomers'. The preparation of green elastomers consumes much less energy and is ecologically sustainable. Because of the profuse

A. Jayan · V. Bijina · K. Abhitha (✉)
Department of Polymer Science and Rubber Technology, CUSAT, Kochi 682022, India
e-mail: abhithak80@cusat.ac.in

A. Jayan
e-mail: aparnajayakumar1996@gmail.com

K. Abhitha
Inter University Centre for Nanomaterials and Devices, CUSAT, Kochi 682022, India

Visakh P. M. *Rubber Based Bionanocomposites*, Advanced Structured Materials 210,
https://doi.org/10.1007/978-981-10-2978-3_6

availability and the excellent thermomechanical properties offered to the composites, starch makes a good replacement for the typical petroleum-based fillers like carbon black in rubber composites [1]. The primary intention behind developing biocomposites is the reduction of toxic synthetic substances that are otherwise used in common. For the fabrication of green composites, either the filler or the matrix or both have to be of biological origin. The green composites can also be called as sustainable composites or environmentally benign composites or biocomposites. In order to enhance the filler-rubber compatibility that would result in better reinforcement, either the filler or the matrix is made to undergo modifications. For instance, modifications are usually done on either starch or natural rubber latex to improve the compatibility between the hydrophilic starch particles and hydrophobic latex [2].

Nanomaterials form a very vital category of fillers, which reinforces rubber. The main reason for their significance when compared to conventional fillers is the high surface area they possess, in addition to the high strength and aspect ratio. The efficiency of the performance of these nanofillers is dependent on the filler-filler interaction, rubber-filler interaction, ease of dispersion in the matrix, efficiency of stress transfer at the interface, the entanglement and the amount of polymer which is bound within the neighbourhood of the filler and so on.

The solitary carbon black filler derived from depleting fossil fuels and several mineral fillers in rubber composites impart severe disruption to our environment in different ways. The usage of renewable as well as sustainable materials in rubber composites such as biocomposites is a possible solution to cut down this environmental pollution, climate change and the energy crisis. Among several biocomposites, starch-based rubber composites gain much attention during the past few years since it is abundantly available in nature as well as biodegradable. Literature points out that starch nanoparticles should have at least one of their dimensions lower than 1000 nm; some specify that at least one of their dimensions should be less than 300 nm [3, 4]. This book chapter will explore the current state of research on starch-based rubber nanocomposites, including synthesis of nanostarch and characterisation, properties, processing methods, and potential applications of the starch-based rubber nanocomposites.

Starch belongs to the family of polysaccharides that can be extracted from natural resources like potatoes, corn, rice, cassava, barley, wheat and so on. The structural composition of starch includes amylose, which is a water-soluble linear unbranched chain constituting nearly 30% of the starch, and amylopectin having a branched structure made of α-D-glucose units comprising nearly 70% of starch, the details of which will be mentioned in the following sections. The cyclic structure of the starch along with its hydrogen bonding imparts more crystallinity and rigidity favouring it for further structural applications. The incorporation of starch in the rubber matrix possesses severe difficulty due to the polarity incompatibility. The presence of hydroxyl groups on the surface of starch molecules makes it difficult to disperse with the non-polar rubber matrix but simultaneously provides abundant possibilities for chemical modifications and functionalisation [5]. Through several modifications, starch-based rubber nanocomposites offer good reinforcement properties and superior biodegradation rates. High shear mixing of polysaccharides in rubber matrix

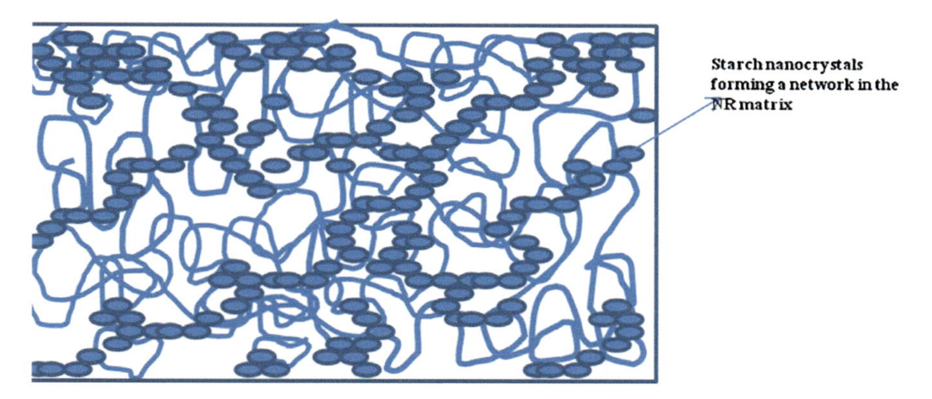

Fig. 6.1 Network formation between rubber and starch nanocrystals [8]

produces uniform dispersion of the composite resulting in enhanced mechanical properties. Similarly incorporating 10 phr of cassava starch directly into the rubber compound also provides enough strength properties to the composites [6]. Misman et al. studied the effect of sulphur pre-vulcanisation and latex co-coagulation method for the preparation of NR-starch composites [7]. They have suggested that polarity incompatibility can be overcome by several methods like matrix modification, usage of interfacial coupling agents, nanostarch modification, etc. Rajisha et al. prepared the nanocomposites consisting of natural rubber filled with potato starch nanocrystals. Strong interactions such as hydrogen bonding between the filler-polymer matrix result in the formation of a rigid three-dimensional networking imposing reinforcing characteristics to the composites [8] (shown in Fig. 6.1).

6.1.1 Structural Details of Starch

Starch is one of the three most prominent biorenewable resources on earth, the other two being sucrose and cellulose. On an estimate, 60–70% of the human caloric intake comes from starch. Starch is a naturally occurring carbohydrate formed in plants by photosynthesis and is present abundance in plant parts like roots, seeds and stalk. Depending on the biological function, starch is categorised into two—storage starch and transitory starch, both of which promote growth and respiration at times when photosynthesis is unlikely to occur. Transient or transitory starch synthesised in daytime accumulates in organelles called chloroplast and is transported at night and metabolised for energy. On the other hand, storage starch is stocked in amyloplasts of non-photosynthetic tissues like endosperm from where they are later taken for fuelling germination and sprouting [9]. Biosynthesis of starch is a highly complicated and synchronised process involving enzymes like adenosine diphosphate—glucose pyrophosphorylase, starch synthase, starch branching and de-branching enzymes, each playing irreplaceable role in the biosynthetic pathway [10].

Starch is a polymer of α-D-glucose and has the basic formula $(C_6H_{10}O_5)_n$. The structural details of starch were confirmed by analysing the residue left from its total acid hydrolysis. Two types of α-glucans are present in starch—amylose and amylopectin. Amylose is the linear form whereas amylopectin is branched and constitutes the major weight of starch. These two glucans make almost 98–99% of the dry weight of starch [11]. In amylopectin, the monomeric α-D-glucose molecules are joined by 1,4-glycosidic linkages that are in addition, linked by 1,6-glycosidic bonds giving rise to branched structures. On the contrary, in amylose, glucose molecules are bonded in a linear fashion via 1,4-glycosidic links, with no or rare 1,6 connections (structure of amylose and amylopectin shown in Fig. 6.2). The amylopectin fraction contributes to the semi-crystalline nature of starch; the adjacent chains orient in a double helix like conformation and pack themselves into an ordered or crystalline lamellae. The branch points in the amylopectin constitute the amorphous segment. The amylose units make up less than 35% and rest in the amorphous region and is not found to be essential in forming the semicrystalline matrix [12].

The weight ratio of amylose and amylopectin in starch varies according to the source. For instance, the weight percentage of amylose in rice starch and corn starch is approximately 23 and 25, respectively. The ratio of amylose and amylopectin differs not only on the basis of plant source but also is found to be different even within a single plant species or organ and depends even on the growth conditions and environment of the plant. The difference in this ratio in turn affects the physical and chemical properties of starch as well as its interaction with other molecules. This invokes difference in its swelling properties, microscopic properties, barrier properties as well as the appearance and stability of starch-based products. For determining the amylose content in starch, several methods are employed, spectrophotometry being the most widely used. Calorimetric, enzymatic and chromatographic techniques are also being used for the purpose as per literature [13]. Most of these methods have certain disadvantages like overestimation of amylose (in spectrophotometry) since iodine is capable of binding with amylopectin as well, tedious sample preparation strategies and lack of cost-effectiveness. Thermogravimetry has been reported as a successful method for amylose content estimation in potato, rice and wheat starch samples. Besides, thermogravimetry is of great practical utility since it needs no sample processing steps and much less time-consuming [13].

The amylose and amylopectin fragments are indeed very different in their properties. Amylose shows a tendency to retrograde and generally forms strong films and hard gels. Amylopectin if dispersed in water is much more stable and forms soft gels and films. Amylopectin and amylose fragments are entangled and contain lipids and phospholipids associated with them. The unique properties of potato starch are attributed to the repulsion of charge between the covalently bonded phosphate derivatives [14].

Starch is synthesised in a granular form and its biosynthesis starts in the hilum. The starch granules are found to exhibit a variety of shapes like ellipsoid, platelets, polygons and tubule. Diameter of the granules varies from 0.1 to 200 microns, and

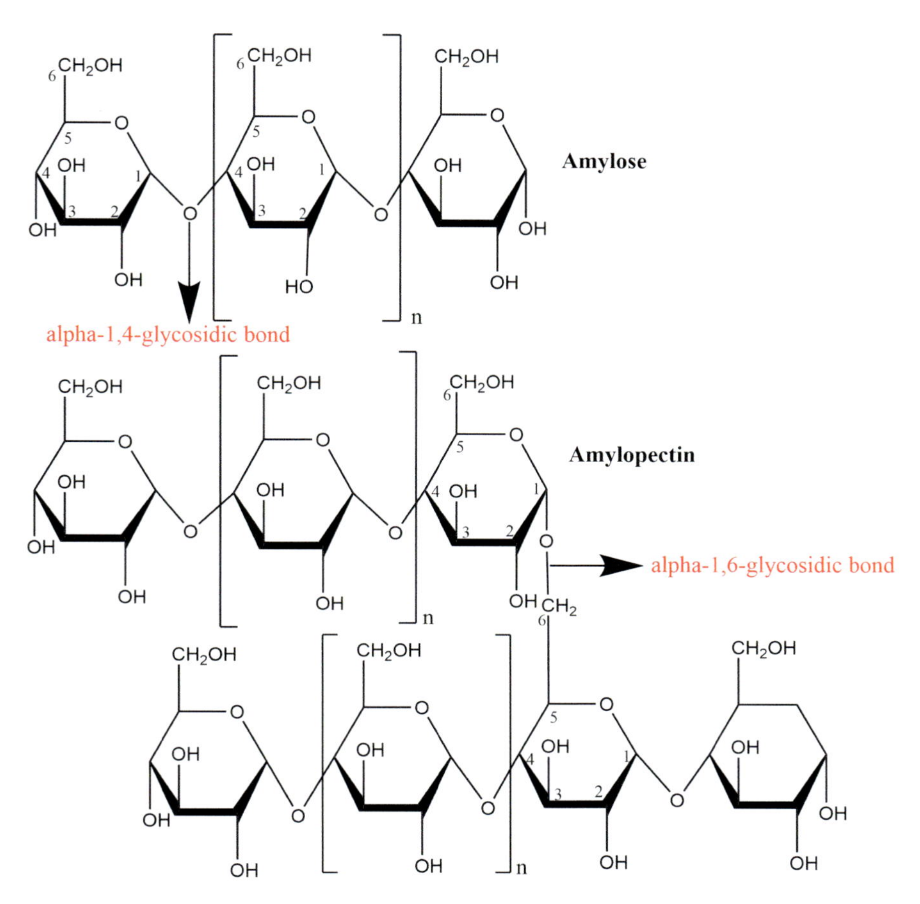

Fig. 6.2 Structure of amylose and amylopectin fragments of starch depicting the glycosidic linkages

this depends on the plant source. These granules are semi-crystalline with density of approximately 1.5 g/cm^3. The percentage crystallinity varies from 15 to 45 [14, 15]. On heating starch with sufficient water, swelling occurs and crystalline nature is disrupted. This gelatinisation can be detected by Differential Scanning Calorimetry (DSC). The endothermic peak in DSC curve accounts for the loss of double helical pattern of amylopectin; a higher temperature and greater energy suggest the presence of greater crystallinity and molecular order. The stage of gelatinisation is followed by agglomeration of starch molecules, which is generally referred to as retrogradation [16].

Starch granules exhibit a multiscale structure (shown in Fig. 6.3). The starch granule has growth rings inside of which further made up of blocks (20–50 nm) composed of crystalline and amorphous lamellae.

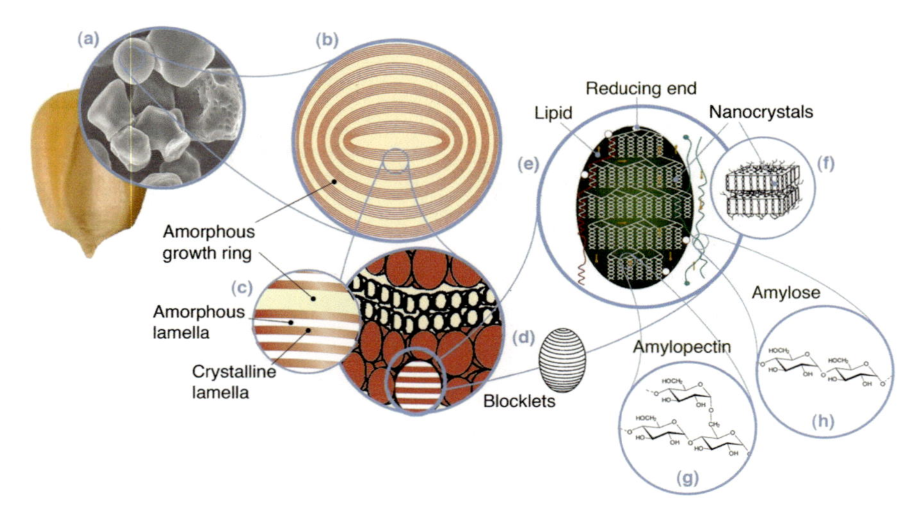

Fig. 6.3 Multiscale structure of starch (**a**) granules of starch taken from maize, (**b**) patterns of growth rings, (**c**) two types of lamellae—amorphous and crystalline; detailed image of crystalline lamellae, (**d**) blocklets constituting unit of growth ring, (**e**) double helices of amylopectin which makes up the crystalline lamellae, (**f**) nanocrystals which is the representation of crystalline lamellae separated by acid hydrolysis, (**g**) molecular structure of amylopectin, (**h**) molecular structure of amylose [17]

The amylopectin fraction mostly constitutes the crystalline part and amylose is present in the amorphous region [17]. Many methodologies are available for evaluating the properties of starch and among them microscopy has gained significant importance [18]. Atomic Force Microscopy (AFM) and Scanning Electron Microscopy (SEM) are some of the prominent tools for examining size and morphological details of starch granules. However due to the complexity and greater operating expenses, optical microscopy and its variants have been used more frequently. Polarised light microscopy (PLM) is a technique commonly used in starch analysis where the sample is illuminated by polarised light; a perpendicular polariser blocks the transmitted light which results in detection of depolarised light [19]. Microscopic evaluation of starch reveals that the carbohydrate shows birefringence in polarised light. The radial orientation of the molecules in the crystalline region is evident from the 'Maltese Cross'-like structures observed in the refraction pattern (shown in Fig. 6.4).

The onion-like structure of the starch granules composed of near concentric circles corresponds to alternating hard crystalline and soft but less ordered regions with hilum at the centre. Irrespective of the botanical origin, the thickness of the crystalline and amorphous regions in combination is found to be 9 nm [21]. PLM technique can identify the source of starch from the typical Maltese cross as well as the shape, size and form of granule and positioning of the hilum [22, 23]. However, using this technique, starch granules appear to have a more smoothened form and the morphology and size of granules obtained might not be accurate. Staining of sample

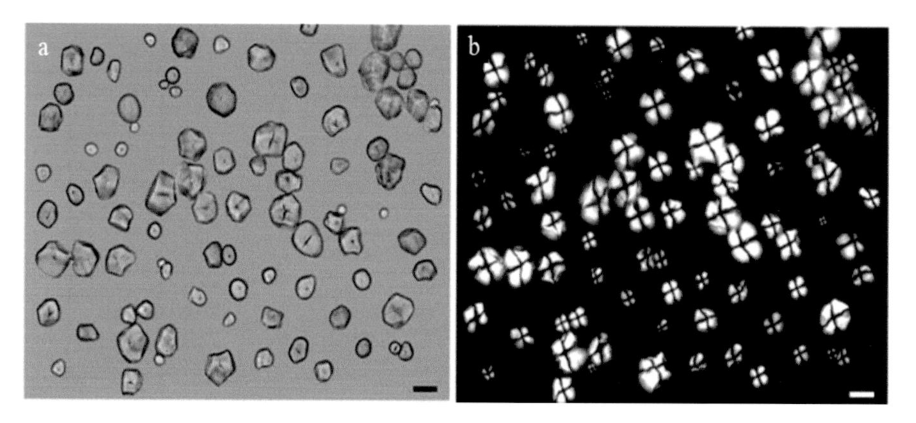

Fig. 6.4 Maltese cross-pattern of maize starch observed under **a** normal light and **b** polarised light [20]

and several complicated algorithms are often employed to yield images with better contrast [24, 25].

6.1.2 Types of Starch

Metabolism of starch in the gut is carried out by the amylolytic enzymes. Hydrolysis of starch generates glucose which after digestion is absorbed by the small intestine. On the basis of rates at which it is digested, starches are classified into Resistant Starch, Slowly Digesting Starch and Rapidly Digesting Starch. The rate of digestion of starch is determined by the percentage of amylose and amylopectin present in it. Resistant starch is defined as that which remains in the small intestine without getting digested and is composed mostly of retrograded amylose. This starch cannot be digested by the amylolytic enzymes present our body. Digestion of starch is initiated by salivary amylase. In large intestine, the microbes present in the alimentary canal break down resistant starch into short chain fatty acids, alcohols and organic acids. A greater percentage of the dietary starch get digested fast. Starches which when subjected to enzymatic digestion convert into glucose within few minutes are categorised as rapidly digestible starch (RDS). The presence of high amount of RDS in food can be of some harm to health because of the sudden increment of glucose level in bloodstream. Then there is the category called slowly digestible starch (SDS), which takes as long as 2 h in the small intestine to get converted into glucose [26, 27].

Another important methodology to analyse the semi-crystalline nature of starch is X-Ray Diffraction (XRD). Starch granules based on their botanical origin exhibit different XRD patterns. Native starch granules, depending on the type of XRD pattern they display, are grouped into A, B and C types. A-type XRD is exhibited by cereal starches where the glucose helices are packed tightly. B-type pattern is shown by

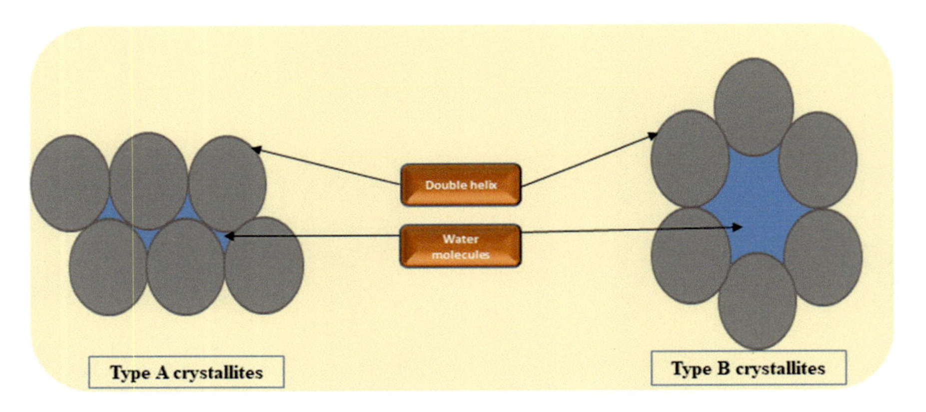

Fig. 6.5 Arrangement of double helices and water molecules in type A and B crystallites

tubers where the double helices are found to be less tightly packed, lodging water molecules in the empty spaces. Legumes are composed of both A and B crystallites and the type of XRD they present is the C pattern. C-type structures have B-type crystallites at the centre and type A crystallites in the surroundings. The source of the starch can thus be identified from its crystallinity pattern [27, 28]. Structural representation of A and B types of starch crystals is shown in Fig. 6.5.

Native starch refers to the starch that is extracted from plant sources directly. Starch can be made to undergo several physical, chemical and enzymatic treatments and starch after treatment is known by the name 'modified starch' [29]. Native starch has its limitations in industrial applications since native starch granules easily take up water, swell, cracks easily and forms a paste after losing viscosity. Physical treatment includes application of heat and pressure in a controlled manner to bring about enhancement in properties. These treatments roughen the surface of granules, increase the amylose content and contribute to lowering the swell percentage and solubility index [30].

Modification can also be done chemically and this introduces new functional groups onto starch molecules. Acid modification is executed by treating starch below its gelatinisation temperature in aqueous acid solutions wherein hydrolysis reduces the size of starch granules and paste viscosity. It is also observed that the acid treatment favours the depolymerisation of amylopectin and enhances the amount of linear chain fragments which are similar to amylose. Enzymatic modifications involve treating starch with different enzymes like hydrolysing enzymes, which enables the formation of derivatives of starch containing several functional groups. An example is pullulanase enzyme, which liberates maltotriose oligomers by acting upon the linear chain glucans as well as hydrolyses 1,6-glycosidic linkages in amylopectin. Isoamylase is a similar enzyme that leads to complete hydrolysis of 1,6-glycosidic bonds [31, 32].

6.1.3 Synthesis Strategies of Nanostarch

The concentric multiscale and semicrystalline nature of starch makes the synthesis of nanoscale structures possible either by acid hydrolysis which forms nanocrystals by enabling the disruption of amorphous realms in starch or by treating gelatinised starch to produce nanoparticles [33].

The strategies available for the synthesis of starch can be broadly grouped into two: top-down and bottom-up approaches.

Nanoprecipitation is a synthetic strategy in which one can precipitate nanoscale polymer particles by adding a dilute solution of the polymer into a solvent which leads to precipitation. Najafi et al. precipitated nanostarch particles by dropwise addition of aqueous phase starch to acetone. They carried out the synthesis using acetylated corn starch, and the final size of the particles was in the range of 221–324 nm [34]. Ethanol is yet another solvent used for carrying out nanoprecipitation. Putro et al. compared acid hydrolysis and ethanol precipitation methods for synthesising nanostarch. Acid hydrolysis yielded particles with much higher crystallinity while ethanol precipitation produced particles with lower crystallinity. Also, acid hydrolysis yielded starch particles of lesser size compared to those obtained by precipitation. This is due to hydrolysis which causes fragmentation of microparticles into smaller sized particles and conversion of some amorphous segments into simple sugar fragments, which resulted in the formation of nanosized particles. Flakes were obtained from acid hydrolysis against spherical nanoparticles, which resulted from precipitation. In the study, further modifications of starch nanoparticles were performed with anionic, cationic and non-ionic surfactants. Ionic surfactants did not seem to cause any impact on the surface morphology of the nanoparticles while the non-ionic surfactant (Tween-20) caused significant changes in the morphology [35].

Then, there is the emulsion technique for synthesising nanoscale starch particles. The technique uses an aqueous solution of the polymer and another solvent which is immiscible or partially miscible with the aqueous phase. Strong agitation of this mixture results in the formation of an emulsion, which is stabilised by the use of surfactants. The droplets act as reaction units favouring the precipitation of size-controlled particles within the drops. The general synthetic route for preparing nanostarch particles using emulsion strategy is shown in Fig. 6.6. Ismail et al. used a microemulsion nanoprecipitation technique for synthesising starch nanoparticles in which they dissolved starch in an alkaline urea solution. Microemulsion solution was a vigorously stirred solution of ethanol, Tween 20 and sunflower oil. Starch solution was dripped gently into the emulsion solution and the formed particles were recovered by centrifugation. The synthesised particles were found to be uniform in size and shape due to the homogeneity of the solution [36].

Recently, Gonde et al. reported a new cavitation-assisted method for the synthesis of starch nanoparticles. They synthesised starch nanoparticles from rice using a cavitation technique under alkaline conditions followed by precipitation.

They have employed both acoustic and hydrodynamic cavitation strategies. The cavitation technique generates micro and nanocavitations which collapse, generating

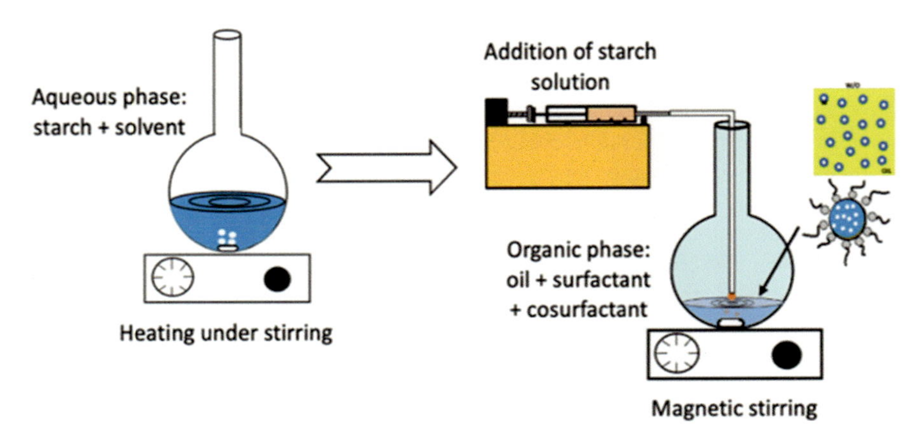

Fig. 6.6 Synthetic route for preparing nanostarch using emulsion technique [37]

high temperature, pressure and high shear microjets locally. When such collapses occur near starch particles in the reaction system, the micro-sized particles break apart into nanosized ones [38]. The synthetic route and the SEM images of the particles and their respective size are detailed in Fig. 6.7.

Starch nanoparticles assume a number of morphologies like nanocrystals, nanospheres, vesicles, nanofibres and micelles, and the formed morphology is largely determined by the preparation strategy. The synthesis of starch by acid hydrolysis consumes great time, and several modifications have been suggested to enhance

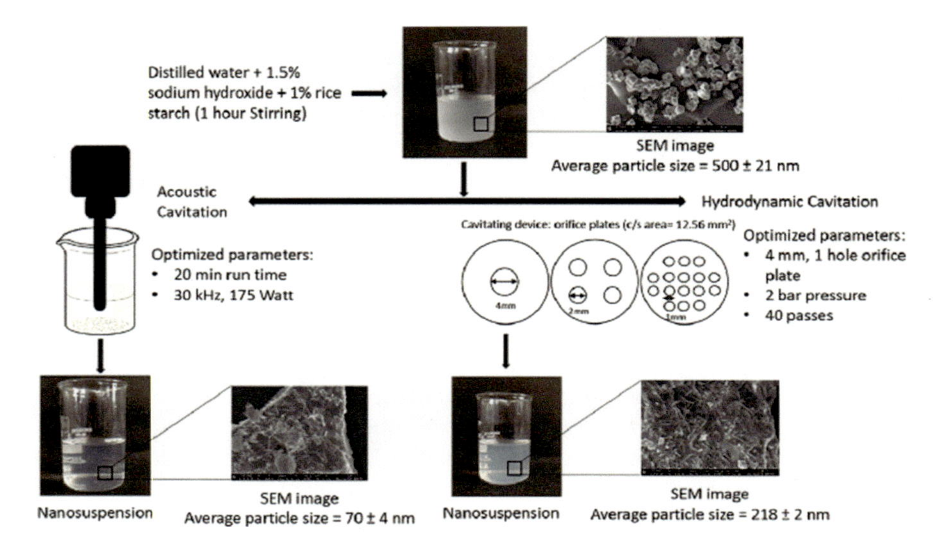

Fig. 6.7 A combined approach utilising cavitation and precipitation for synthesising starch nanoparticles [38]

the efficiency of preparation of starch nanoparticles. One of the approaches is the enzyme-assisted synthesis, which shortens the preparation time effectively. α-amylase, β-amylase and glucoamylase are the enzymes most commonly used for the synthesis of starch nanoparticles [39]. Pre-treatment of waxy corn starch by glucoamylase for 2 h reduced the acid hydrolysis time from 24 to 6 h. Such particles also formed better dispersions. Another strategy to enhance the rate of formation of starch nanoparticles was ball milling. As a result of a combination of ball milling and 3 days of sulphuric acid hydrolysis, the researchers were able to produce starch nanocrystals (SNCs) with a yield of 19.3 wt% and a shape that was comparable to that of SNCs produced by the standard method after 5 days of sulphuric acid hydrolysis. SNCs with round edges and an average diameter of about 31 nm were formed by this combined approach [40].

Under the same H^+ ion concentration, HCl and HNO_3 are found to have the most efficiency in bringing about the hydrolysis of starch, while sulphuric acid and phosphoric acid are the least efficient. However, due to the higher H^+ ion availability, lesser time is usually needed for sulphuric acid-assisted hydrolysis than that taken for HCl-assisted hydrolysis. From the mechanism of acid hydrolysis of starch, more concentrated acids would furnish greater amounts of hydronium ions which would lead to increased α (1–4) glycosidic bond cleavages in starch, leading to formation of shorter chains. Sai Li, et al. used a mixture of sulphuric acid and hydrochloric acid for carrying out the acid hydrolysis of starch and compared the results with the conventional strategy of hydrolysis using sulphuric acid. No differences were observed in the particle morphology, group structure and surface elements in the nanoparticles prepared by the two methods. Compared to SNC (starch nanoparticles synthesised using sulphuric acid), HSNC (nanoparticles synthesised using mixture of acids) had a lower yield. However, the time required to make HSNC was significantly reduced to 1 h, taking only 0.83% of the time (5 days) required to prepare SNC using sulfuric acid. HSNC also displayed higher relative crystallinity and smaller sizes than those of SNC. In addition, the crystal pattern of HSNC (when the ratio of sulfuric acid to hydrochloric acid was 1:1) changed to V-type (a starch crystal pattern, not so common, observed when amylose, in the presence of a suitable ligand, adopts a single helical conformation[41]), while SNC had an A-type crystalline pattern [42].

6.1.4 Various Modifications of Starch Nanoparticles

Starch nanoparticles (SNPs) have been made to undergo several modifications through additional strategies in order to broaden their use or enhance their functionality. Chemical and physical treatments and grafting are some of the modifications commonly carried out. The common modifications of starch nanoparticles and their applications are shown in Fig. 6.8. Particularly, methods of modification that lessen the hydrophilic property of SNPs have attracted a lot of attention. For instance, SNPs have undergone graft modification to improve their hydrophobicity, which is better for interfacial adsorption in pickering emulsions.

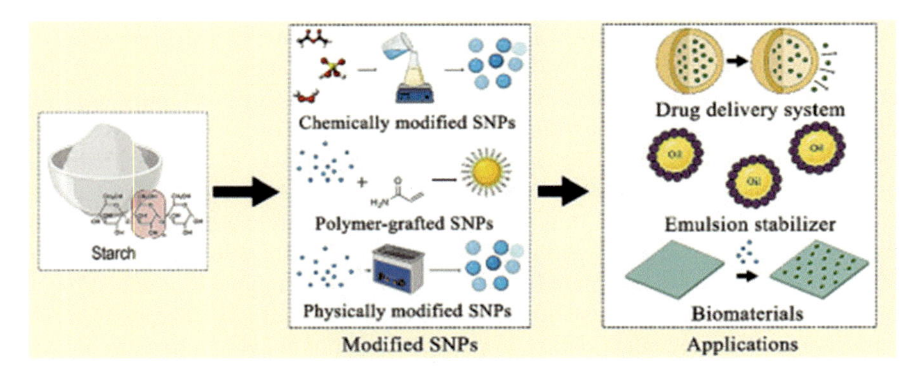

Fig. 6.8 Various techniques for modifying starch [39]

Core–shell SNPs, on the other hand, are useful fillers to increase the durability of weak materials and are created using emulsion polymerisation [43, 44]. Core–shell particle synthesis is a way to enhance the properties of materials since a synergistic action of both the core and shell could be observed in the synthesised nanoparticles. The preparation of starch-PLA nanocomposites is not an easy task due to the thermodynamic immiscibility between hydrophobic PLA and hydrophilic starch. This incompatibility leads to weak mechanical properties of the composites. In a study conducted by Yang Wang et al., stearyl chloride and acryloyl chloride were used to esterify starch. The esterified starch demonstrated an exceptional emulsifying property and might show satisfactory graft efficiency in the grafting procedure due to the addition of a long alkyl chain and double bond. Then, using a soap-free emulsion copolymerisation to create a layer of poly ethyl acrylate (PEA), the core–shell starch-based nanoparticles (CSS NPs) were synthesised. After that, melt-blend technology was used to create PLA/CSS nanocomposites. The starch-induced stiffness and the PEA phase's induced elastic characteristics will be combined in the CSS NPs. The addition of CSS NPs to a PLA matrix improved the matrix's toughness and impact characteristics while retaining a respectably high tensile strength through synergistic action [43].

Starch can be physically altered by microwave irradiation, which uses electromagnetic waves with a frequency between 300 GHz and 300 MHz. In terms of absorption, reflection and transmission of microwave energy, microwave irradiation of starch alters its relative complex permittivity. In-vitro digestibility, morphological, crystalline, pasting, and rheological characteristics of the underused talipot starch were examined by Basheer et al. in relation to the effects of microwave irradiation and chemical treatments, such as oxidation, esterification and cross-linking. They conducted this study to see the impact of microwave radiation as a pre-treatment strategy to chemical modification of starch. The reduced retrogradation chances and enhanced resistant starch content in the modified starch indicate their potential in frozen products and low calorie foods [45].

Starch nanoparticles (SNPs), produced by recrystallising debranched waxy maize starch, were modified using oxygen and ammonia vacuum cold plasma (CP). Fourier transform infrared spectroscopy measurements of the modified SNPs revealed a new carbonyl or carboxyl signal at 1720 cm^{-1}. SNPs changed by oxygen CP treatment exhibit negative charges (between 21.6 and 15.1 mV). Modified SNPs with diameters ranging from 75.94 to 159.72 nm showed good dispersibility and minimal aggregation. Modified SNPs' relative crystallinity fell from 44.13 to 33.80%. Moreover, modified SNPs exhibited substantial absorption of tea polyphenols, indicating that as nanocarriers, they can carry more cargo molecules than primary SNPs. SNPs can be easily, sustainably, and cheaply modified using CP. Modified SNPs can be utilised as nanocarriers to deliver medicine or food components in the pharmaceutical and food industries [45]. Jaimie et al. reported the modification of starch by nitroxide mediated polymerisation (NMP) to modify the surface of starch nanoparticles by grafting onto it poly (methyl methacrylate-co-styrene) (P(MMA-co-S)), poly methyl acrylate (PMA) and poly acrylic acid (PAA) with different monomer loadings.

The grafting was carried out using a three-step procedure as shown in Fig. 6.9. The success of the grafting syntheses was verified using FTIR and ^{1}H NMR. To confirm the grafting copolymerisation of SNP with synthetic polymers, thermogravimetric analysis (TGA) and elemental analysis (EA) were used. To separate the grafted P(MMA-co-S) chains from the SNP-g-P(MMA-co-S) hybrid materials and enable their analysis by gel permeation chromatography (GPC), acid hydrolysis was performed. The modification of SNP via grafting using NMP method shown in this work gives the opportunity for grafting an enormous range of monomers for a broad range of applications. These brand-new hybrid materials made of starch might be used in polymer latex paints, coatings for paper and adhesives [46].

6.1.5 Starch-Based Rubber Nanocomposites

A family of two-phase materials known as 'nanocomposite' have at least one dimension that is less than 100 nm in size. Clay, silica beads, carbon nanotubes and other dispersible colloidal particles are common reinforcing agents. These materials have attracted the most economic attention as fillers for nanocomposites. The potential to obtain enhanced properties even with small fractions of the reinforcing phase has significantly increased the interest in the development of nanocomposites throughout time. Despite the development of carbon nanotube and graphene composites in recent years, the usage of bio nanofillers like nanostarch, nanochitin and nanocellulose in polymer composites has grown as a result of their good physical qualities and environmentally friendly nature and low cost [47]. These biofillers are abundantly available on earth and are renewable materials since they are isolated from natural sources and are thus excellent replacements for the petroleum-derived fillers.

The preparation and the various characterisation of starch-based rubber nanocomposites will be discussed from hereon.

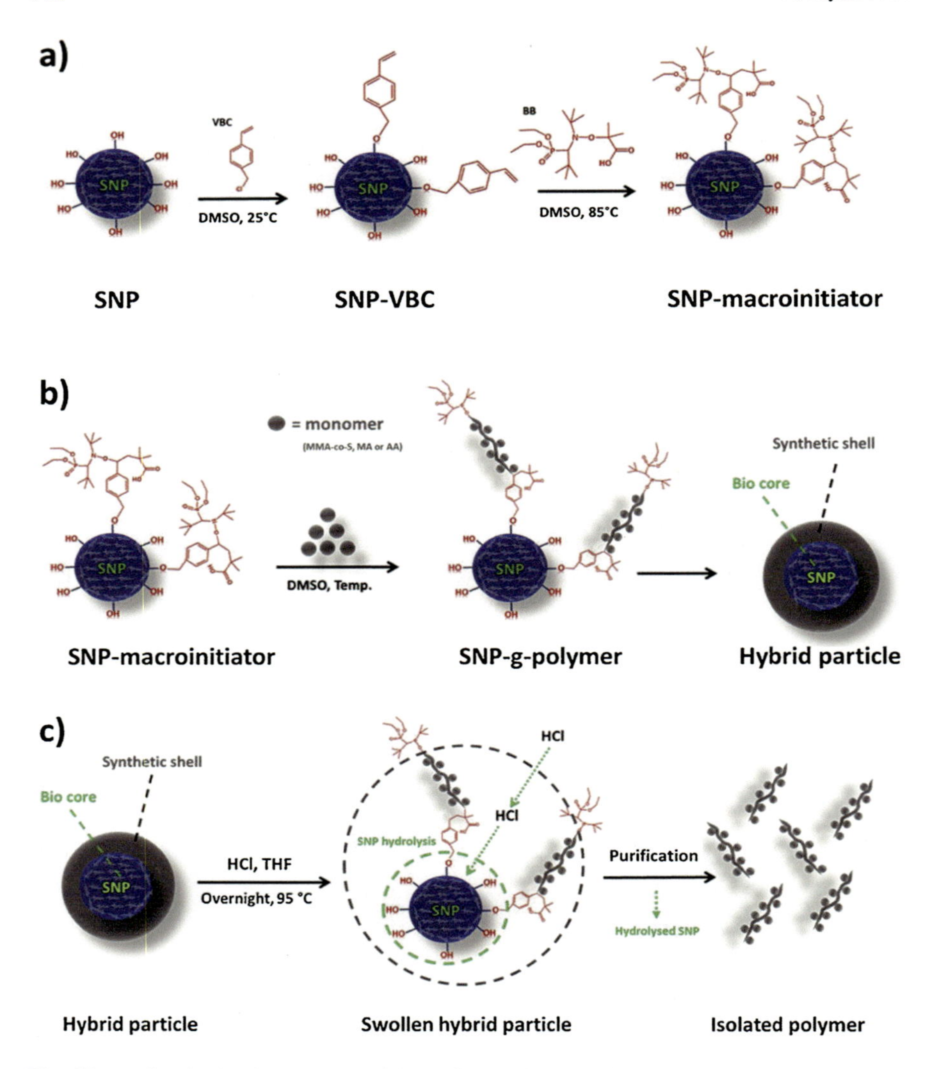

Fig. 6.9 (**a**) Synthesis of SNP macroinitiator, (**b**) grafting, (**c**) acid hydrolysis [46]

6.1.6 Preparation of Starch-Based Rubber Nanocomposites

Several processing methods have been used to prepare starch-based rubber nanocomposites, including melt blending, solution casting and extrusion. Starch rubber nanocomposites can also be prepared by incorporating starch and a nanosized filler like clay into the rubber matrix. Melt blending involves mixing the starch and rubber components in a molten state, followed by the addition of nanoparticles and subsequent cooling to form a solid material. Solution casting involves dissolving the starch and rubber components in a solvent, adding nanoparticles, and then casting the solution into a mould to solidify. Extrusion involves passing the starch, rubber and nanoparticles through an extruder to produce a continuous solid material. Each of these methods has its own advantages and disadvantages, and the choice of method depends on the desired properties and application of the final product.

Casting Method

Film casting is one of the simplest methods available for preparing the rubber latex-based composites. Deborah et al. synthesised starch nanocrystals from various botanical sources and then prepared natural rubber nanocomposites with these synthesised starch nanoparticles. The method they adopted for the synthesis was casting or evaporation method. For preparing the starch nanocrystals, a litre of previously prepared, diluted (3.16 M) sulfuric acid was combined with 147 g of starch. For 5 days, the suspension was kept under 100 rpm mechanical stirring in a silicon bath at 40 °C. The finished suspensions were redispersed with Ultra Turrax for 5 min at 13,000 rpm to break up aggregates after being rinsed repeatedly with distilled water until they reached neutral pH. On a filter fabric, the resulting suspensions were filtered. Before being stored at 4 °C, sodium azide was added to the suspensions to prevent microbial growth. The aqueous suspensions of SNC and NR latex were mixed in different quantities, and the resulting materials were cast and evaporated at 40 °C to create the SNC/NR nanocomposites [48].

Dun Lop Procedure for Starch Rubber Foams

For the preparation of rubber foams with starch, the Dun lop procedure has been adopted. To reduce the content of ammonia, NR latex was stirred in a blender; this was followed by addition of potassium oleate and stirred to enhance the volume. The blending speed was reduced and various compounding ingredients like sulphur, ZDEC, ZMBT and antioxidant Lowinox CPL were added; then rice starch and activated charcoal were added. Finally, ZnO and DPG were added and blended for 2 min.

This was further followed by the addition of sodium silicofluoride (SSF) that was the gelling agent, and mixing was carried out for another 35 s. This mixture was then transferred into the mould and kept at room temperature overnight. Then, the curing was carried out at 100 °C for 1 h in a hot air oven and the cured foam was rinsed with water and dried.

The manufactured rubber foams found application in food packaging, with specific application in the adsorption of ethylene released by postharvest bananas [49]. The pictorial representation of the whole process of manufacture of foam is given in Fig. 6.10.

Rubber Starch Hydrogels

Hydrogels are an interesting class of materials having great utility in the biomedical fields due to their capability of taking up considerable amounts of water and biological fluids without getting dissolved. The mechanical strength and toughness could be improved by preparing interpenetrating network (IPN) hydrogels, by blending two

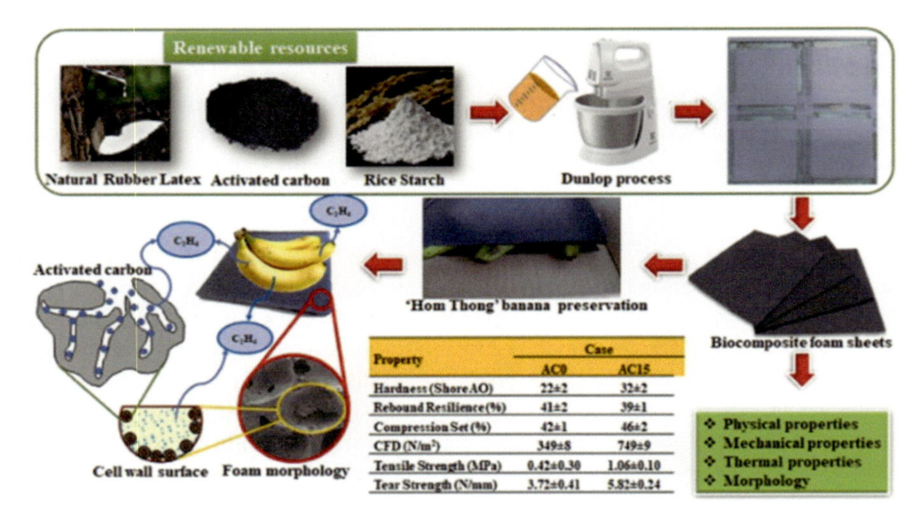

Property	Case	
	AC0	AC15
Hardness (Shore AO)	22±2	32±2
Rebound Resilience (%)	41±2	39±1
Compression Set (%)	42±1	46±2
CFD (N/m²)	349±8	749±9
Tensile Strength (MPa)	0.42±0.30	1.06±0.10
Tear Strength (N/mm)	3.72±0.41	5.82±0.24

Fig. 6.10 Development of starch-activated carbon/NR biocomposite foam sheets for food packaging [49]

or more polymers. Most hydrogels are based on petroleum-based polymers that are non-renewable, expensive and nondegradable. So, the preparation of bio-based hydrogels has an ever-greater demand these days. Starch-based hydrogels of natural rubber are getting greater interest, but phase separations are usually seen due to poor compatibility. Vudjung et al. synthesised cross-linked natural rubber (XNR) and cassava starch (CSt) by IPN method. To a round bottom reactor, latex, KOH and stabiliser were added and stirred followed by the addition of potassium persulphate. Then methylene bisacrylamide (MBA) cross-linker was added to mixture, heated and stirred and cooled to room temperature to obtain the XNR latex. Starch was gelatinised and mixed with this latex at 50:50 wt%; 2% maleic acid was added, and the mixture was cast on a glass mould and dried in an oven to obtain IPN/XNR/CSt [50].

Mixing

For the preparation of dry rubber-based composites, mixing is the most common strategy. In a study conducted by Xueyu et al., modification of porous starch was carried out by dodecenyl succinic anhydride (modified starch is abbreviated as DDSA-PS). They mixed NR latex with the modified starch and coagulated it by adding acetic acid. Then, carbon black and vulcanisation ingredients were incorporated into the coagulated rubber filler compound by mixing in an internal mixer and then the different compounds were vulcanised at 145° to prepare the composite [51]. Open mill mixing using two roll mills is yet another strategy to prepare nanocomposites of rubber. Melt blending can also be performed for preparing rubber starch composites. This can be done in an internal mixer or on a two-roll mill by applying temperatures as high as 140 °C [52].

Extrusion

Extrusion is yet another method for preparing nanocomposites based on rubber. Mondragon et al. prepared thermoplastic starch/natural rubber/montmorillonite clay nanocomposites using twin screw extrusion. This was followed by injection moulding to finally prepare the test samples. The powdered formulations of clay and starch were first prepared and then liquid formulations were made by mixing glycerol with natural rubber. Two forms of natural rubber were taken: one was the unmodified form and the other was modified using dimethylaminoethyl methacrylate. Both these formulations were extruded and pelletised using an extruder at different temperatures and then the pellets were dried and injection moulded [53].

6.1.7 Potential Applications of Starch-Based Rubber Nanocomposites

Starch-based rubber nanocomposites have potential applications in a wide range of fields, including packaging, automotive, construction and biomedical industries. In the packaging industry, these materials can be used to produce biodegradable and compostable materials that are environmentally friendly. In the automotive industry, starch-based rubber nanocomposites can be used to produce lightweight and fuel-efficient parts, while in the construction industry, they can be used to produce materials with improved insulation and fire resistance. In the biomedical industry, these materials can be used to produce biocompatible materials for use in implants and medical devices [33, 54].

6.2 Characterisation Techniques

Starch-based rubber nanocomposites offer several advantages over traditional rubber compounds, including improved mechanical properties, increased thermal stability and reduced environmental impact. The addition of nanoscale particles to the starch matrix enhances the stiffness and strength of the material, while also reducing its weight and increasing its resistance to heat and wear. The type and concentration of nanoparticles used in these can also affect the properties of the composite, with different materials and concentrations leading to varying levels of reinforcement. The characterisation techniques used for starch-rubber composites will be discussed in the following section.

6.2.1 Tensile Testing

In a study conducted, the authors developed rubber-based nanocomposites of starch using starch nanocrystals, synthesised from various botanical sources like potato, maize and wheat. The evaluation of tensile properties of the various starch films suggested that all the nanocomposites showed an elasto-plastic behaviour in the non-linear range of the stress–strain curve, which is typical of amorphous rubber materials. They also reported that for all composites, irrespective of their botanical origin, Young's modulus and tensile strength showed great enhancement when compared with the properties obtained for neat NR. They have also suggested that the composites have become reinforced since the elongation at break values obtained were rather stable. The relative Young's modulus values were found to be higher for the composites where the starch loading was 30 wt% and this might be due to the possibility of formation of network through hydrogen bonding [48].

At normal temperature, the tensile behaviour of potato starch nanocrystal reinforced NR latex nanocomposite sheets was examined. The starch-natural rubber composite's stress–strain curves with increasing starch content are shown in Fig. 6.11. a. Stress is shown to rise steadily with strain until the sample breaks. Tensile strength and modulus both noticeably rise as filler loading is increased. The reinforcement effect of starch nanocrystal is demonstrated by the significant rise in tensile strength of the NR nanocomposites with the addition of the filler.

When 20 wt% of the starch nanocrystals were added, the elongation at break of the nanocomposites was found to be reduced from a value of 1351 for the neat rubber to 536 (as shown in Fig. 6.11d). With filler loading, tensile modulus of the composites also dramatically increased. As a function of starch concentration, the tensile modulus of the NR/potato starch nanocrystal composites is shown in Fig. 6.11c. Tensile modulus similarly rises in an almost linear direction, as seen in the graph. The reinforcing effect of starch nanoparticles seems to be dependent on the formation of strong interactions like hydrogen bonds within the sample resulting in the formation of a three-dimensional network. With increase in the filler loading, reinforcement becomes stronger due to the greater possibilities of polymer-filler interaction [8].

6.2.2 Differential Scanning Calorimetry

Differential scanning calorimetry (DSC) is a technique commonly used to know the response of different polymers when subjected to heating. This is primarily used to determine the melting behaviour and glass transition, in addition to knowing the various thermal transitions happening in the polymer.

Starch particles were modified by sodium isobutyl xanthate (SIBX) and used as a vulcanisation accelerator for natural rubber (NR-SSX) and the properties compared with the vulcanizate prepared using 2-mercapto benzothiazole (NR-M). DSC technique was used to determine the glass transmission temperature of the rubber

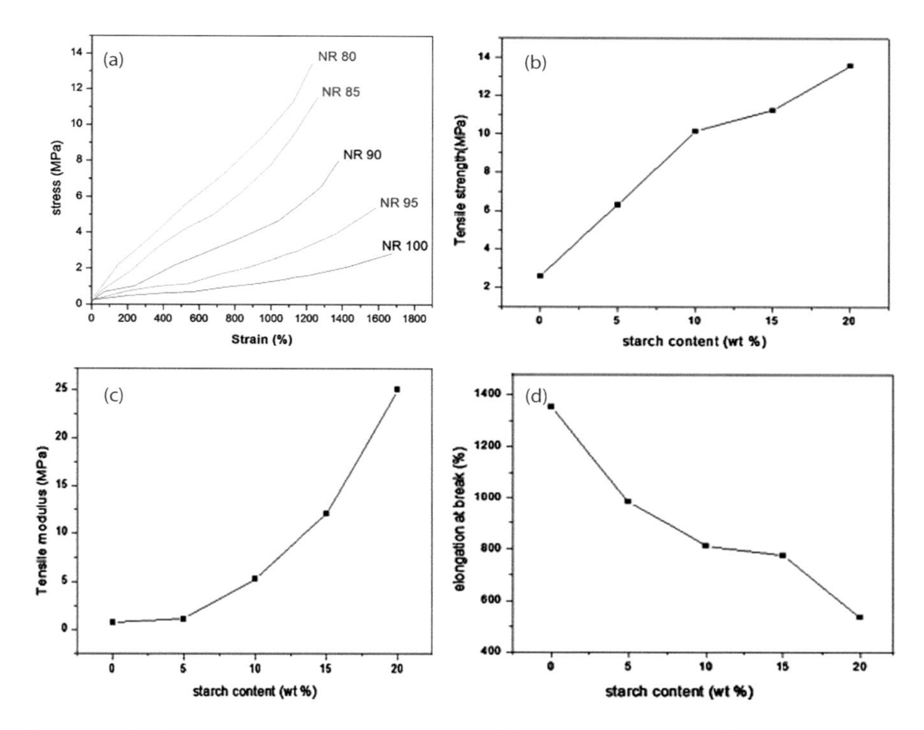

Fig. 6.11 Mechanical properties of potato starch-NR nanocomposites. (**a**) Stress–strain curves of nanocomposites with various starch loadings, (**b**) tensile strength, (**c**) tensile modulus and (**d**) elongation at break of various starch-rubber nanocomposites [8]

compounds. The DSC curves are shown in the Fig. 6.12a. The mid-point of the glass transition region corresponded to T_g of the rubber compounds. The glass transition temperatures of the rubber compounds vulcanised by the conventional system and the modified starch were obtained as −60.32 and −62.65 °C, respectively. T_g of NR-SSX showed an obvious decrease when compared to NR-M, which might be possibly due to the softer molecular segments and higher molecular weight. Weight fraction of the immobilised rubber (χ_{im}) and the change in the heat capacity (ΔC_{pn}) were also found out by referring literature. The lower values of ΔC_{pn} obtained for NR-SSX suggested that the rubber immobilised on SSX particles greater in amount which in turn indicated that rubber chains were cross-linked directly by SSX. The glass transition temperatures were determined from tan delta peak obtained from DMA (Fig. 6.12b) and the results obtained were in agreement with the value obtained from DSC. The modified starch exhibited a better dispersion as schematically shown in Fig. 6.12c (which was further confirmed by SEM analysis of the vulcanizate) and thus a stronger interfacial interaction [55].

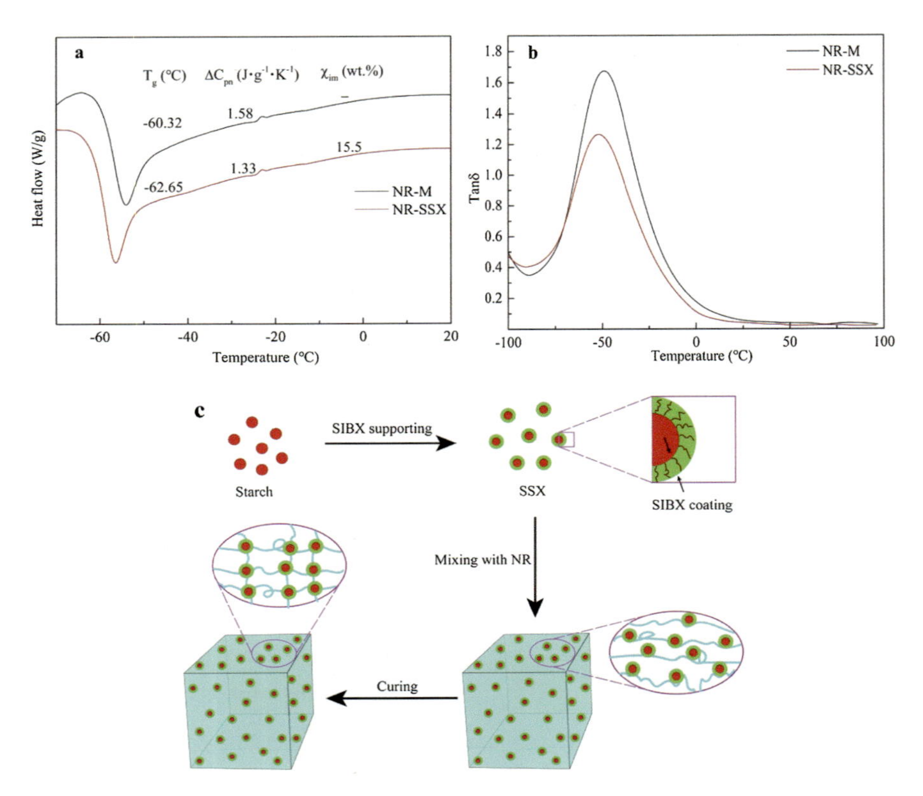

Fig. 6.12 (**a**) DSC curves of the various vulcanizates, (**b**) tanδ from DMA and (**c**) vulcanisation procedure with the modified starch [55]

6.2.3 Dynamic Mechanical Analysis

DMA analysis is carried out to study the viscoelastic response of the polymer. DMA analysis of the PMMA-modified starch-SBR composites prepared by Mei-Chun et al. was conducted to examine the interaction of the fillers and rubber and the plot obtained is shown in Fig. 6.13.

The storage modulus of the MDI (rubber compound with coupling agent methylene diphenyl isocyanate) was found to be greater than that obtained for the control (rubber composite without coupling agent). This suggested a better confinement of the modified starch particles in the rubber matrix. Glass transition temperature (T_g) was obtained from the peak of the tan delta curve and for the composite with coupling agent, the T_g was enhanced to −27.4 °C from the value of −34.4 °C obtained for the control. Furthermore, the peak of tan delta curve obtained for the composite was much broader compared to the control and this might be due to the restricted mobility of the rubber chains. The MDI modification has resulted in a strong interfacial adhesion between the rubber and the filler [56].

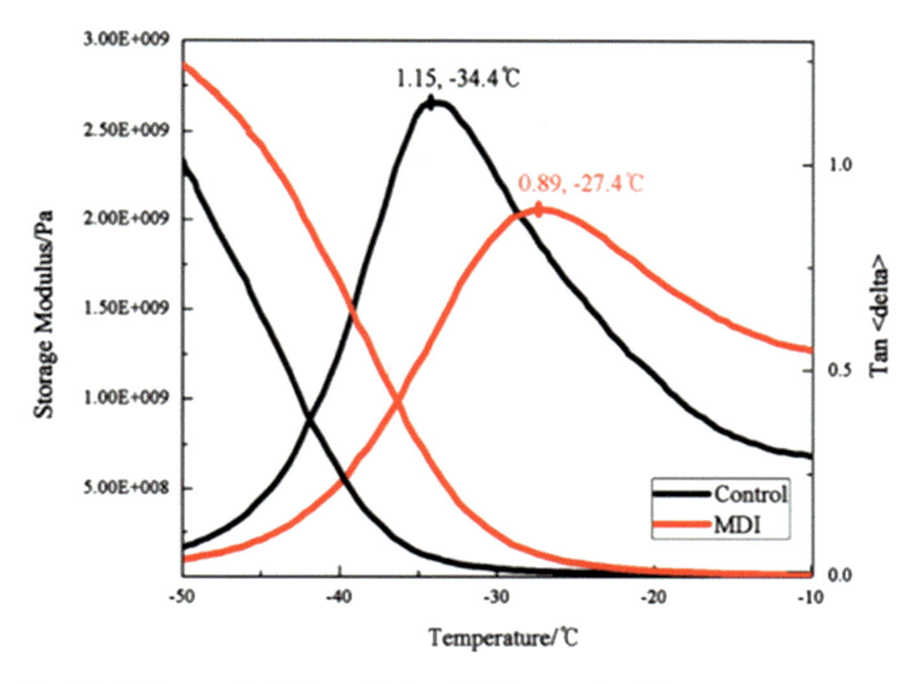

Fig. 6.13 DMA curve of PMMA-modified starch-SBR composites [56]

Porous starch was modified by dodecyl succinate anhydride and mixed natural rubber latex and coagulated with acetic acid at various concentrations (0, 10 and 20 phr). DMA analysis was carried out and tanδ values at 60 and 80 °C were found. The glass transition temperature shifted to lower values with the addition of modified starch. The lower tanδ values at 60° and 80° are indicative of lower rolling resistance and lower heat generation for the vulcanised rubber. This might be the result of internal plasticisation of NR by modified starch which in turn enhances the activity of NR. This might again contribute in reducing the friction between rubber and starch molecules [57].

6.2.4 Thermogravimetric Analysis

Thermogravimetric analysis evaluates the thermal stability of the polymer composites.

Styrene butadiene latex and PMMA-modified starch emulsion were stirred vigorously followed by drying in an oven; the further mixing of coupling agents and vulcanisation ingredients was carried out in a two-roll mixing mill and vulcanised at 150°. The composites prepared without and with coupling agents (marked as control and MDI, respectively) were evaluated for their thermal stability. The addition of

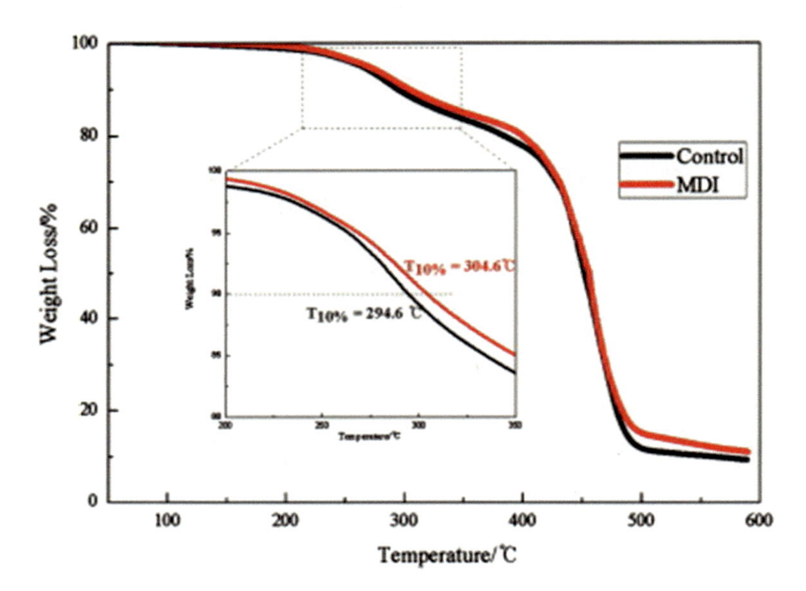

Fig. 6.14 TGA curves of the PMMA modified starch-SBR composites [56]

the coupling agent MDI has resulted in enhanced interaction between the starch and the rubber matrix and hence, the initial decomposition temperature, as evaluated from the TGA curve (shown in Fig. 6.14), of the composite increased by 10° when compared to the control sample [56].

In a study to prepare starch-based natural rubber hydrogels, thermal degradation of natural rubber (NR), cassava starch (CSt), NR-CSt blend and IPN XNR/CSt hydrogels were studied. They found that the thermal stability of starch was very low when compared to NR. The neat NR showed a single-step decomposition with the initial and final decomposition temperatures at 353° and 440°, while cassava starch exhibited a two-step degradation; the first in the range of 38°–150° corresponded to the loss of water while the one in the range 285°–327° corresponded to the degradation of amylose and amylopectin fragments in the starch. On the other hand, IPN XNR/CSt showed a three-step decomposition: first two stages (100–140 and 270–315 °C) corresponding to the thermal decomposition of starch and the last one corresponding to the degradation of rubber (440 °C). On increasing the MBA content (cross-linking agent), the decomposition temperatures also increased correspondingly. With increase in the MBA content, the cross-link density of the rubber and the gel fraction of the hydrogel improved and this might have contributed to the enhanced thermal stability of IPN XNR/CSt [50].

6.2.5 SEM Analysis

Scanning electron microscopy is an effective method to analyse the surface topology of the sample using a focused beam of electrons. A study was conducted to prepare natural rubber starch-based nanocomposites using starch nanocrystals (SNC) obtained from different botanical sources by the casting method. The morphology of the prepared nanocomposites was analysed using SEM analysis. The unfilled NR matrix's fracture surface was obviously smooth and homogeneous. In contrast, SNC and SNC concentration were visible as white dots in the filled NR composites, making them simple to detect. The fact that SNC can be seen in these micrographs gives an indication that although the distribution of SNC inside the matrix appears to be quite uniform, sections of the SNC are rather 'homogeneously aggregated' throughout the matrix [48].

The FE-SEM images of the NR-potato starch nanocrystal composites prepared by casting method were recorded. A uniform dispersion of starch nanocrystals, which resemble white dots, could be seen in the FE-SEM micrograph. For the compositions, a consistent distribution of the nanocrystal within the matrix was seen. To achieve the best properties, the filler must be distributed throughout the matrix in an even and uniform manner. The nanocrystals have a tendency to aggregate as the filler loading is increased, as shown in the image. The nanocrystal's particle size was found to be between 12 and 20 nm [8].

Starch was modified by carbon disulphide to form starch xanthates and natural rubber composites were prepared with the modified and unmodified starch and compared by SEM analysis. SEM analysis of the starch particles as well as the tensile-fracture surface of the composites were done and is shown in Fig. 6.15. The size of the unmodified starch granules was in the range of 1000–5000 nm; after modification the size of the starch particles was reduced to less than 200 nm. For the composites with unmodified starch, holes were present on the fracture surface suggesting a weak interfacial interaction between the rubber and the filler. As for the composites with modified starch, with increase in the dosage of the modified starch, the fracture surface was found to become more coarse, suggesting increased resistance of the composite to external force [58].

6.2.6 TEM Analysis

Transmission Electron Microscopy is a powerful tool to determine the morphology of rubber-based nanocomposites. TEM images of unmodified and modified natural rubber nanocomposites (NR modified by grafting dimethyl aminoethyl methacrylate) with thermoplastic starch and montmorillonite clay (TPS/uNR/MMT) and (TPS/mNR/MMT) have been analysed (given in Fig. 6.16). The white area, as can be seen from the figure, denotes the themopalstic starch and the dark zone represents the NR domain. The montmorillonite seems to be dispersed in the NR region and intercalated

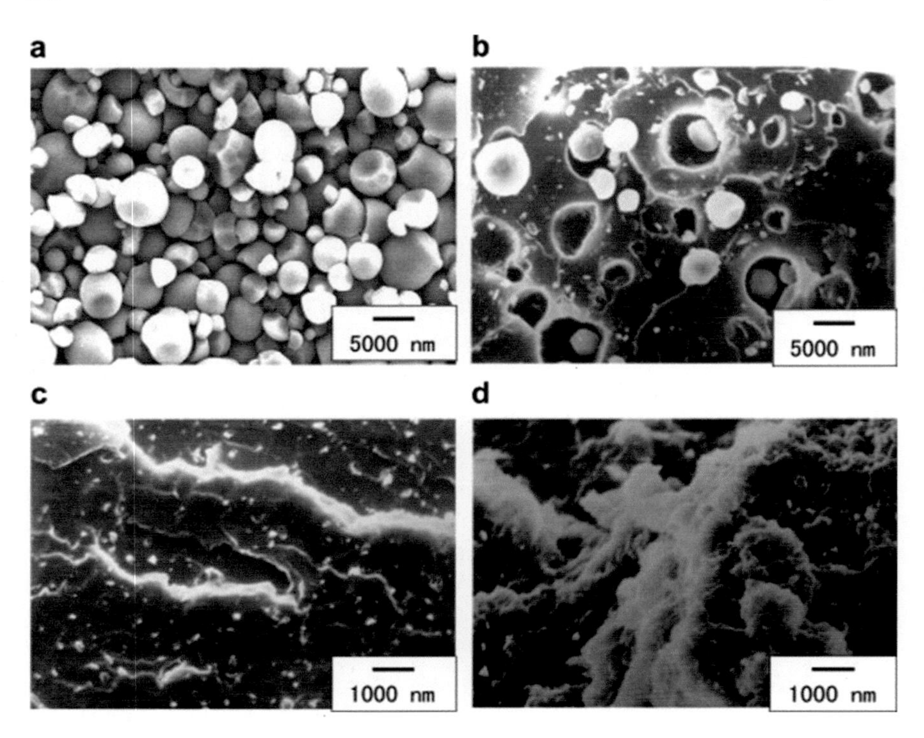

Fig. 6.15 SEM images of (**a**) unmodified starch, (**b**) unmodified starch(15phr)-natural rubber composite, (**c**) modified starch(5phr)-natural rubber composite, (**d**) modified starch(15phr)-natural rubber composite [58]

structures have been formed. This distribution seems to be quite peculiar since starch particles have higher degree of polarity when compared to rubber matrix.

High shearing force generated during the melt processing as well as the presence of polar moieties like proteins and lipids on the natural rubber surface may have resulted in greater intercalation of the polymer chains into the silicate layers [53].

Xu et al. prepared nanosized starch particles by means of a reverse titration strategy. Ethylene-co-vinyl acetate (EVM) rubber-starch composites were prepared using starch particles of different sizes (Mst: microsized starch and Nst: nanosized starch), and agglomeration phenomenon was found to be higher for Mst particles, leading to an uneven distribution in the rubber matrix. Epoxy functionalised EVM (GMA/EVM) was employed as the compatibiliser. In the absence of compatibiliser, the nanostarch particles were found to be exposed on the surface of the rubber due to lack of compatibility between rubber and filler. The nanoparticles were found to be agglomerated and were of the size between 350 and 1500 nm. But for the composites prepared using compatibiliser, the number of exposed surfaces of nanoparticles reduced significantly and a better dispersion and filler network formation could be observed. The incorporation of a silane coupling agent 3-(2-aminoethylamino)

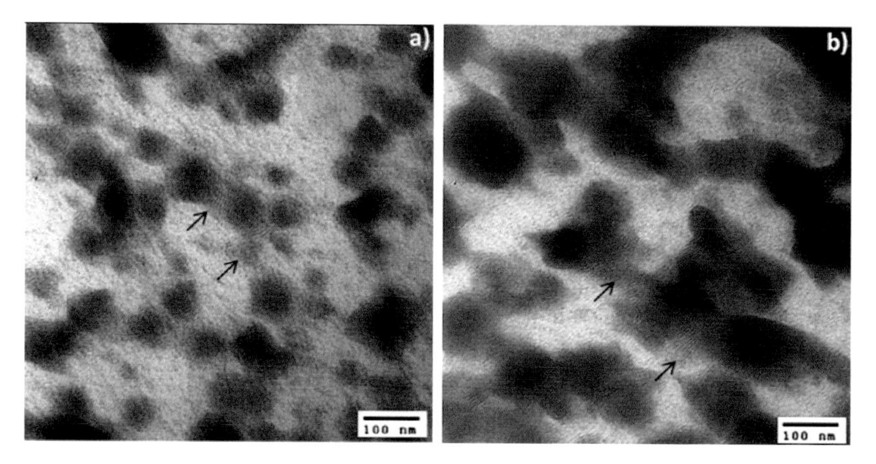

Fig. 6.16 TEM images of nanocomposites of (**a**) unmodified and (**b**) modified NR with montmorillonite clay and starch [53]

propyl dimethoxy methyl silane (KH602) further improved the compatibilisation effect which was confirmed by morphology analysis [59].

6.2.7 IR Spectrum

Fourier Transform Infrared Spectroscopy (FTIR) and Attenuated Total Reflection (ATR) are very useful tools used for analysing the incorporation and interaction of fillers in polymer-based composites by monitoring the peaks corresponding to various functional groups.

Zhi et al. esterified starch using carbon disulphide to produce starch Xanthates and natural rubber-modified starch composites were prepared. FTIR analysis of the natural rubber, modified starch and the NR-modified starch composite were done and the plots are shown in Fig. 6.17. The stretching vibrations of methyl and methene groups present in NR results in peaks at 2960, 2917 and 2851 cm^{-1} as seen from the curve a. The peaks at 1662 and 1547 cm^{-1} correspond to the vibrational frequencies of amides of proteins present in NR.

The peaks at 1447 and 1375 cm^{-1} result from bending vibrations of methene and methyl group from NR. The band at 1036 cm^{-1} is due to the N–H groups of amino acids in natural rubber. From the spectrum of modified starch, the stretching vibrations of the hydroxyl group have resulted in a peak at 3340 cm^{-1}. Absorption at 1638 cm^{-1} is C–O vibrations in the chelate aldehyde group. The unsymmetric stretching vibrations of the C–O–C bond and C–O bonds correspond to the peaks observed at 1154 and 1045 cm^{-1}, respectively. The peak at 879 cm^{-1} in the modified starch arises from the methene vibration.

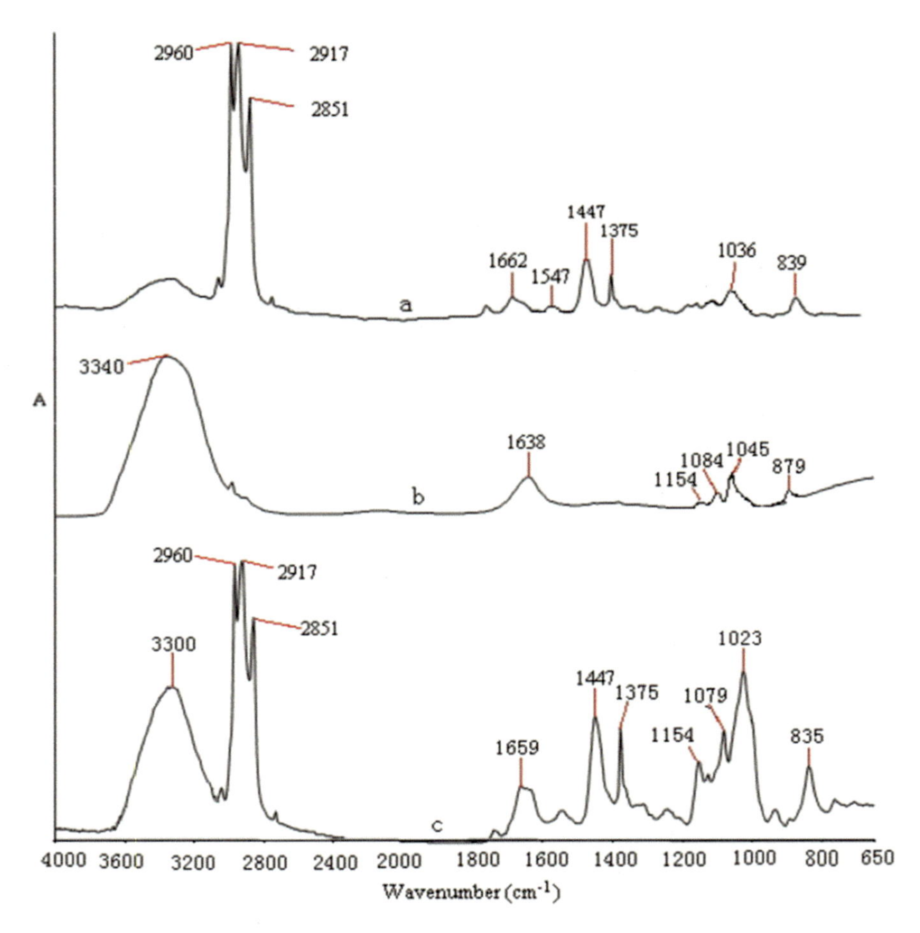

Fig. 6.17 FTIR spectrum of (**a**) NR, (**b**) modified starch and (**c**) modified starch-NR composite [58]

From the IR spectrum obtained for the composite, it can be seen that there is a strong and broad peak at 1659 cm^{-1} that arises from the vibrations of amide group in NR (1662 cm^{-1}) and C–O group from the chelate aldehyde group present in the starch (1638 cm^{-1}). This is a strong indication of the formation of hydrogen bonds between starch and NR.

The peak at 1023 cm^{-1} corresponds to N–H vibrations (1036 cm^{-1}) in NR as well C–O stretches from the primary alcohol (1045 cm^{-1}) present in starch. The formation of hydrogen bond between starch and rubber is further confirmed by the reduction in the stretching frequencies of the N–H group in NR and C–O group in starch and the formation of stronger and broader peak corresponding to amide moiety.

The IR studies indicate the possibility of formation of strong hydrogen bonds between the amide group present in the protein in NR and hydroxyl group in starch [58].

6.2.8 XRD Analysis

X-ray diffraction studies are carried out on polymer composites to analyse solid-state structural data like degree of crystallinity. Rajisha et al. synthesised potato starch nanocrystals (SNCs) from native potato starch powder using sulphuric acid hydrolysis. Then, using these SNCs, they prepared natural rubber latex nanocomposites by a casting method, the schematic procedure for which is shown in Fig. 6.18. The prepared nanocomposites were characterised using X-ray diffraction method. XRD graphs, corresponding to the different compositions of nanocomposites prepared by varying the weight percentage of starch, were obtained.

For the sake of having a comparison, XRD graphs of the pristine natural rubber and starch nanocrystals were also taken. The frozen nanocrystals of starch were pressed to form a film, whose XRD pattern corresponded with that of A-type amylose allomorph. The characteristic peaks obtained were two weak peaks at $2\theta = 10.1°$ and, a strong peak at 18.2°, and a strong peak at 23.5°. The natural rubber film shows the characteristic pattern of a typical amorphous material and evidently a broad peak was observed at $2\theta = 18°$. The addition of nanocrystals made the peak corresponding to the A-amylose stronger.

This suggests an increment in the total crystallinity of the composite with an increase in the starch content. Moreover, the XRD analysis confirmed that the casting and evaporation techniques carried out at 40 °C had no impact on the crystallinity of the starch nanoparticles [8].

Yang et al. synthesised oxidised starch (OST) with varying carboxyl content using hydrogen peroxide as green oxidising agent. This was then added to carboxyl nitril rubber (XNBR) latex and stirred, and then air dried; ZnO was mixed with the dry rubber using a two-roll mill and the composites were compression moulded; in a

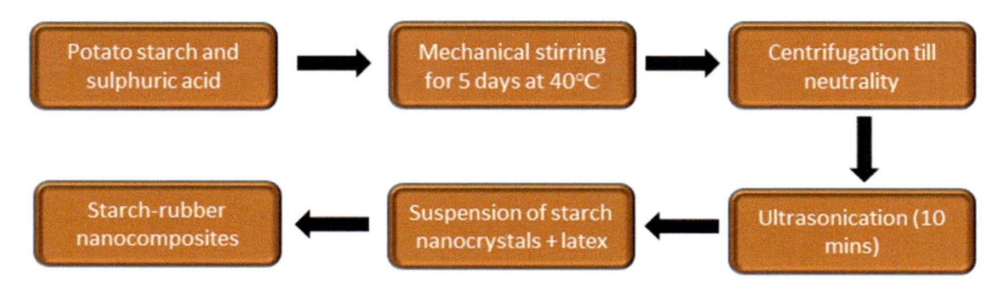

Fig. 6.18. Preparation strategy of potato starch nanocrystals and starch-NR nanocomposites

similar manner, composites were prepared with native starch for the sake of comparison. XRD patterns of composites labelled XNBR/5Z (5 phr ZnO), XNBR/15OST-13/5Z (15 phr oxidised starch with carboxyl content of 13% and 5 phr ZnO) and XNBR/15OST-57/5Z (15 phr oxidised starch with carboxyl content of 57% and 5 phr ZnO) before and after hot pressing were analysed and the plots are provided in Fig. 6.19. The XNBR/5Z and XNBR/15OST-13/5Z curves showed the normal zinc oxide peaks at $2\theta = 32.3°$, 34.3°, 36.1° and 56.5°. The peak strength of the ZnO in the composites obtained after the hot-pressing got decreased in comparison to the composites formed without the heating process, showing that the ZnO has not entirely reacted in these composites. When compared to curve d, the peak intensities at $2\theta = 32.3°$ and 36.1° on curve f are noticeably lower. This suggests that the heating process is the primary catalyst for the interfacial neutralisation reaction between XNBR, OST, and ZnO. When the curve e is employed as the control, however, the intensity of the distinctive ZnO diffraction peaks in XNBR/15OST-57/5Z (curve f) has decreased with increment in the carboxyl content. Since both the XNBR and ZnO concentrations are constant, the results suggest that OSTs with high carboxyl content can utilise more ZnO to produce more Zn^{2+} salt bondings [60].

Angellier et al. have done extensive work on natural rubber-starch nanocomposites. They prepared waxy maize starch nanocrystals by hydrolysing native starch granules using sulphuric acid. The obtained particles were of the shape of platelet with a thickness of 6–8 nm, length of 40–60 nm and a width of 15–30 nm. The NR-nanostarch films were prepared by varying the weight percentage of starch from 0 to 50. They also modified the starch using succinic anhydride and phenyl isocyanate. Composite films of rubber were prepared using the modified starch as well. The typical peaks of the A-type amylose allomorph can be seen in the diffraction pattern recorded for a film of pure waxy maize starch nanocrystals produced by pressing freeze-dried nanocrystals.

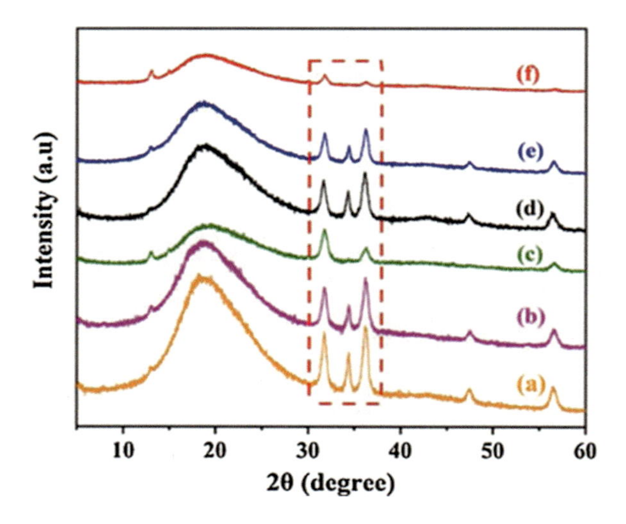

Fig. 6.19 XRD patterns of (**a**, **d**) XNBR/5Z, (**b**, **e**) XNBR/15OST-13/5Z and (**c**, **f**) XNBR/15OST-57/5Z; (**a**–**c**) before and (**d**–**f**) after a hot press [60]

It has two strong peaks at 23.5° and 15.3°, two weak peaks at 10.1° and 11.5° and double strong peak at 17.1° and 18.2°. The natural rubber film exhibits the typical characteristics of a completely amorphous polymer. It is distinguished by a wide hump at about 18°. The simple mixing formula of the diffractograms of the two pure parent components, $I = WsI_{WN} + (1 - Ws)\, I_{L100}$, which states that I is the diffracted intensity at each angle for a specific composite film, does not exactly match the diffraction patterns of the varied NR/starch nanocrystals. (L_{100} corresponds to natural rubber film, Ws corresponds to weight fraction of starch, WN corresponds to waxy maize starch nanocrystals). The experimental double central peak is indeed more powerful than the expected one. This is most likely caused by the various starch nanocrystal orientations found in the films made from solely starch and the composite. In contrast to the reference starch nanoparticle films, which are made of in-plane oriented filler, the distribution of nanocrystals in the composite is likely random [61].

6.2.9 *Rheology Measurements*

Rheology is the evaluation of deformation and flow properties of rubber compounds. Kuncai et al. modified starch using sodium isobutyl xanthate (SIBX) to prepare a novel environment-friendly vulcanisation accelerator for rubber. Comparison was made between natural rubber compounds vulcanised with starch modified by isobutyl xanthate (NR-SSX) and with 2-mercaptobenzothiazole (NR-M). The cure curves of both the compounds (Fig. 6.20) exhibited the same pattern and showed three periods: induction period, curing period and overcure period. Optimum cure time for NR-SSX was slightly lower than that for NR-M. The maximum torque developed for both the composites was in the range of 5.6 dNm. The cure properties indicate that SSX can be an effective replacement for the traditional accelerator [55].

Porous starch was modified by dodecenyl succinic anhydride (DDSA), and this was used as filler for natural rubber in combination with carbon black. The impact of different ratios of modified starch-carbon black on the cure properties was studied using a multifunction rheometer (curves given in Fig. 6.21).

When compared to the composite where carbon black alone is used as filler, the combination was found to cause a drastic reduction in the cure time. The 5/55 combination resulted in enhancement in minimum torque (M_L), maximum torque (M_H) and $M_H - M_L$. All these values were found to decrease with further increase in the loading ratio and this might be due to attenuation of filler-filler and rubber-filler interactions at higher loadings.

The storage modulus of all the composites were found to first increase with strain, but after 1%, the modulus values went down with increasing strain (shown in Fig. 6.21). This could be explained in terms of Payne effect. Initially, the stress applied might have resulted in the formation of a network, but on further increasing strain, the destruction of the network dominated over its formation and resulted in reduction in storage modulus. When carbon black alone was present, the higher

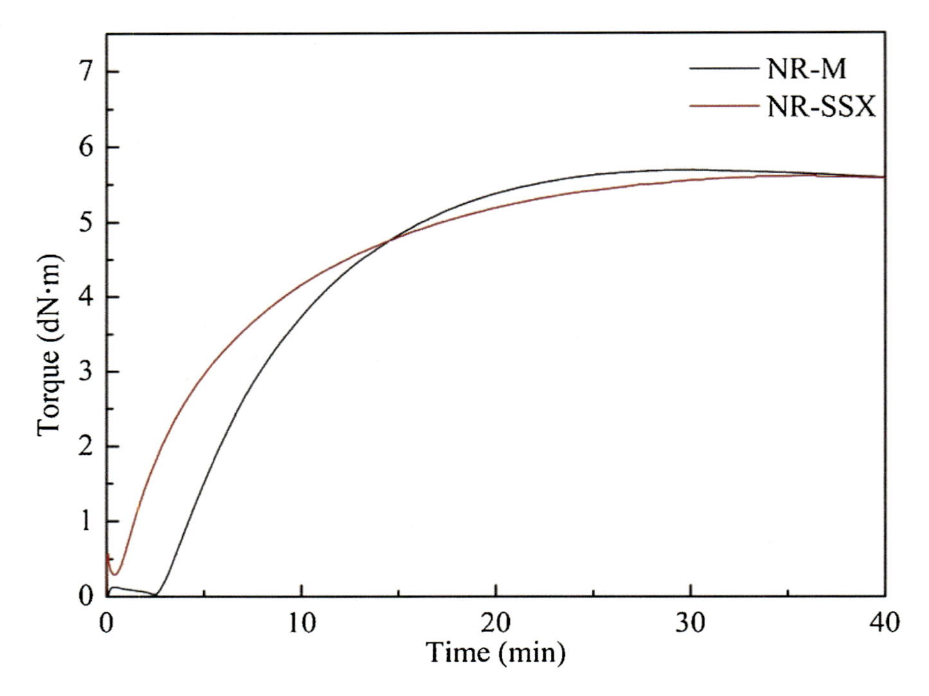

Fig. 6.20 Cure curves of NR-SSX and NR-M compounds [55]

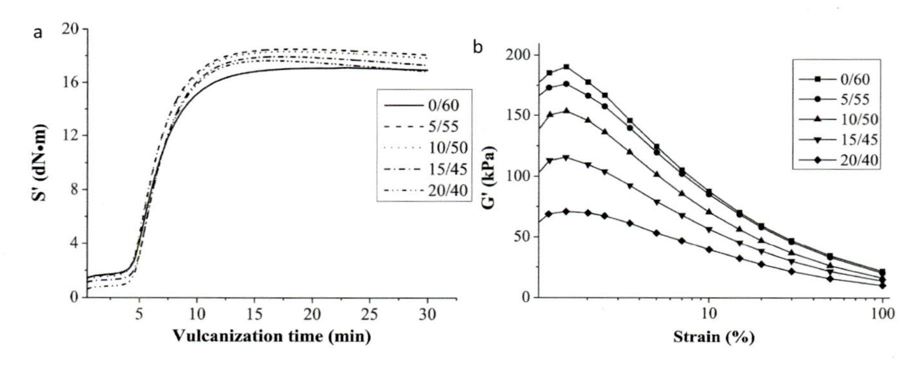

Fig. 6.21 (**a**) Torque versus time curves and (**b**) storage modulus versus strain curves of various composites with different DDSA-PS/CB ratios [51]

surface area promoted the network formation and thus higher storage modulus was obtained for the composite. With incorporation of the modified starch, the resultant higher size and lower surface area weakened the possibility of network formation and correspondingly storage modulus reduced with increase in the amount of DDSA-PS [51].

6.3 Conclusion

Starch is one of the most promising fillers of biological origin that can effectively replace the conventional non-renewable fillers used in rubber compounding like carbon black. The structural details of starch, its properties, techniques for synthesising nanostarch, its various modifications, preparation of starch rubber composites and their various characterisations have been described in detail. Performing various kinds of modifications is quite essential in improving the compatibility between hydrophilic starch and hydrophobic rubber. Modified nanostarch particles have great potential in replacing conventional fillers and can impart good mechanical properties to the composite at much lower filler loading.

References

1. Jang SM, Li MC, Lim JH, Cho UR (2013) Study on properties of natural rubber compound using starch as filler. Asian J Chem 25:5221–5225. https://doi.org/10.14233/ajchem.2013.f22
2. Thomas SK, Parameswaranpillai J, Krishnasamy S et al (2021) A comprehensive review on cellulose, chitin, and starch as fillers in natural rubber biocomposites. Carbohydr Polym Technol Appl 2:100095. https://doi.org/10.1016/j.carpta.2021.100095
3. Sun Q, Li G, Dai L et al (2014) Green preparation and characterisation of waxy maize starch nanoparticles through enzymolysis and recrystallisation. Food Chem 162:223–228. https://doi.org/10.1016/j.foodchem.2014.04.068
4. Campelo PH, Sant'Ana AS, Pedrosa Silva Clerici MT (2020) Starch nanoparticles: production methods, structure, and properties for food applications. Curr Opin Food Sci 33:136–140. https://doi.org/10.1016/j.cofs.2020.04.007
5. Mélé P, Angellier-Coussy H, Molina-Boisseau S, Dufresne A (2011) Reinforcing mechanisms of starch nanocrystals in a nonvulcanized natural rubber matrix. Biomacromolecules 12:1487–1493. https://doi.org/10.1021/bm101443a
6. Duy NQ, Rashid AA, Ismail H (2012) Effects of filler loading and different preparation methods on properties of cassava starch/natural rubber composites. Polym Plast Technol Eng 51:940–944. https://doi.org/10.1080/03602559.2012.679377
7. Misman MA, Rashid AA, Yahya SR (2018) Modification and application of starch in natural rubber latex composites. Rubber Chem Technol 91:184–204. https://doi.org/10.5254/rct-18-82604
8. Rajisha KR, Maria HJ, Pothan LA et al (2014) Preparation and characterization of potato starch nanocrystal reinforced natural rubber nanocomposites. Int J Biol Macromol 67:147–153. https://doi.org/10.1016/j.ijbiomac.2014.03.013
9. Streb S, Eicke S, Zeeman SC (2012) The simultaneous abolition of three starch hydrolases blocks transient starch breakdown in arabidopsis. J Biol Chem 287:41745–41756. https://doi.org/10.1074/jbc.M112.395244
10. Qu J, Xu S, Zhang Z et al (2018) Evolutionary, structural and expression analysis of core genes involved in starch synthesis. Sci Rep 8:1–16. https://doi.org/10.1038/s41598-018-30411-y
11. Tester RF, Karkalas J, Qi X (2004) Starch—composition, fine structure and architecture. J Cereal Sci 39:151–165. https://doi.org/10.1016/j.jcs.2003.12.001
12. Seung D (2020) Amylose in starch: towards an understanding of biosynthesis, structure and function. New Phytol 228:1490–1504. https://doi.org/10.1111/nph.16858
13. Lemos PVF, Barbosa LS, Ramos IG et al (2019) Characterization of amylose and amylopectin fractions separated from potato, banana, corn, and cassava starches. Int J Biol Macromol 132:32–42. https://doi.org/10.1016/j.ijbiomac.2019.03.086

14. Pérez S, Bertoft E (2010) The molecular structures of starch components and their contribution to the architecture of starch granules: a comprehensive review. Starch/Staerke 62:389–420. https://doi.org/10.1002/star.201000013
15. Hoover R (2001) Composition , molecular structure, and physicochemical properties of tuber and root starches: a review. 45
16. Srichuwong S, Candra T, Mishima T et al (2005) Starches from different botanical sources I: contribution of amylopectin fine structure to thermal properties and enzyme digestibility. 60:529–538. https://doi.org/10.1016/j.carbpol.2005.03.004
17. Dufresne A (2014) Crystalline starch based nanoparticles. Curr Opin Colloid Interface Sci 19:397–408. https://doi.org/10.1016/j.cocis.2014.06.001
18. Chakraborty I, Pallen S, Shetty Y et al (2020) Advanced microscopy techniques for revealing molecular structure of starch granules. 105–122
19. Xiao H, Wang S, Xu W et al (2020) The study on starch granules by using dark field and polarized light microscopy. J Food Compos Anal 92:103576. https://doi.org/10.1016/j.jfca.2020.103576
20. Cai C, Zhao L, Huang J et al (2014) Morphology, structure and gelatinization properties of heterogeneous starch granules from high-amylose maize. Carbohydr Polym 102:606–614. https://doi.org/10.1016/j.carbpol.2013.12.010
21. Dufresne A (2014) Crystalline starch based nanoparticles. Curr Opin Colloid Interface Sci 19:397–408. https://doi.org/10.1016/j.cocis.2014.06.001
22. SNYDER EM (1984) Industrial microscopy of starches. In: Starch: chemistry and technology, 2nd edn. Science Direct
23. Ghiasi K, Hoseney RC, Marston EV (1982) Geletinization of wheat starch. III. Comparison by differential scanning calorimetry and light microscopy. Cereal Chem 59:258–262
24. Langenaeken NA, De SCF, De SDP, Courtin CM (2019) Different gelatinization characteristics of small and large barley starch granules impact their enzymatic hydrolysis and sugar production during mashing. Food Chem 295:138–146. https://doi.org/10.1016/j.foodchem.2019.05.045
25. Tao J, Huang J, Yu L et al (2018) A new methodology combining microscopy observation with Artificial Neural Networks for the study of starch gelatinization. Food Hydrocoll 74:151–158. https://doi.org/10.1016/j.foodhyd.2017.07.037
26. Raigond P, Ezekiel R, Raigond B (2015) Resistant starch in food: a review. J Sci Food Agric 95:1968–1978. https://doi.org/10.1002/jsfa.6966
27. Govindaraju I, Chakraborty I, Baruah VJ et al (2021) Structure and morphological properties of starch macromolecule using biophysical techniques. Starch/Staerke 73:1–12. https://doi.org/10.1002/star.202000030
28. Wang J, Ren F, Yu J et al (2019) Toward a better understanding of different dissolution behavior of starches in aqueous ionic liquids at room temperature. 4–11. https://doi.org/10.1021/acsomega.9b00962
29. Amani NG, Kamenan A, Rolland-Sabaté A, Colonna P (2005) Stability of yam starch gels during processing. Afr J Biotechnol 4:94–101
30. Goel C, Semwal AD, Khan A et al (2020) Physical modification of starch: changes in glycemic index, starch fractions, physicochemical and functional properties of heat-moisture treated buckwheat starch. J Food Sci Technol 57:2941–2948. https://doi.org/10.1007/s13197-020-04326-4
31. Ulbrich M, Natan C, Flöter E (2014) Acid modification of wheat, potato, and pea starch applying gentle conditions—impacts on starch properties. Starch/Staerke 66:903–913. https://doi.org/10.1002/star.201400089
32. Chavan P, Sinhmar A, Nehra M et al (2021) Impact on various properties of native starch after synthesis of starch nanoparticles: a review. Food Chem 364:130416. https://doi.org/10.1016/j.foodchem.2021.130416
33. Le Corre D, Bras J, Dufresne A (2010) Starch nanoparticles: a review. Biomacromolecules 11:1139–1153. https://doi.org/10.1021/bm901428y
34. Mahmoudi Najafi SH, Baghaie M, Ashori A (2016) Preparation and characterization of acetylated starch nanoparticles as drug carrier: ciprofloxacin as a model. Int J Biol Macromol 87:48–54. https://doi.org/10.1016/j.ijbiomac.2016.02.030

35. Putro JN, Ismadji S, Gunarto C et al (2020) A study of anionic, cationic, and nonionic surfactants modified starch nanoparticles for hydrophobic drug loading and release. J Mol Liq 298:112034. https://doi.org/10.1016/j.molliq.2019.112034
36. Ismail NS, Gopinath SCB (2017) Enhanced antibacterial effect by antibiotic loaded starch nanoparticle. J Assoc Arab Univ Basic Appl Sci 24:136–140. https://doi.org/10.1016/j.jaubas.2016.10.005
37. Morán D, Gutiérrez G, Blanco-López MC et al (2021) Synthesis of starch nanoparticles and their applications for bioactive compound encapsulation. Appl Sci 11. https://doi.org/10.3390/app11104547
38. Gonde S, Badve M, Jain B (2022) Cavitation assisted novel method for synthesis of starch nanoparticles. Chem Eng Process Process Intensif 175. https://doi.org/10.1016/j.cep.2022.108935
39. Lecorre D, Vahanian E, Dufresne A, Bras J (2012) Enzymatic pretreatment for preparing starch nanocrystals. Biomacromolecules 13:132–137. https://doi.org/10.1021/bm201333k
40. Dai L, Li C, Zhang J, Cheng F (2018) Preparation and characterization of starch nanocrystals combining ball milling with acid hydrolysis. Carbohydr Polym 180:122–127. https://doi.org/10.1016/j.carbpol.2017.10.015
41. Dries DM, Gomand SV, Delcour JA, Goderis B (2016) V-type crystal formation in starch by aqueous ethanol treatment: the effect of amylose degree of polymerization. Food Hydrocoll 61:649–661. https://doi.org/10.1016/j.foodhyd.2016.06.026
42. Li S, Zhou W, Huang C et al (2023) Rapid preparation of starch nanocrystals by the mixed acid of sulfuric acid and hydrochloric acid. Int J Biol Macromol 232. https://doi.org/10.1016/j.ijbiomac.2023.123402
43. Wang Y, Hu Q, Li T et al (2018) Core-shell starch nanoparticles and their toughening of polylactide. Ind Eng Chem Res 57:13048–13054. https://doi.org/10.1021/acs.iecr.8b02695
44. Torres FG, De-la-Torre GE (2022) Synthesis, characteristics, and applications of modified starch nanoparticles: a review. Int J Biol Macromol 194:289–305. https://doi.org/10.1016/j.ijbiomac.2021.11.187
45. Aaliya B, Sunooj KV, John NE et al (2022) Impact of microwave irradiation on chemically modified talipot starches: a characterization study on heterogeneous dual modifications. Int J Biol Macromol 209:1943–1955. https://doi.org/10.1016/j.ijbiomac.2022.04.172
46. Cazotti JC, Fritz AT, Garcia-Valdez O et al (2020) Graft modification of starch nanoparticles using nitroxide-mediated polymerization and the grafting from approach. Carbohydr Polym 228:115384. https://doi.org/10.1016/j.carbpol.2019.115384
47. Julie JC, Bipinbal PK, Renju VS et al (2022) Bionanocomposites based on natural rubber and cellulose nanofibrils from arecanut husk: rheological, mechanical and thermal characterizations. J Polym Res 29. https://doi.org/10.1007/s10965-022-03069-4
48. Lecorre DS, Bras J, Dufresne A (2012) Influence of the botanic origin of starch nanocrystals on the morphological and mechanical properties of natural rubber nanocomposites. Macromol Mater Eng 297:969–978. https://doi.org/10.1002/mame.201100317
49. Nooun P, Chueangchayaphan N, Ummarat N, Chueangchayaphan W (2023) Fabrication and properties of natural rubber/rice starch/activated carbon biocomposite-based packing foam sheets and their application to shelf life extension of 'Hom Thong' banana. Ind Crops Prod 195:116409. https://doi.org/10.1016/j.indcrop.2023.116409
50. Vudjung C, Saengsuwan S (2017) Synthesis and properties of biodegradable hydrogels based on cross-linked natural rubber and cassava starch. J Elastomers Plast 49:574–594. https://doi.org/10.1177/0095244316676868
51. Du X, Zhang Y, Pan X et al (2019) Preparation and properties of modified porous starch/carbon black/natural rubber composites. Compos Part B Eng 156:1–7. https://doi.org/10.1016/j.compositesb.2018.08.033
52. Jantanasakulwong K, Leksawasdi N, Seesuriyachan P et al (2016) Reactive blending of thermoplastic starch, epoxidized natural rubber and chitosan. Eur Polym J 84:292–299. https://doi.org/10.1016/j.eurpolymj.2016.09.035

53. Mondragón M, Hernández EM, Rivera-Armenta JL, Rodríguez-González FJ (2009) Injection molded thermoplastic starch/natural rubber/clay nanocomposites: morphology and mechanical properties. Carbohydr Polym 77:80–86. https://doi.org/10.1016/j.carbpol.2008.12.008
54. Ogunsona E, Ojogbo E, Mekonnen T (2018) Advanced material applications of starch and its derivatives. Eur Polym J 108:570–581. https://doi.org/10.1016/j.eurpolymj.2018.09.039
55. Li K, You J, Liu Y et al (2020) Functionalized starch as a novel eco-friendly vulcanization accelerator enhancing mechanical properties of natural rubber. Carbohydr Polym 231:115705. https://doi.org/10.1016/j.carbpol.2019.115705
56. Li MC, Cho UR (2013) Effectiveness of coupling agents in the poly (methyl methacrylate)-modified starch/styrene-butadiene rubber interfaces. Mater Lett 92:132–135. https://doi.org/10.1016/j.matlet.2012.10.050
57. Wu J, Li K, Pan X et al (2018) Preparation and physical properties of porous starch/natural rubber composites. Starch/Staerke 70:1–8. https://doi.org/10.1002/star.201700296
58. Wang ZF, Peng Z, Li SD et al (2009) The impact of esterification on the properties of starch/natural rubber composite. Compos Sci Technol 69:1797–1803. https://doi.org/10.1016/j.compscitech.2009.04.018
59. Xu P, Zhao X, Niu D et al (2018) Superior reinforcement of ethyl-co-vinyl acetate rubber composites by using nano-sized starch filler: the role of particle size and reactive compatibilization. Eur Polym J 105:107–114. https://doi.org/10.1016/j.eurpolymj.2018.05.019
60. Yang J, Gao B, Zhang S, Chen Y (2021) Improved antibacterial and mechanical performances of carboxylated nitrile butadiene rubber via interface reaction of oxidized starch. Carbohydr Polym 259:117739. https://doi.org/10.1016/j.carbpol.2021.117739
61. Angellier H, Molina-Boisseau S, Dufresne A (2005) Mechanical properties of waxy maize starch nanocrystal reinforced natural rubber. Macromolecules 38:9161–9170. https://doi.org/10.1021/ma0512399

Chapter 7
Bacterial Cellulose (BC) Based Rubber Nanocomposites

Alvina Augusthy, Harinand Satheesan, Reshma Varghese, Sreejith Puthuvalsthalath Madhusudhanan, and Jayalatha Gopalakrishnan

Abstract Carbon black and silica are the prominent fillers used in the rubber industry. However, these fillers can seriously pollute the environment, need to be added in large quantities and carbon black, being produced from fossil fuels demands an alternative filler that is renewable, ubiquitous, less expensive and biodegradable. Though nanocellulose, especially, from plant sources has been explored as reinforcing fillers to a great extent in rubber matrices, they rely on climatic and geographical conditions, require energy-intensive processing, etc. Bacterial cellulose exhibits superior features such as purity, high crystallinity, high modulus and strength making it a potential candidate for reinforcing rubbers. This chapter discusses the production of bacterial cellulose, factors affecting the production, characterization and its influence on the mechanical and viscoelastic properties of natural and synthetic rubbers.

Keywords Nanocellulose · Whiskers · Viscoelastic properties · DMA · Stimuli-responsive behaviour · Elastomers

7.1 Introduction

Rubbers or their blends are widely utilized in the manufacture of products such as tyres, gaskets, seals, conveyor belts and hoses due to their remarkable elasticity, flex resistance, durability and wear resistance. For such applications, it is common to use reinforcing fillers, as rubbers inherently have low mechanical strength and modulus. The conventional reinforcing fillers that dominate the rubber industry include carbon black and silica. Nevertheless, the processing of these fillers generates environmental pollution and substantial energy consumption. Furthermore, the high density of these

A. Augusthy · H. Satheesan · R. Varghese · S. P. Madhusudhanan · J. Gopalakrishnan (✉)
Department of Polymer Science and Rubber Technology, CUSAT, Kochi 682022, India
e-mail: jayalatha@cusat.ac.in

J. Gopalakrishnan
Interuniversity Centre for Nanomaterials and Devices, CUSAT, Kochi 682022, India

Visakh P. M. *Rubber Based Bionanocomposites*, Advanced Structured Materials 210,
https://doi.org/10.1007/978-981-10-2978-3_7

fillers imparts high density to the ultimate filled rubber product [1–3]. During the past decades, there has been considerable interest in the development of nanocomposites using nanofillers such as nanoclay, carbon nanotube (CNT) and graphene. The intriguing characteristics of nanofillers such as high surface area and interaction with the matrix can significantly enhance the mechanical properties of nanocomposites even with a small fraction of fillers. However, there was an increase in demand for biobased fillers that are lightweight, carbon neutral and have a low impact on the environment. Cellulose has witnessed an expeditious advancement due to its renewable nature, low density, high mechanical strength, ease of availability, diverse resources, etc. [4].

7.1.1 Cellulose

Cellulose, the most prominent biopolymer on earth is made up of repeating glucose units $(C_6H_{10}O_5)_n$ linked through β-1,4-glucosidic bonds. It is the primary structural element of plant cell walls and plays a vital role in the paper, textile and pulp industries. Cellulose can be easily isolated from plants via the elimination of lignin and hemicellulose. Furthermore, cellulose can be biosynthesized by a variety of microorganisms such as algae, fungi or bacteria. Some sea organisms (tunicates) also biosynthesized cellulose and are hence named tunicin. Besides, other routes that are of scientific importance include in vitro enzymatic synthesis using cellulase as a catalyst and chemosynthesis from glucose via cationic ring-opening polymerization of benzylated-pivaloylated glucose derivatives. Figure 7.1 displays the various pathways for the synthesis of cellulose [5, 6]. The hydroxyl groups on C2, C3 and C6 of glucose units render it hydrophilic with the formation of inter- and intramolecular hydrogen bonds which hinders its ability to dissolve in water as well as in most organic solvents [7].

Based on the plant species, nearly 18–36 individual cellulose chains assemble via hydrogen bonds to produce elementary fibrils. These fibrils gather to form microfibrils and subsequently bundled macrofibrils. Furthermore, these micro- and macrofibrils in conjunction with lignin and hemicellulose build the cell wall, which eventually interact with other substances such as proteins, inorganic materials to form the plants. Each fibril is made up of both ordered (crystalline) domains and disordered (amorphous) domains with cross-sectional dimensions ranging from 2 to 20 nm. A single cellulose chain transits through many crystalline and amorphous regions. Within the crystalline region of fibrils, the chains of cellulose are remarkably aligned [8, 9].

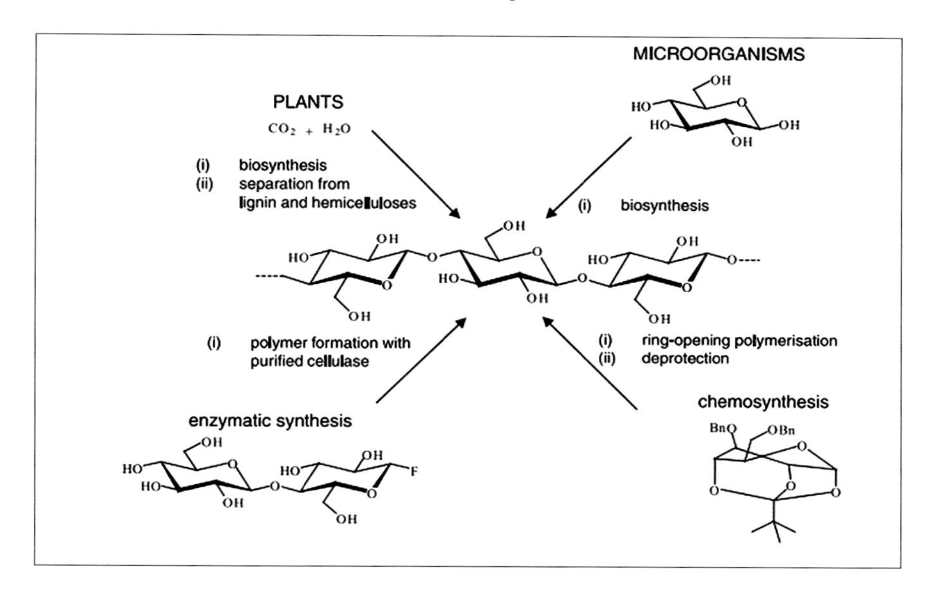

Fig. 7.1 Different synthesis modes of cellulose (Reprinted with permission from Progress in Polymer Science, 2001, 26(9), 1561–1603, Copyright (2001) Elsevier)

7.1.2 Bacterial Cellulose (BC)

Brown, in 1886, first discovered bacterial cellulose (BC) as a gelatinous translucent membrane on the surface of acetic acid fermentation broth [10]. Bacterial cellulose, which is chemically identical to plant cellulose is produced by the fermentation of bacteria, the most noteworthy being the non-pathogenic species of *Komagateibacter* such as *K. xylinus*, formerly known as *Acetobacter* or *Gluconacetobacter*. During its production, the gelatinous BC pellicle can be shaped into a variety of sizes. The nanofibrils of BC have cross-sectional dimensions in the nanometer range and can aggregate to form ribbon-like microfibrils, which have dimensions of 50–80 nm in width and 3–8 nm in thickness (Fig. 7.2). BC microfibrils are organized into a 3-D network structure, forms a membrane that never dries out and maintains up to 99 wt. % of water, making it less vulnerable to microbial attack [11]. In contrast to plant cellulose, BC is free of lignin, hemicellulose, pectin and other components of lignocellulosic thus rendering high purity. It has a unique 3D nanofiber network structure with fibres having a diameter nearly 100 times thinner than plant-based fibres [12]. BC possesses high surface area with striking features such as high crystallinity, high mechanical strength and modulus, high sorption ability, biodegradability, non-toxicity and hydrophilicity [13–15]. Further, BC is independent of regional and climatic conditions. It has been reported that the elastic modulus of single BC microfibrils is about 78 GPa which is higher compared to natural fibres [16].

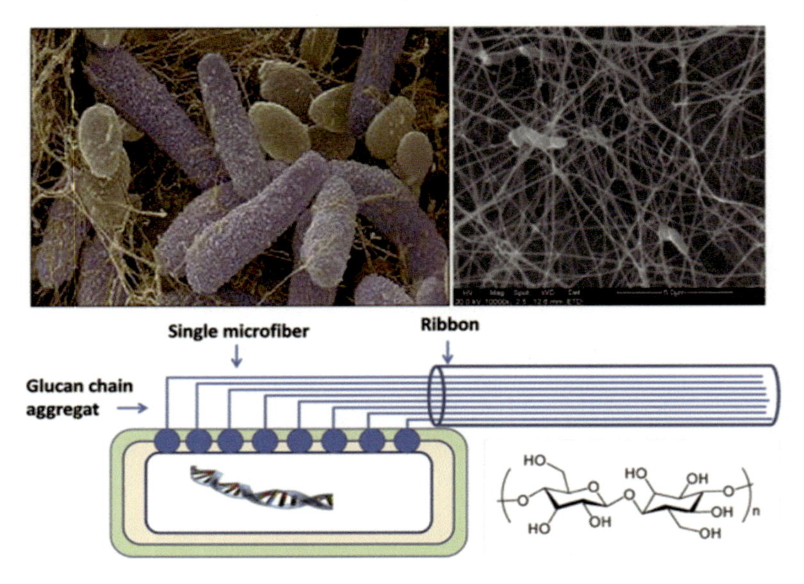

Fig. 7.2 SEM images of *Acetobacter xylinus* and formation of Bacterial cellulose (Reprinted with permission from Food Hydrocolloids, 2014, 35, 539–545, Copyright (2014) Elsevier)

7.1.3 Bacterial Cellulose Whiskers (BCW)

Like plant cellulose, BC can be processed into cellulose nanocrystals of diameter 10–50 nm and length 100–300 nm via a top-down approach. The most common method is acid hydrolysis in conjunction with ultrasonication. Each microfibril consists of strings of cellulose crystals that are linked with disordered or paracrystalline regions. Treatment with acids, e.g. sulphuric acid leads to the elimination of regions of low lateral order resulting in the formation of rod-like cellulose nanocrystals, also referred to as whiskers [17–21]. The limitations associated with acid hydrolysis include the formation of sulphate groups on the nanocellulose crystals that can diminish the thermal stability and significantly reduce the degree of polymerization (DP) which can adversely affect the reinforcing properties in nanocomposites [22]. Another approach for producing BC nanocrystals that can retain the structural properties of cellulose is enzymatic hydrolysis, employing the use of commercially available cellulase enzyme. This multicomponent enzyme hydrolyzes the cellulose into low molecular weight fragments, cellobiose and glucose [23]. The cellulase complex contains endoglucanase which promotes the hydrolysis of amorphous domains that act as structural defects causing the random scission of the chains to smaller fragments [24]. Nonetheless, the crystalline regions are less vulnerable to hydrolysis as a result of the strong hydrogen bonding that exists between them. The enzymatic hydrolysis depends on various factors such as duration of enzyme activity, concentration of enzyme, surface area of cellulose, temperature and pH. Higher concentration of enzyme enhances its adsorption on cellulose fibrils resulting in a rapid fragmentation of cellulose. However, beyond a certain concentration, there is no further increase in the enzyme activity due to the saturation of the available surface. It has been reported that a low concentration of enzyme has a better control on the rate of hydrolysis. Further, a prolonged enzyme action may adversely break down the compact crystalline regions. A cellulase concentration of 0.004% (v/v) in acetate buffer at pH 5 and temperature of 50 °C for a duration of 12 h has been observed as the optimum condition for the enzymatic hydrolysis. Moreover, the process is eco-friendly and cleaner in comparison to conventional acid hydrolysis [20].

7.2 Production of Bacterial Cellulose

Bacterial strains such as *Acetobacter, Sarcina, Achromobacter, Aerobacter, Alcaligenes, Agrobacterium, Pseudomonas, Zoogloea* and *Rhizobium* along with carbon source, oxygen and nitrogen source produce cellulose with high purity. The pathway for the biosynthesis of BC (Fig. 7.3) comprises the following steps:

(1) Phosphorylation of glucose to glucose-6-phosphate by glucokinase,
(2) Isomerization of glucose-6-phosphate by phosphoglucomutase to glucose-1-phosphate,
(3) Conversion of glucose-1-phosphate by UDPG pyrophosphorylase to uridine diphosphate glucose (UDPG) via UDPG pyrophosphorylase and

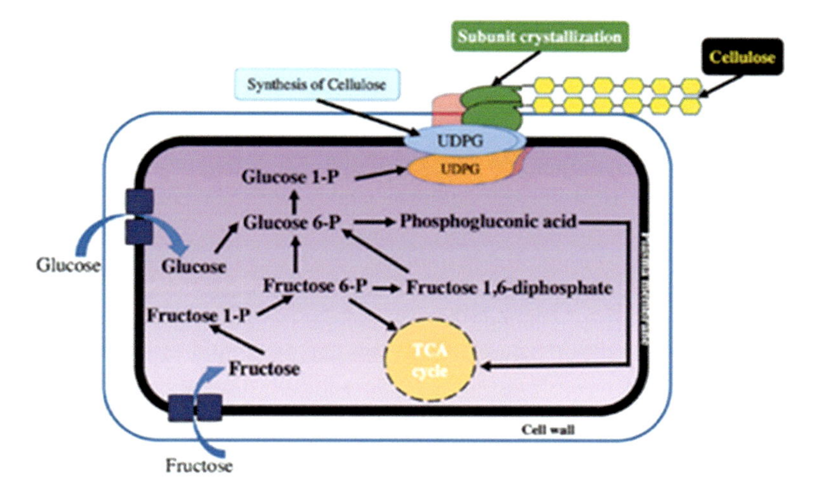

Fig. 7.3 The bottom-up biosynthesis of BC in the cells of microorganisms (Reprinted with permission from ACS Sustainable Chemistry & Engineering, 2020, 8(20), 7536–7562, Copyright (2020) American Chemical Society)

(4) Expansion of UDPG by cellulose synthase to form cellulose chain coupled with crystallization.

The BC is synthesized in the form of a ribbon composed of 10–100 microfibrils that protrude from the pores of bacterial rod. The rate at which the ribbon elongates is 2 μ/min. There are nearly 50 sites arranged in rows along the longitudinal axis of the outer envelope of the bacterial cell that can produce BC microfibrils [6, 17, 25].

Compared to plant cellulose, cellulose production via microbial fermentation has numerous benefits:

(i) BC is not reliant on climate and geographical factors,
(ii) The growth rate of the microorganisms may be regulated to generate BC in the required amounts within the timeframe,
(iii) Energy-intensive plant cellulose processing may be eliminated because of the absence of lignin or other impurities in the cellulose generated by microbes,
(iv) Genetically altered microbes can manufacture BC with the necessary characteristics and
(v) Both agricultural and industrial wastes can be utilized as a growing medium.

The species and genetic mutation of the utilized bacteria, the kind of feedstock, its composition and the type of reactor for the production process must be considered to guarantee optimal production yields and regulated costs [11].

7.2.1 Bacterial Strains and Its Genetic Modification

Apart from *Gluconacetobacter xylinus*, other gram-negative bacterial strains such as *Rhizobium, Agrobacterium, Achromobacter, Pseudomonas, Aerobacter, Azotobacter, Salmonella, Escherichia* have also produced BC as part of their metabolic activity. However, *Gluconacetobacter* is the only species that is capable of producing BC at the commercial level. It has been reported that a gram-positive strain *Sarcina ventriculi* can also produce BC but lacks cellulase activity [16]. Gluconic acid produced during the synthesis can reduce the pH of the medium and thereby decline the production of BC. Genetically engineered mutant bacteria have been found to improve the yield of BC. UV mutagenesis strategy has been utilized by De Wulf et al. to obtain ketogluconate non-producing mutant from its parent strain with an increased productivity of 3.3 g/L from 1.8 g/L after 10 days of shaking culture [26]. Further, when cultivated in a stirred reactor, the mutant can enhance the production by 36% in comparison to its parent strain [27]. The BC yield of *Gluconacetobacter xylinus* (*G. xylinus*) has been found to increase from 106 to 158 g/L by mutagenesis, induced via high hydrostatic pressure treatment [28]. Glutamate dehydrogenase deficient mutant of *G. xylinus* BPR strain produces BC twice as that of its parent strain [29]. It produces BC with more porosity, low film thickness and low density than that of *G. xylinus*.

Moreover, supercritically dried BC from *G. europaeus* exhibits better water absorption capability, higher Young's modulus and higher hardness in comparison to room temperature and freeze-dried BC [6, 30].

7.2.2 Feedstock and Culture Medium

Though Hestrin-Schramm (HS) medium has been considered the standard medium for BC cultivation, it is expensive and needs supplements for effectual cultivation. The prime components essential for the growth of microbes are carbon sources (maltose, fructose, starch, glycerol) and nitrogen source. When fructose and sucrose were used as dual carbon sources, the BC yield was found to be 8.79 g/L [31]. Industrial and agricultural wastes have been identified as the cost-effective feedstock for the growth of bacteria [32–35]. For instance, tea infusions [36], corn steep liquor [37], fruit and vegetable juices [38], molasses [39, 40], wheat straw acid hydrolysate [41], waste cotton fabrics [42], rice bark [43], konjac powder hydrolysate [44], etc. have been extensively investigated for optimizing BC production. It has been reported that sucrose, glucose and mannitol are the carbon sources that can consistently produce high yields of BC. The culture medium accounts for 30% of the production cost of BC. Wastewater from rice wine distillery referred to as thin stillage used to replace distilled water in HS medium can produce 2.5 fold BC (10.38 g/L) after 10 days of static cultivation [45]. Further, nitrogen sources such as peptone and yeast are generally used for supplying nitrogen and growth factors [46]. Inorganic nitrogen has been found to inhibit the growth of bacteria [47]. Moreover, vitamins such as nicotinic acid, pyridoxine, biotin and p-amino benzoic acid can further enhance the metabolism and growth of bacteria unlike riboflavin and pantothenate [48]. Furthermore, culture conditions such as variation in temperature, pH, presence of ions in the culture medium, supply of oxygen and additives such as nicotinamide, ethanol, organic acid and sodium alginate have a major influence on BC productivity (Fig. 7.4) [11, 25, 49, 50].

7.2.3 Temperature

Temperature is a crucial parameter that not only influences the homeostatic physiology of an organism but also regulates the adaption pattern in its localized niche. Zahan et al. reported that 28 °C is the optimum temperature for maximum BC production while no production was observed at 5° and 40 °C. High temperature develops a denaturing environment for the bacterial strain thereby affecting their metabolic components. Lower temperature adversely affects BC production by reducing the energy supply that is mandated for cell development and conversion of glucose to cellulose [51].

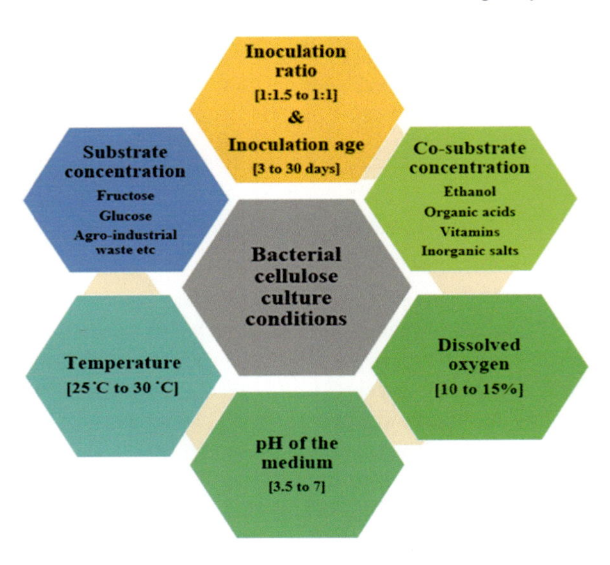

Fig. 7.4 Factors that influence BC production (Reprinted with permission from International Journal of Biological Macromolecules, (2020)164, 2598–2611, Copyright (2020) Elsevier)

7.2.4 pH

An acidic or near-neutral pH during fermentation is essential for BC production. Further, the production of secondary metabolites such as acetic acid, gluconic acid and lactic acid can shift the pH. Thus, a pH of around 4–6 is ideal for BC production [52]. However, the optimal pH may vary with the carbon source.

7.2.5 Oxygen Level

Cellulose-secreting bacteria are highly aerobic and hence it is necessary to control the aeration rate in the culture medium. Oxygen content is also crucial for the metabolism of cells. Higher content of dissolved oxygen in the medium favours the production of BC along with gluconic acid formation while lower oxygen content prevents the growth of bacteria. Oxygen content may be scanty in static cultivation leading to lower BC yield [53–55]. The O_2 transfer rate, which continues to decrease with an increase in the viscosity of the broth, is another factor that affects the net BC productivity. The problem can be alleviated by using a two-stage cultivation approach where the dissolved oxygen level is increased in the culture during the first stage to support cell growth before it reaches the exponential phase of the growth curve and at the second stage to maintain a hypoxic condition in the stationary phase to promote BC production [56].

7.2.6 *Agitation Rate*

The agitation rate has a major impact on the characteristics, morphology and net productivity of BC. At lower agitation rates (100 rpm), the cellulose attains a ball-shaped morphology with a size range of 0.5–1 cm. With increasing agitation rates (150–200 rpm), the size drastically reduces and at 300 rpm, clusters of irregular size and shape are formed. A moderate rpm can result in paramount productivity. Rigorous agitation of the culture medium induces shear stresses which strengthens the formation of cellulose-negative mutants and affects productivity. Moreover, the agitation rate varies considerably for different bacterial strains and different culture broths and hence needs to be optimized for mass BC production.

7.2.7 *Fermentation Techniques*

The strategies for BC fermentation using microorganisms include static, agitated and bioreactor fermentation. In the static method which involves no shear, the culture medium is set in a shallow tray, inoculated and incubated for 5–20 days at temperatures of 28–30 °C. A thick gelatinous BC pellicle forms at the air–liquid interface and restricts the oxygen beneath the pellicle exposed to the medium [57]. Further, the floating nature of the pellicle is due to the trapped CO_2 bubbles formed during the metabolism of bacteria. The low productivity due to large cultivation area, long culture time and intensive labour limits the industrial application [49]. A modified system referred to as fed-batch strategy utilizing the waste from beer fermentation broth has been found to produce around 750 g of BC sheet with a thickness of 2.5 cm after 30 days of cultivation. The fibrils were crowded and thinner in contrast to the conventional system [58]. Utilizing this strategy, a novel bioreactor was developed by Kralisch et al. named as horizontal lift reactor, which produces fleeces of adjustable length and height. Continuous harvesting under static conditions and significant cost reduction are advantages in contrast to static cultivation [59].

Under agitated fermentation, BC in the form of pellets or threads is produced in less time. The low production costs result from the continuous mixing of oxygen into the medium along with less space and workforce requirements. Though several attempts have been made to increase the production by elevating the shear rates, it not only promotes the turbulence force of the medium but also induces mutation of the bacterial strains inhibiting the synthesis of BC. In contrast to a stirred tank reactor, airlift reactor is advantageous as it is energy efficient, involves low shear stress thereby controlling the BC-producing strains and continuously transfers oxygen from the bottom of the reactor to the medium facilitating an aerobic atmosphere to the microbes [53]. Further with a stirred tank reactor and air lift reactor, BC adheres to the shaft of the reactor making it difficult to collect the product and clean up the reactor. Rotating disk bioreactor is a better option for BC production as it augments the oxygen supply, provides a high surface-volume ratio and reduces shear force. It

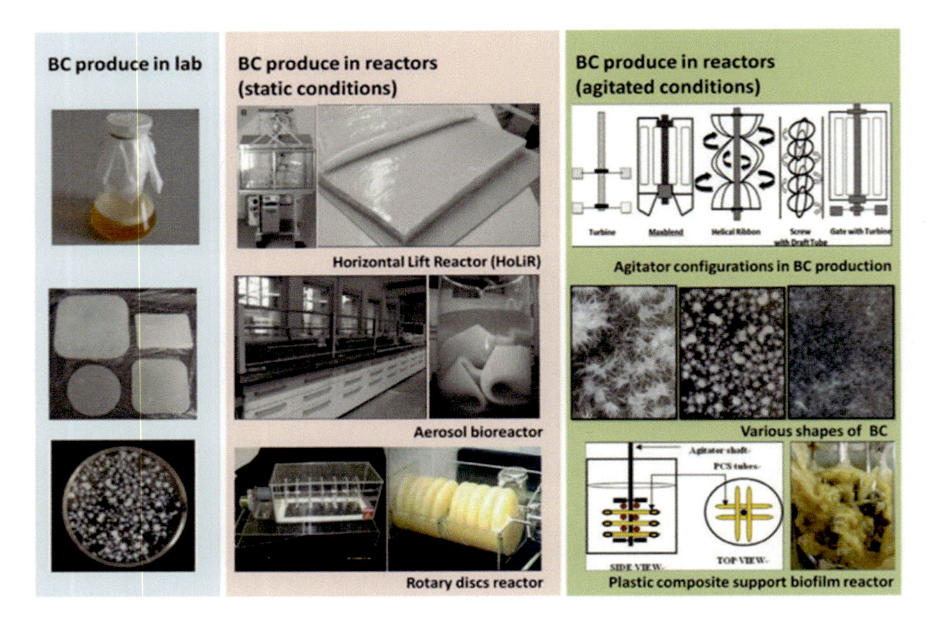

Fig. 7.5 Bacterial cellulose produced under static and agitated conditions (Reprinted with permission from Food Hydrocolloids, 2014, 35, 539–545, Copyright (2014) Elsevier)

consists of numerous circular discs mounted on a central rotating shaft and an inlet for inoculation. Half the area of the disc is submerged in the medium while the other half is exposed to the atmosphere. During rotation, the disk surface is alternately located between the medium and atmosphere. Moreover, the BC produced will adhere to the disk and restore its strength [11, 16, 49, 55, 60–62]. BC produced under static and agitated reactors is displayed in Fig. 7.5.

7.3 Characterization of BC

A. xylinium produces two forms of cellulose: ribbon-like polymer, Cellulose I and amorphous form which is more thermodynamically stable, Cellulose II. The β-1,4 glucan chains in Cellulose I are arranged parallel via van der Waals forces while the chains in Cellulose II are antiparallel and with a large number of hydrogen bonds which makes it more stable [63, 64]. Cellulose I structure, naturally, is found in two allomorphic forms, I_{α} or I_{β} depending on the arrangement of chains between each other. BC exhibits a higher percentage of I_{α} structure in contrast to plant cellulose which shows a higher percentage of I_{β} structure. Various factors that can affect the characteristics of BC include fermentation method, culture medium and carbon sources. Bacterial cellulose can have various morphologies like pellicle, gel, hollow tube, fibre, sphere, etc. based on the bacterial strain, culture and growth medium conditions. Further, static and agitated culture can produce BC with different morphologies which is attributed to varying degrees of crystallinity, crystallite size and I_α content [65].

The FESEM images of BC shown in Fig. 7.6 indicate a reticulated structure comprising ultrafine cellulose fibrils. The chains of BC aggregate to form subfibrils which crystallize into microfibrils and get bundled to form a ribbon-like structure [66]. The BCWs formed after acid hydrolysis appears as slender rod-like shapes in the TEM image (Fig. 6b). The whiskers aggregate together as a result of strong hydrogen bonds between the hydroxyl groups present on the surface [67].

The XRD pattern of BC (Fig. 7.7) reveals three characteristic peaks at $2\theta = 14.5°, 16.4°$ and $22.5°$ corresponding to (101), (10) and (002) crystallographic planes, respectively, indicating that cellulose exhibits an I_α structure. The crystallinity index can be calculated using the following equation [68]

$$CI(percentage) = A_{crystalline}/A_{total} \times 100 \tag{7.1}$$

where $A_{crystalline}$ – sum of the areas relevant to crystalline peaks

A_{total} – sum of the areas under the diffraction peaks

Further, the crystallite sizes can be estimated from the peaks of (101), (10) and (002) planes using the equation

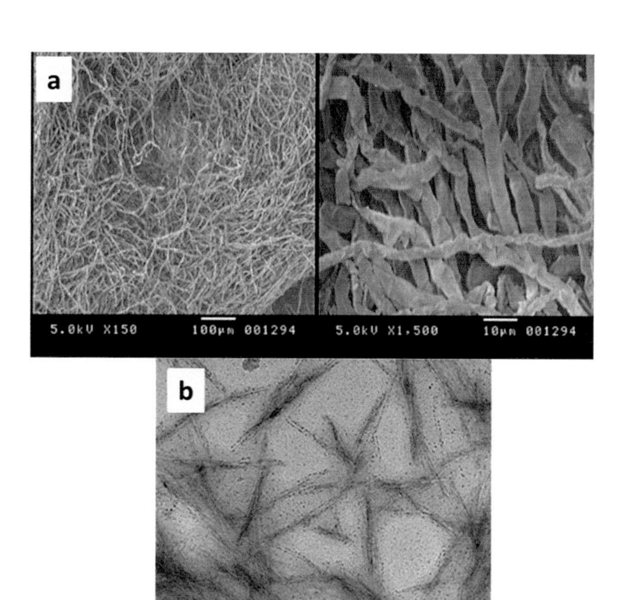

Fig. 7.6 FE-SEM micrograph of BC (**a**) (Reprinted with permission from Carbohydrate polymers, 2014, 106, 132–141, Copyright (2014) Elsevier) and TEM of BCW (**b**) (Reprinted with permission from Biomacromolecules, 2013, 14(4), 1078–1084, Copyright (2013) American Chemical Society)

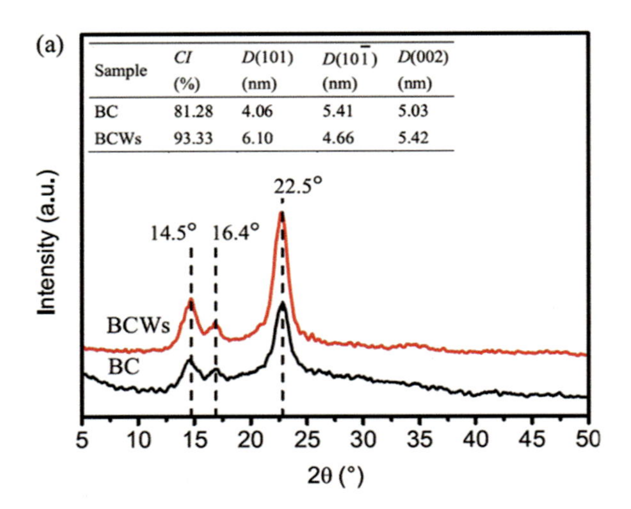

Fig. 7.7 X-ray diffraction pattern of BC and BCW (Reprinted with permission from industrial crops and products, 2018,113, 240–248. Copyright (2018) Elsevier)

$$D_{hkl} = k\lambda/\beta \cos\theta \tag{7.2}$$

BCWs, in comparison to BC exhibit a higher crystallinity index (CI) on account of the elimination of amorphous domains [67]. The crystallite size related to (101) and (002) planes of BCW are slightly higher than BC indicating that the strong acid hydrolysis may also digest the smaller as well as more defective crystals. The atomic force microscopy (AFM) images of dilute suspensions of BCWs revealed that the nanowhiskers had diameters in the range of 3–30 nm. Nevertheless, the shape of the whiskers appears different than that observed in TEM images which may be due to the tip broadening effect of AFM [69]. The FTIR spectra of BC (Fig. 7.8) exhibit the characteristic peaks at 3350 and 2960 cm^{-1} attributed to the O–H and C–H stretching vibrations, respectively. The absorption signal at 1650 and 1060 cm^{-1} corresponds to the deformation vibration of the adsorbed water molecules and C–O–C stretching of glucose. Two additional bands at 1430 and 1370 cm^{-1} are assigned to the symmetric deformation and bending vibration of C–H, respectively [70–74].

Bacterial cellulose shows high thermal stability and can withstand temperatures up to 220 °C without degradation. TGA studies have revealed that BC shows a one-step degradation process and ~ 65% weight loss occurs at 280 °C with a maximum degradation rate at 300 °C. The purity of BC, high crystallinity and orientation of cellulose chains within fibrils induce high thermal stability. It exhibits a T_g of nearly 40 °C and T_m of 85 °C [75]. It is able to exhibit thermal conductivity as low as 0.03 W/mK, which is much lower than some of the common insulating materials like fiberglass [76]. It also shows unique phase transition and thermal expansion. Bacterial cellulose does not expand or contract significantly with changes in temperature. The phase transition of bacterial cellulose can be controlled by varying the degree of polymerization. It exhibits a mechanical strength of around 9.51 MPa. Using AFM, the elastic modulus of BC is found to be 78 GPa which is substantially higher in comparison to natural fibres. It is capable of retention of large amount of water

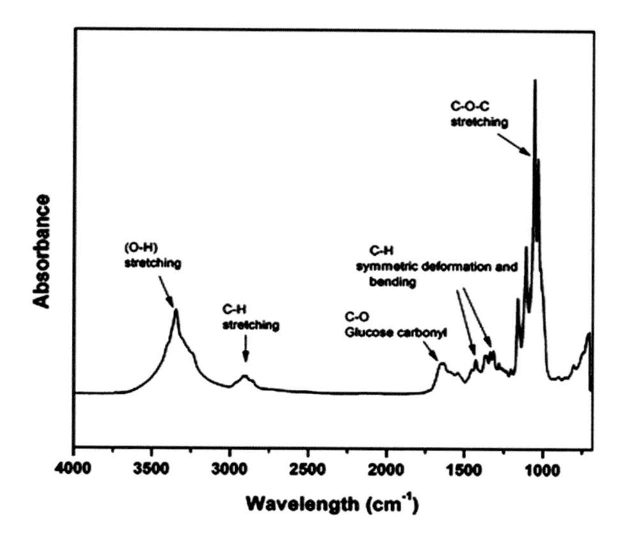

Fig. 7.8 FT-IR spectra of BC (Reprinted from RSC advances, 2017, *7*(23), 13,678–13,688)

which makes it hydrophilic due to numerous hydroxyl groups. The surface area of bacterial cellulose is reported to be in the range of 20–200 m^2g^{-1}. It has a high surface area because of its unique nanostructure which consists of interconnected nanofibrils [77].

The following section discusses the influence of bacterial cellulose on the morphology, mechanical and viscoelastic properties of rubber nanocomposites.

7.3.1 Rubber/Bacterial Cellulose Composites

7.3.2 Natural Rubber/Bacterial Cellulose Nanocomposite

Natural rubber is a sustainable polymer that is endowed with high elasticity and film-forming capability. Nevertheless, it exhibits low abrasion resistance and mechanical strength which needs to be tailored for specific performances in products [78]. Carbon black and silica are the most common nanoscale fillers that make NR suitable for automobile and truck tyre applications [79]. Although nylon, aramid, glass fibres have been used as reinforcing fillers [80], the exigency for a renewable green filler has promoted the use of nanocellulose as a viable alternative to conventional fillers. With regard to the chemical structure of cellulose, the –OH functionality, being polar can pose the problem of poor interfacial compatibility with the non-polar NR matrix. This can result in non-homogeneous distribution of cellulose in the NR matrix. The issue can be mitigated by specific chemical modifications of cellulose [81] which aids to reduce the hydrophilic nature or preparing NR composites via an aqueous latex microdispersion process.

In situ surface modification of nanocellulose in their natural aqueous medium is a suitable technique as it eliminates the laborious operations involved in other modification strategies, which utilize organic solvents. Trovatti et al. have modified the surface of bacterial cellulose nanofibers (BC) via in situ micellar polymerization of styrene using hexadecyl pyridinium chloride as the surfactant referred to as BCPS. The process involves the adsorption of a double layer of cationic surfactant onto the surface of BC nanofibres followed by adsolubilization of the styrene monomer into the surfactant layer. The use of a water-soluble initiator promotes the polymerization of styrene in the aqueous medium. The adsorption of cationic surfactant on nanocellulose is favoured by the electrostatic interaction with the negative charges present on nanocellulose. With regard to BCPS, polystyrene (PS) gets homogeneously distributed as a thin layer on the surface with occasional bulging (Fig. 7.9). NR composite films are produced by the incorporation of BC and BCPS fibres into NR latex containing vulcanizing agents, stirring, drying followed by vulcanization. The composites containing 7 wt. % of BCPS exhibited a tensile strength and modulus of 6 and 0.4 MPa, respectively. The enhanced mechanical properties are due to the effective stress transfer from matrix to fibre where the –OH groups are masked by PS leading to high compatibility and homogeneous dispersion. The modulus of a rubbery plateau for both BC and BCPS showed a substantial increment in contrast to the vulcanized NR [82].

Another simple strategy involves the incorporation of BC nanofibres in the form of dilute slurry in aqueous NR latex which enables higher loading of BC (20–80 wt. %). The films can be fabricated by casting the suspension in trays and drying. At lower loadings of BC, the nanofibrils are homogenously dispersed while BC loading above 40 wt.% results in a layered structure. With 80 wt.% of BC nanofibers, the tensile strength improved from 0.8 to ~75 MPa (Fig. 7.10). The high modulus of the composite may be due to the rigidity imparted by the crystalline BC fibres in the NR matrix. The network of BC fibres restricted the mobility of the chains leading to enhancement in mechanical properties [83]. Further, above 50 wt. % of BC nanofibers, T_g decreased slightly as a result of the segmental mobility of the NR chain within the multilayered porous structure. The thermal decomposition of the composites containing 20 wt.% BC shifted to a higher temperature in comparison to NR and BC. Higher loadings of BC can cause agglomeration of the fillers leading to slightly lower stability [84].

The reinforcing effect of BCW in rubbers has received great attention on account of its high crystallinity, high aspect ratio, outstanding mechanical properties, etc. The extent of the filler network, dispersion of fillers in rubber and the physical/chemical interfacial interaction are the prime factors that influence the mechanical properties of the nanocomposites. Yin et al. have developed NR/BCW composites by mixing aqueous suspensions of BCW with NR latex via ultrasonication followed by casting and drying. The XRD pattern of natural rubber exhibits a broad peak at $2\theta = 18°$ owing to its amorphous nature. The nanocomposite combines the peaks of NR with the characteristic peaks of BCW. With increasing content of BCW, the peak corresponding to NR apparently weakens while that of BCW displays an enhanced intensity. The characteristic peaks of NR comprise the asymmetric and

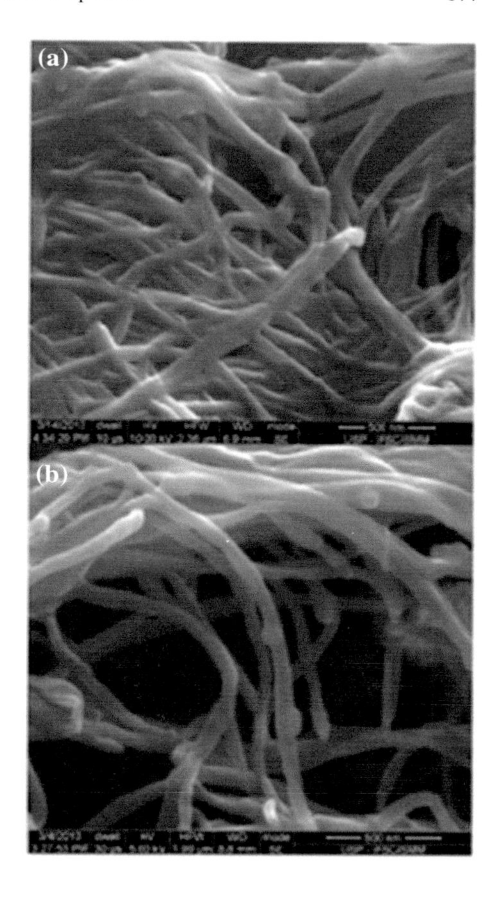

Fig. 7.9 SEM micrographs of **a** BC and **b** BCPS fibres (Reprinted with permission from Biomacromolecules, 2013, 14(8), 2667–2674. Copyright (2013), American Chemical Society)

symmetric stretching vibrations of $-CH_2$ (2921, 2851 cm^{-1}), asymmetric and deformation vibrations of $-CH_3$ (2957, 1375 cm^{-1}), asymmetric stretching vibration of C=C (1657 cm^{-1}) and out-of-plane bending vibration of C=C–H (832 cm^{-1}) [85]. The addition of BCW caused a rougher texture to the nanocomposite than the neat NR which presented a homogenously striped texture. Higher loadings of BCW (>10 phr) led to agglomerate formation of irregular shapes. The DMA studies (Fig. 7.11) revealed three distinct regions in the storage modulus (E′) vs temperature plots of the nanocomposites viz. a glassy region followed by a transition region and rubbery plateau. Storage modulus is related to the stiffness and load-bearing capacity of the material. At low temperatures (~ −60 °C), there is only a moderate increase in storage modulus with an increase in BCW content, i.e. from 1918 MPa for neat NR to 5878 MPa for NR/BCW20. This gradual increase may be due to the fact that NR being in the glassy state in the nanocomposite may not result in a significant change in modulus between the rubber and the filler. A sharp decline in E′ occurs in the glass transition region in NR in contrast to the nanocomposite which is ascribed to the relaxation phenomenon in rubber. In the rubbery plateau region, E′ is significantly

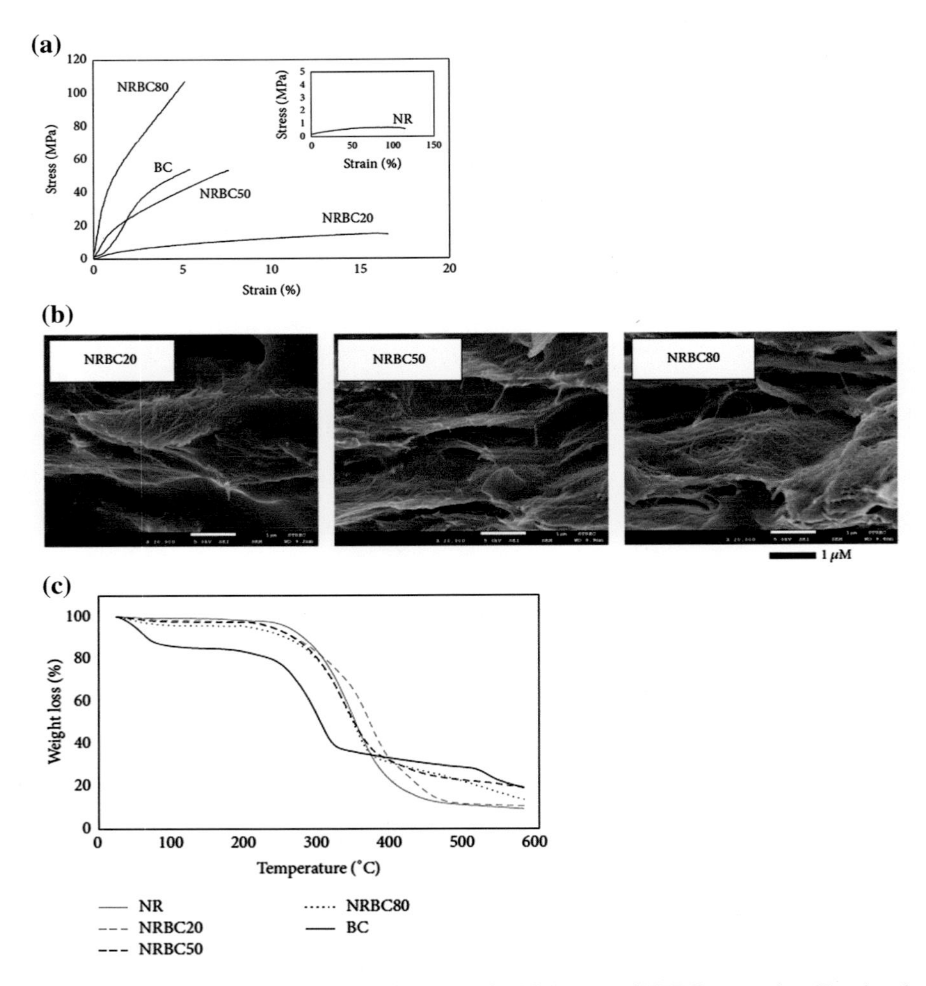

Fig. 7.10 **a** Stress–strain curves, **b** SEM images and **c** TGA curve of NRBC composites (Reprinted with permission from Journal of Nanomaterials, 2017)

higher in the case of NR/BCW20 (635 MPa) compared to NR (1.3 MPa). The tremendous increase may be due to the 3D network of BCWs that are linked via hydrogen bonding and the numerous immobilized chains around the surfaces of BCW [86]. Further, the T_g increased to higher temperatures due to immobilization of the rubber chains in the proximity of BCW. Additionally, the intensity of the tan δ reduced for the nanocomposite as a result of a drop in the volume fraction of mobile polymer segments [87]. Moreover, the storage modulus was triggered by water intake and its removal thereby exhibiting the behaviour of a shape memory polymer. In the presence of water, the modulus of NR/BCW film decreased drastically as the BCW promoted water intake and disrupted self-interaction via competitive hydrogen bonding with

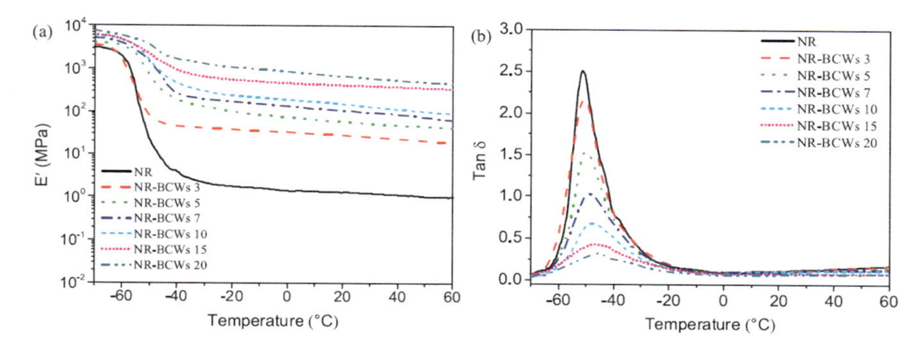

Fig. 7.11 Temperature dependence of **a** storage modulus and **b** tan δ curves of NR/BCW nanocomposites with varying BCW content (Reprinted with permission from Industrial crops and products, 2018,113, 240–248.] Copyright (2018) Elsevier)

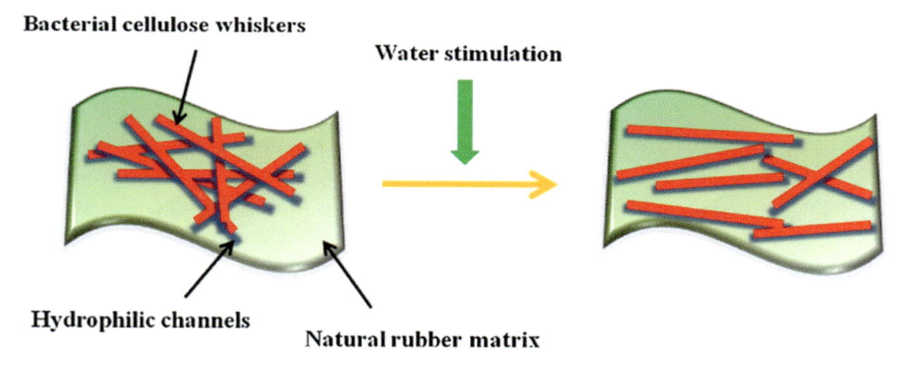

Fig. 7.12 The architecture of NR/BCW nanocomposite before and after immersion in water (Reprinted with permission from Industrial crops and products, 2018,113, 240–248. Copyright (2018) Elsevier)

water (Fig. 7.12). While in the dry condition, the hydrogen bonding between BCW in NR was recovered. Such shape memory materials may be suitable for applications such as self-retractable and movable stents, adaptive implants and self-tightening sutures [69].

7.3.3 Styrene Butadiene Rubber/Bacterial Cellulose Nanocomposite

Styrene butadiene rubber (SBR) is a non-polar rubber with intriguing properties such as abrasion resistance and ageing resistance. It is widely used in tyre industries, which makes use of carbon black as the reinforcing filler at very high loadings.

Though there are reports on the utilization of cellulose nanofibers from plant sources as reinforcements in SBR [88, 89], the usage of bacterial cellulose is scarce.

BCW, the derivative of BC has a high aspect ratio and can impart a high reinforcing effect. Yin et al. have developed SBR/ BCW nanocomposites by mixing SBR latex with BCW suspensions and subsequently coagulated using aqueous sodium chloride (NaCl) solution. The dried coagulum is melt mixed with compounding ingredients and compression moulded. The XRD pattern of amorphous SBR exhibits a broad peak around $2\theta = 20°$. Although low loadings of BCW do not display the characteristic peak of BCW, the peaks gradually emerge for the vulcanizates beyond 0.5 phr manifesting the increase in crystallinity of the nanocomposites. The FTIR spectra of neat SBR and SBR/BCW vulcanizates present the characteristic peaks of SBR. The peaks at 2919, 2846, 1454 and 1395 cm^{-1} reflect the strong asymmetrical and symmetrical stretching vibrations of C–H in methylene, scissoring vibrations of CH_2 and C–H bending vibrations, respectively. The bands at 960, 907, 756 and 704 cm^{-1} are attributed to the trans –CH=CH- group vibration of butadiene, out-of-plane bending vibration of C–H of a vinyl group, out-of-plane bending vibration of aromatic =C–H group and C=C groups of styrene units, respectively [90]. The peaks at 1645 and 1056 cm^{-1} relevant to BCW are predominant in the spectra of nanocomposites.

At low loadings <2.5 phr, the whiskers get well dispersed in the matrix, while higher loadings lead to the formation of aggregates. A loading of 2.5 phr of BCW increases the tensile strength significantly to 9.9 MPa in comparison to neat SBR (2.3 MPa). The high aspect ratio of BCW favours a higher extent of load sharing with the rubber matrix. Furthermore, BCW enables crack-bridging in the rubber matrix thus restricting the failure under stress. Agglomerate formation at higher loadings impedes the stress transfer and thereby reduces the strength. On comparing with the SBR matrix, the tear strength and modulus at 300% enhanced to a value of 35.2 kN/m and 4 MPa. The elongation at break improved substantially up to 2 phr of BCW as a result of the synergistic effect of whisker orientation and chain slippage. Under stretch, the physical cross-links between SBR and BCW permit the reorientation of whiskers along with the elongation of SBR in the stress direction. Nonetheless, the higher loadings of BCW form a stiff network in the composite and restrict the extension of the rubber chain [91, 92]. The cross-link density of nanocomposite increases up to 2 phr BCW content revealing its role as a physical cross-linker. Further, the storage modulus increases with BCW content both in the glassy region as well as rubbery region [93].

Akin to NR/BCW composite films, SBR/BCW films exhibit the water-stimuli responsive dynamic mechanical behaviour where the films are produced by mixing of SBR latex and BCW suspension followed by evaporation. The storage modulus increases rapidly with BCW content which is attributed to the three-dimensional network of rigid whiskers connected via hydrogen bonding and immobilization of numerous rubber chains around the whisker surface. In the rubbery region, there is augmentation of the storage modulus (E′) from 2.8 to 1161 MPa at 20 phr of BCW. After equilibrium swelling in water, the E′ decreases to a value of 8.9 MPa due to competitive hydrogen bonding with water thereby disrupting the whisker network [15].

7.3.4 Carboxylated Acrylonitrile Butadiene Rubber/Bacterial Cellulose Nanocomposite

Carboxylated nitrile rubber (XNBR) is a polar polymer composed of acrylonitrile, butadiene and monomer containing carboxylic groups such as methacrylic or acrylic acid. XNBR exhibits improved physico-mechanical properties such as tensile strength, oil resistance, ageing and abrasion resistance in contrast to conventional NBR rubber [94]. XNBR are extensively used in applications such as high-pressure seals, medical gloves, oil and wear-resistant rubber parts. XNBR are extensively used in applications such as high-pressure seals, medical gloves, oil and wear-resistant rubber parts. Rather than the classical vulcanizing agents such as sulphur or dicumyl peroxide used to cross-link the double bonds in XNBR, divalent metal oxides (ZnO) can also be used to react with the carboxyl groups to form metal-carboxylate cross-linked structure. Numerous fillers such as carbon black, silica, clay, calcium carbonate, and graphene oxide have been incorporated in XNBR not only to reinforce rubber but also to enhance the thermal stability and thermal conductivity [95]. Only a few studies have been reported on the effect of sustainable fillers such as BC on XNBR.

XNBR/BCW nanocomposites can be prepared by combining suspensions of BCW and XNBR latex followed by casting into films or by coagulating using an aqueous NaCl solution (6.5%). The coagulated rubber is dried, further compounded and vulcanized. Wang et al. have studied the effect of varying content of BCW on XNBR nanocomposites that have been vulcanized by dicumyl peroxide. It was noticed that BCW content of 1 phr enhanced the vulcanization rate of XNBR. Higher loadings of BCW resulted in a drop in the vulcanization rate due to the tendency of the whisker in scavenging the free radicals during peroxide curing. The XNBR, being amorphous in nature presents a unique broad diffraction peak at around 18.7° in X-ray diffractogram (XRD) [96]. The XNBR/BCW vulcanizates comprise the amorphous peak of XNBR along with the crystalline peaks of BCW at 14.5° and 22.5° revealing that the crystal structure of BCW is retained after processing. In the FTIR spectrum of BCW, the peak at 3345 cm^{-1} corresponds to the hydrogen-bonded O–H stretching. For XNBR, the peak at 3468 cm^{-1} is ascribed to the O–H group stretching vibration of the carboxylic groups (COOH) while the peaks at 1732 and 1697 cm^{-1} are assigned to the stretching vibration of free and hydrogen bonded C=O groups, respectively [97]. The extent of hydrogen bonding formed between the OH group of BCW and COOH group of XNBR can be determined by curve fitting of the spectra using the Lorentz function.

The fraction of hydrogen-bonded C=O groups F_{H-CO} is calculated using the equation [98].

$$F_{H-CO} = (A_H/r)/(A_H/r + A_f) \quad (7.3)$$

where A_f – peak area of free C=O groups

A_H – peak area of H-bonded C=O groups and

r – conversion constant, is the specific absorption ratio of free to H-bonded C=O groups (r =1.5).

An increase of BCW content (XNBR) enhances the $F_{H\text{-}CO}$ from 0.42 to 0.49 (13 phr BCW) revealing the formation of H-bonding between the surface –OH groups of BCW and –COOH groups of XNBR. The tensile strength increased with BCW content showing an increment from 2.9 to 12.1 MPa at 13 phr loading. The Young's modulus and tear strength also increased from 1.7 to 4.3 MPa and 14.6 to 36.7 N/mm, respectively. The augmentation can be attributed to the formation of a percolating network of BCW which could easily transfer the stress through hydrogen bonds [99]. Increasing the content of BCW enhances the storage modulus of XNBR nanocomposites not only in the glassy region but also in the rubbery region [67].

Chen et al. have fabricated water-stimuli responsive XNBR nanocomposites via mixing XNBR latex and BCW suspension followed by casting films. The storage modulus at −60 °C increased from 2108 to 5683 MPa for the composite containing 20 phr BCW. Above the glass transition region, i.e. at 25 °C, E′ enhanced by two orders of magnitude from 3.1 to 570 MPa. BCW acts as physical cross-link points due to interfacial hydrogen bonds with XNBR and thereby intensifies the modulus. Moreover, the interaction of hydrogen bonds promotes the internal plasticization up to 7 phr of BCW leading to a shift of T_g to lower temperatures (−10 °C) of composites in contrast to neat XNBR (T_g −3 °C). Higher loadings of BCW form a rigid network structure and impede the segmental motion of rubber resulting in high T_g. The tan δ peak drops continuously with increasing whisker content [100]. The interfacial interactions between the filler and rubber can be predicted by an interaction parameter B which is calculated using the equation.

$$tan\delta_f = tan\delta_m/(1 + 1.5B\phi) \tag{7.4}$$

where tan δ_f and tan δ_m are the peak values of tan δ for neat rubber and filled rubber, respectively, and Φ represents the volume fraction of filler [101]. The parameter B rises with an increase in whisker content revealing the enhanced interfacial interaction between rubber and filler.

The hydrophilic nature of BCW enhances the water uptake behaviour of the composite with increasing BCW content. The storage modulus (E′) of the nanocomposite containing 20 phr of BCW on equilibrium swelling in water drops from a value of 570 to 11.1 MPa. The whiskers act as channels for water uptake, disentangle the hydrogen-bonded percolation network and switch to hydrogen bonding with water indicating the water-induced mechanically adaptive behaviour of the nanocomposite [14].

7.4 Conclusions

Bacterial cellulose has received a great deal of attention as a renewable material in the scenario of sustainable development as it could play a key role in the reduction of carbon footprint. They are natively pure, biodegradable, non-toxic, biocompatible along with high water-holding capacity. Though it has been explored for a plethora of applications such as burn and wound dressings, scaffolds for tissue engineering, drug delivery, medical implants and waste water treatment, there are only a few reports on its use as reinforcing filler in rubber matrices. The characterization results from the literature review reveal that BC is a promising alternative filler to petroleum-based ones. BC can significantly enhance the tensile properties and viscoelastic properties of NR, SBR and XNBR composites. Moreover, its water uptake behaviour within rubber matrices can make it an attractive candidate for water-induced mechanically adaptive materials. However, its use as reinforcing filler in other elastomers along with modifications to enhance the physico-mechanical properties are yet to be explored to enable commercialization in rubber industries.

References

1. Zhang C, Dan Y, Peng J, Turng LS, Sabo R, Clemons C (2014) Thermal and mechanical properties of natural rubber composites reinforced with cellulose nanocrystals from southern pine. Adv Polym Technol 33(S1)
2. CS JC, PK B, VS R, Raman V, TK BS, Sasi S, Antony JV (2022) Bionanocomposites based on natural rubber and cellulose nanofibrils from arecanut husk: rheological, mechanical and thermal characterizations. J Polym Res 29(6):217
3. Kumagai A, Tajima N, Iwamoto S, Morimoto T, Nagatani A, Okazaki T, Endo T (2019) Properties of natural rubber reinforced with cellulose nanofibers based on fiber diameter distribution as estimated by differential centrifugal sedimentation. Int J Biol Macromol 121:989–995
4. PM V, Thomas S, Oksman K, Mathew AP (2012) Effect of cellulose nanofibers isolated from bamboo pulp residue on vulcanized natural rubber. Bioresources 7(2):2156–2168
5. Klemm D, Schumann D, Udhardt U, Marsch S (2001) Bacterial synthesized cellulose—artificial blood vessels for microsurgery. Prog Polym Sci 26(9):1561–1603
6. Ullah H, Wahid F, Santos HA, Khan T (2016) Advances in biomedical and pharmaceutical applications of functional bacterial cellulose-based nanocomposites. Carbohyd Polym 150:330–352
7. Zainal SH, Mohd NH, Suhaili N, Anuar FH, Lazim AM, Othaman R (2021) Preparation of cellulose-based hydrogel: a review. J Market Res 10:935–952
8. Liu K, Du H, Zheng T, Liu H, Zhang M, Zhang R, … Si C (2021) Recent advances in cellulose and its derivatives for oilfield applications. Carbohyd Polym 259:117740
9. Seddiqi H, Oliaei E, Honarkar H, Jin J, Geonzon LC, Bacabac RG, Klein-Nulend J (2021) Cellulose and its derivatives: towards biomedical applications. Cellulose 28(4):1893–1931
10. Brown AJ (1886) XLIII.—On an acetic ferment which forms cellulose. J Chem Soc Trans 49:432–439
11. Shi Z, Zhang Y, Phillips GO, Yang G (2014) Utilization of bacterial cellulose in food. Food Hydrocolloids 35:539–545
12. Portela R, Leal CR, Almeida PL, Sobral RG (2019) Bacterial cellulose: a versatile biopolymer for wound dressing applications. Microb Biotechnol 12(4):586–610

13. Sintharm P, Nimpaiboon A, Liao YC, Phisalaphong M (2022) Bacterial cellulose reinforced with skim/fresh natural rubber latex for improved mechanical, chemical and dielectric properties. Cellulose 29(3):1739–1758
14. Chen Y, Wang L, Yin Q, Jia H, Wang D, Li G, … Xu Z (2019) Water-induced mechanically adaptive behavior of carboxylated acrylonitrile-butadiene rubber reinforced by bacterial cellulose whiskers. Polym Eng Sci 59(1):58–65
15. Chen Y, Li G, Yin Q, Jia H, Ji Q, Wang L, … Yin B (2018) Stimuli-responsive polymer nanocomposites based on styrene-butadiene rubber and bacterial cellulose whiskers. Polym Adv Technol 29(5):1507–1517
16. Pandit A, Kumar R (2021) A review on production, characterization and application of bacterial cellulose and its biocomposites. J Polym Environ 29(9):2738–2755
17. Azizi Samir MAS, Alloin F, Dufresne A (2005) Review of recent research into cellulosic whiskers, their properties and their application in nanocomposite field. Biomacromol 6(2):612–626
18. Hirai A, Inui O, Horii F, Tsuji M (2009) Phase separation behavior in aqueous suspensions of bacterial cellulose nanocrystals prepared by sulfuric acid treatment. Langmuir 25(1):497–502
19. Habibi Y, Lucia LA, Rojas OJ (2010) Cellulose nanocrystals: chemistry, self-assembly, and applications. Chem Rev 110(6):3479–3500
20. George J, Ramana KV, Bawa AS (2011) Bacterial cellulose nanocrystals exhibiting high thermal stability and their polymer nanocomposites. Int J Biol Macromol 48(1):50–57
21. George J (2012) High performance edible nanocomposite films containing bacterial cellulose nanocrystals. Carbohyd Polym 87(3):2031–2037
22. Roman M, Winter WT (2004) Effect of sulfate groups from sulfuric acid hydrolysis on the thermal degradation behavior of bacterial cellulose. Biomacromol 5(5):1671–1677
23. Tebeka IR, Silva AG, Petri DF (2009) Hydrolytic activity of free and immobilized cellulase. Langmuir 25(3):1582–1587
24. Ahola S, Turon X, Osterberg M, Laine J, Rojas OJ (2008) Enzymatic hydrolysis of native cellulose nanofibrils and other cellulose model films: effect of surface structure. Langmuir 24(20):11592–11599
25. Liu W, Du H, Zhang M, Liu K, Liu H, Xie H, … Si C (2020) Bacterial cellulose-based composite scaffolds for biomedical applications: a review. ACS Sustain Chem Eng 8(20):7536–7562
26. De Wulf P, Joris K, Vandamme EJ (1996) Improved cellulose formation by an Acetobacter xylinum mutant limited in (keto) gluconate synthesis. J Chem Technol Biotechnol: Int Res Process, Environ Clean Technol 67(4):376–380
27. Bae SO, Sugano Y, Ohi K, Shoda M (2004) Features of bacterial cellulose synthesis in a mutant generated by disruption of the diguanylate cyclase 1 gene of Acetobacter xylinum BPR 2001. Appl Microbiol Biotechnol 65:315–322
28. Wu RQ, Li ZX, Yang JP, Xing XH, Shao DY, Xing KL (2010) Mutagenesis induced by high hydrostatic pressure treatment: a useful method to improve the bacterial cellulose yield of a Gluconoacetobacter xylinus strain. Cellulose 17:399–405
29. Shigematsu T, Takamine K, Kitazato M, Morita T, Naritomi T, Morimura S, Kida K (2005) Cellulose production from glucose using a glucose dehydrogenase gene (gdh)-deficient mutant of Gluconacetobacter xylinus and its use for bioconversion of sweet potato pulp. J Biosci Bioeng 99(4):415–422
30. Zeng M, Laromaine A, Roig A (2014) Bacterial cellulose films: influence of bacterial strain and drying route on film properties. Cellulose 21:4455–4469
31. Hungund B, Prabhu S, Shetty C, Acharya S, Prabhu V, Gupta SG (2013) Production of bacterial cellulose from Gluconacetobacter persimmonis GH-2 using dual and cheaper carbon sources. J Microb Biochem Technol 5(2):31–33
32. Ogrizek L, Lamovšek J, Čuš F, Leskovšek M, Gorjanc M (2021) Properties of bacterial cellulose produced using white and red grape bagasse as a nutrient source. Processes 9(7):1088
33. Saleh AK, El-Gendi H, Ray JB, Taha TH (2021) A low-cost effective media from starch kitchen waste for bacterial cellulose production and its application as simultaneous absorbance for methylene blue dye removal. Biomass Convers Biorefinery, 1–13

34. Singh O, Panesar PS, Chopra HK (2017) Response surface optimization for cellulose production from agro industrial waste by using new bacterial isolate Gluconacetobacter xylinus C18. Food Sci Biotechnol 26:1019–1028
35. Carreira P, Mendes JA, Trovatti E, Serafim LS, Freire CS, Silvestre AJ, Neto CP (2011) Utilization of residues from agro-forest industries in the production of high value bacterial cellulose. Biores Technol 102(15):7354–7360
36. Fontana JD, Franco VC, De Souza SJ, Lyra IN, De Souza AM (1991) Nature of plant stimulators in the production of Acetobacter xylinum ("tea fungus") biofilm used in skin therapy. Appl Biochem Biotechnol 28:341–351
37. Toyosaki H, Naritomi T, Seto A, Matsuoka M, Tsuchida T, Yoshinaga F (1995) Screening of bacterial cellulose-producing Acetobacter strains suitable for agitated culture. Biosci Biotechnol Biochem 59(8):1498–1502
38. Kurosumi A, Sasaki C, Yamashita Y, Nakamura Y (2009) Utilization of various fruit juices as carbon source for production of bacterial cellulose by Acetobacter xylinum NBRC 13693. Carbohyd Polym 76(2):333–335
39. Keshk S, Sameshima K (2006) The utilization of sugar cane molasses with/without the presence of lignosulfonate for the production of bacterial cellulose. Appl Microbiol Biotechnol 72:291–296
40. Bae S, Shoda M (2004) Bacterial cellulose production by fed-batch fermentation in molasses medium. Biotechnol Prog 20(5):1366–1371
41. Hong F, Zhu YX, Yang G, Yang XX (2011) Wheat straw acid hydrolysate as a potential cost-effective feedstock for production of bacterial cellulose. J Chem Technol Biotechnol 86(5):675–680
42. Hong F, Guo X, Zhang S, Han SF, Yang G, Jönsson LJ (2012) Bacterial cellulose production from cotton-based waste textiles: enzymatic saccharification enhanced by ionic liquid pretreatment. Biores Technol 104:503–508
43. Goelzer FDE, Faria-Tischer PCS, Vitorino JC, Sierakowski MR, Tischer CA (2009) Production and characterization of nanospheres of bacterial cellulose from Acetobacter xylinum from processed rice bark. Mater Sci Eng, C 29(2):546–551
44. Hong F, Qiu K (2008) An alternative carbon source from konjac powder for enhancing production of bacterial cellulose in static cultures by a model strain Acetobacter aceti subsp. xylinus ATCC 23770. Carbohyd Polym 72(3):545–549
45. Wu JM, Liu RH (2012) Thin stillage supplementation greatly enhances bacterial cellulose production by Gluconacetobacter xylinus. Carbohyd Polym 90(1):116–121
46. Rangaswamy BE, Vanitha KP, Hungund BS (2015) Microbial cellulose production from bacteria isolated from rotten fruit. Int J Polym Sci 2015(2):1–8
47. Aswini K, Gopal NO, Uthandi S (2020) Optimized culture conditions for bacterial cellulose production by Acetobacter senegalensis MA1. BMC Biotechnol 20:1–16
48. Matsuoka M, Tsuchida T, Matsushita K, Adachi O, Yoshinaga F (1996) A synthetic medium for bacterial cellulose production by Acetobacter xylinum subsp. sucrofermentans. Biosc Biotechnol Biochem 60(4):575–579
49. Lin SP, Loira Calvar I, Catchmark JM, Liu JR, Demirci A, Cheng KC (2013) Biosynthesis, production and applications of bacterial cellulose. Cellulose 20(5):2191–2219
50. El-Gendi H, Taha TH, Ray JB, Saleh AK (2022) Recent advances in bacterial cellulose: a low-cost effective production media, optimization strategies and applications. Cellulose 29(14):7495–7533
51. Zahan KA, Nordin K, Mustapha M, Mohd Zairi MN (2015) Effect of incubation temperature on growth of Acetobacter xylinum 0416 and bacterial cellulose production. Appl Mech Mater 815:3–8
52. El-Saied H, Basta AH, Gobran RH (2004) Research progress in friendly environmental technology for the production of cellulose products (bacterial cellulose and its application). Polym-Plast Technol Eng 43(3):797–820
53. Wu SC, Li MH (2015) Production of bacterial cellulose membranes in a modified airlift bioreactor by Gluconacetobacter xylinus. J Biosci Bioeng 120(4):444–449

54. Lee KY, Buldum G, Mantalaris A, Bismarck A (2014) More than meets the eye in bacterial cellulose: biosynthesis, bioprocessing, and applications in advanced fiber composites. Macromol Biosci 14(1):10–32
55. Islam MU, Ullah MW, Khan S, Shah N, Park JK (2017) Strategies for cost-effective and enhanced production of bacterial cellulose. Int J Biol Macromol 102:1166–1173
56. Liu M, Li S, Xie Y, Jia S, Hou Y, Zou Y, Zhong C (2018) Enhanced bacterial cellulose production by Gluconacetobacter xylinus via expression of Vitreoscilla hemoglobin and oxygen tension regulation. Appl Microbiol Biotechnol 102:1155–1165
57. Czaja W, Romanovicz D, Brown RM (2004) Structural investigations of microbial cellulose produced in stationary and agitated culture. Cellulose 11:403–411
58. Shezad O, Khan S, Khan T, Park JK (2010) Physicochemical and mechanical characterization of bacterial cellulose produced with an excellent productivity in static conditions using a simple fed-batch cultivation strategy. Carbohyd Polym 82(1):173–180
59. Kralisch D, Hessler N, Klemm D, Erdmann R, Schmidt W (2010) White biotechnology for cellulose manufacturing—the HoLiR concept. Biotechnol Bioeng 105(4):740–747
60. Serafica G, Mormino R, Bungay H (2002) Inclusion of solid particles in bacterial cellulose. Appl Microbiol Biotechnol 58:756–760
61. Fernandes IDAA, Pedro AC, Ribeiro VR, Bortolini DG, Ozaki MSC, Maciel GM, Haminiuk CWI (2020) Bacterial cellulose: from production optimization to new applications. Int J Biol Macromol 164:2598–2611
62. Navya PV, Gayathri V, Samanta D, Sampath S (2022) Bacterial cellulose: a promising biopolymer with interesting properties and applications. Int J Biol Macromol 220:435–461
63. Brown RM Jr (1989) Bacterial cellulose. In: Kennedy P, Williams (eds) Cellulose: structural and functional aspects. Ellis Horwood Ltd., Chichester, UK, pp 145–151
64. Yu X, Atalla RH (1996) Production of cellulose II by Acetobacter xylinum in the presence of 2, 6-dichlorobenzonitrile. Int J Biol Macromol 19(2):145–146
65. Chawla PR, Bajaj IB, Survase SA, Singhal RS (2009) Microbial cellulose: fermentative production and applications. Food Technol Biotechnol 47(2)
66. Park JK, Khan T, Jung J (2009). Ch26. Bacterial cellulose. In: Phillips GO, Williams PA (eds) Handbook of hydrocolloids, 2nd edn. Woodhead Publishing Ltd., Cambridge, UK, pp 724–739
67. Wang J, Jia H, Zhang J, Ding L, Huang Y, Sun D, Gong X (2014) Bacterial cellulose whisker as a reinforcing filler for carboxylated acrylonitrile-butadiene rubber. J Mater Sci 49:6093–6101
68. Martínez-Sanz M, Lopez-Rubio A, Lagaron JM (2011) Optimization of the nanofabrication by acid hydrolysis of bacterial cellulose nanowhiskers. Carbohyd Polym 85(1):228–236
69. Yin Q, Wang D, Jia H, Ji Q, Wang L, Li G, Yin B (2018) Water-induced modulus changes of bio-based uncured nanocomposite film based on natural rubber and bacterial cellulose nanocrystals. Ind Crops Prod 113:240–248
70. Wong SS, Kasapis S, Tan YM (2009) Bacterial and plant cellulose modification using ultrasound irradiation. Carbohyd Polym 77(2):280–287
71. Sheykhnazari S, Tabarsa T, Ashori A, Shakeri A, Golalipour M (2011) Bacterial synthesized cellulose nanofibers; Effects of growth times and culture mediums on the structural characteristics. Carbohyd Polym 86(3):1187–1191
72. Trovatti E, Serafim LS, Freire CS, Silvestre AJ, Neto CP (2011) Gluconacetobacter sacchari: an efficient bacterial cellulose cell-factory. Carbohyd Polym 86(3):1417–1420
73. Yang Y, Jia J, Xing J, Chen J, Lu S (2013) Isolation and characteristics analysis of a novel high bacterial cellulose producing strain Gluconacetobacter intermedius CIs26. Carbohyd Polym 92(2):2012–2017
74. Li G, Nandgaonkar AG, Habibi Y, Krause WE, Wei Q, Lucia LA (2017) An environmentally benign approach to achieving vectorial alignment and high microporosity in bacterial cellulose/chitosan scaffolds. RSC Adv 7(23):13678–13688
75. Mohite BV, Patil SV (2014) Physical, structural, mechanical and thermal characterization of bacterial cellulose by G. hansenii NCIM 2529. Carbohyd Polym 106:132–141

76. Padmanabhan SK, Protopapa C, Licciulli A (2021) Stiff and tough hydrophobic cellulose-silica aerogels from bacterial cellulose and fumed silica. Process Biochem 103:31–38
77. Retegi A, Gabilondo N, Peña C, Zuluaga R, Castro C, Gañán P, ... Mondragon I (2010) Bacterial cellulose films with controlled microstructure–mechanical property relationships. Cellulose 17:661–669
78. Roberts AD (1988) Natural rubber science and technology. Oxford University Press
79. Ooi ZX, Ismail H, Bakar AA (2013) Optimisation of oil palm ash as reinforcement in natural rubber vulcanisation: a comparison between silica and carbon black fillers. Polym Testing 32(4):625–630
80. Zhou Y, Fan M, Chen L, Zhuang J (2015) Lignocellulosic fibre mediated rubber composites: an overview. Compos B Eng 76:180–191
81. Belgacem MN, Gandini A (2008) Surface modification of cellulose fibres. In: Monomers, polymers and composites from renewable resources. Elsevier, pp 385–400
82. Trovatti E, Carvalho AJ, Ribeiro SJ, Gandini A (2013) Simple green approach to reinforce natural rubber with bacterial cellulose nanofibers. Biomacromol 14(8):2667–2674
83. Thomas MG, Abraham E, Jyotishkumar P, Maria HJ, Pothen LA, Thomas S (2015) Nanocelluloses from jute fibers and their nanocomposites with natural rubber: preparation and characterization. Int J Biol Macromol 81:768–777
84. Phomrak S, Phisalaphong M (2017) Reinforcement of natural rubber with bacterial cellulose via a latex aqueous microdispersion process. J Nanomater 2017:7
85. Rolere S, Liengprayoon S, Vaysse L, Sainte-Beuve J, Bonfils F (2015) Investigating natural rubber composition with Fourier Transform Infrared (FT-IR) spectroscopy: rapid and non-destructive method to determine both protein and lipid contents simultaneously. Polym Testing 43:83–93
86. Dagnon KL, Shanmuganathan K, Weder C, Rowan SJ (2012) Water-triggered modulus changes of cellulose nanofiber nanocomposites with hydrophobic polymer matrices. Macromolecules 45(11):4707–4715
87. Liu M, Peng Q, Luo B, Zhou C (2015) The improvement of mechanical performance and water-response of carboxylated SBR by chitin nanocrystals. Eur Polym J 68:190–206
88. Sinclair A, Zhou X, Tangpong S, Bajwa DS, Quadir M, Jiang L (2019) High-performance styrene-butadiene rubber nanocomposites reinforced by surface-modified cellulose nanofibers. ACS Omega 4(8):13189–13199
89. Balachandrakurup V, Gopalakrishnan J (2022) Enhanced performance of cellulose nanofibre reinforced styrene butadiene rubber nanocomposites modified with epoxidised natural rubber. Ind Crops Prod 183:114935
90. Gunasekaran S, Natarajan RK, Kala A (2007) FTIR spectra and mechanical strength analysis of some selected rubber derivatives. Spectrochim Acta Part A Mol Biomol Spectrosc 68(2):323–330
91. Peddini SK, Bosnyak CP, Henderson NM, Ellison CJ, Paul DR (2015) Nanocomposites from styrene–butadiene rubber (SBR) and multiwall carbon nanotubes (MWCNT) part 2: Mechanical properties. Polymer 56:443–451
92. Gatos KG, Sawanis NS, Apostolov AA, Thomann R, Karger-Kocsis J (2004) Nanocomposite formation in hydrogenated nitrile rubber (HNBR)/organo-montmorillonite as a function of the intercalant type. Macromol Mater Eng 289(12):1079–1086
93. Yin B, Li G, Wang D, Wang L, Wang J, Jia H, … Sun D (2018) Enhanced mechanical properties of styrene–butadiene rubber with low content of bacterial cellulose nanowhiskers. Adv Polym Technol 37(5):1323–1334
94. Ai C, Gong G, Zhao X, Liu P (2017) Determination of carboxyl content in carboxylated nitrile butadiene rubber (XNBR) after degradation via olefin cross metathesis. Polym Test 60:250–252
95. Gao B, Yang J, Chen Y, Zhang S (2021) Oxidized cellulose nanocrystal as sustainable crosslinker to fabricate carboxylated nitrile rubber composites with antibiosis, wearing and irradiation aging resistance. Compos B Eng 225:109253

96. Wang J, Jia H, Tang Y, Ji D, Sun Y, Gong X, Ding L (2013) Enhancements of the mechanical properties and thermal conductivity of carboxylated acrylonitrile butadiene rubber with the addition of graphene oxide. J Mater Sci 48:1571–1577
97. Li M, Ren W, Zhang Y, Zhang Y (2012) Study on preparation and properties of gel polymer electrolytes based on comb-like copolymer matrix of poly (ethylene glycol) monomethylether grafted carboxylated butadiene-acrylonitrile rubber. J Polym Res 19:1–7
98. Hameed N, Guo Q, Tay FH, Kazarian SG (2011) Blends of cellulose and poly (3-hydroxybutyrate-co-3-hydroxyvalerate) prepared from the ionic liquid 1-butyl-3-methylimidazolium chloride. Carbohyd Polym 86(1):94–104
99. Capadona JR, Shanmuganathan K, Trittschuh S, Seidel S, Rowan SJ, Weder C (2009) Polymer nanocomposites with nanowhiskers isolated from microcrystalline cellulose. Biomacromol 10(4):712–716
100. Qiao R, Brinson LC (2009) Simulation of interphase percolation and gradients in polymer nanocomposites. Compos Sci Technol 69(3–4):491–499
101. Xue X, Yin Q, Jia H, Zhang X, Wen Y, Ji Q, Xu Z (2017) Enhancing mechanical and thermal properties of styrene-butadiene rubber/carboxylated acrylonitrile butadiene rubber blend by the usage of graphene oxide with diverse oxidation degrees. Appl Surf Sci 423:584–591

Chapter 8
Lignin-Based Rubber Composites and Bionanocomposites

Carlos A. Rodriguez Ramirez, Mirta L. Fascio, Nancy L. García, and Norma B. D'Accorso

Abstract Lately, rubbers or elastomers have acquired a highly important role in daily life. They are part of many inputs of great industrial importance. These rubbers require processing that can modify their mechanical properties. In this aspect, products from biomass are attractive due to their low cost, renewable and that allow the application of environmentally friendly methodologies. In this context, lignin is an opportunity to obtain new materials with specific properties. This chapter presents the main characteristics of this natural product and its application as a stabilizer of rubber nanocomposites, as flame retardants, production of antioxidants, applications in agriculture and also in 3D printing.

8.1 Introduction

Elastomers and rubbers are critical components in modern technology, playing a key role in tyres, seals and gaskets, soft robotics, damping systems, wearable electronics and flexible sensors, footwear soles, 3D and 4D printing materials [1]. They are widely used because of their unique properties such as elasticity and flexibility. However, in order for raw rubber to be used industrially, it must be chemically cross-linked using the well-known vulcanization process. This curing process gives the rubber chemical and thermal stability, mechanical strength and excellent noise and vibration-damping properties, but it also raises environmental concerns due to its non-recyclability. In addition, abrasion, tear, cut and break resistance, stiffness and

C. A. Rodriguez Ramirez · M. L. Fascio (✉) · N. B. D'Accorso
Departamento de Química Orgánica, Facultad de Ciencias Exactas y Naturales, Universidad de Buenos Aires, Buenos Aires, Argentina
e-mail: mfascio@qo.fcen.uba.ar

N. B. D'Accorso
e-mail: norma@qo.fcen.uba.ar

C. A. Rodriguez Ramirez · M. L. Fascio · N. L. García · N. B. D'Accorso
Centro de Investigaciones en Hidratos de Carbono (CIHIDECAR), CONICET–Universidad de Buenos Aires, Buenos Aires, Argentina

Visakh P. M. *Rubber Based Bionanocomposites*, Advanced Structured Materials 210,
https://doi.org/10.1007/978-981-10-2978-3_8

toughness are improved by adding fillers to rubber products. The unique characteristics of vulcanized rubber are primarily determined by reinforcing fillers such as carbon black and silica [2]. Recent research is aimed at improving the recyclability and self-healing properties of engineering rubbers, some of which have been summarized in several reviews [3, 4].

Carbon black is a crucial product in the rubber industry, as it enhances its mechanical and electrical properties. However, processing carbon black into rubbers can present several challenges such as inhomogeneous dispersion, diminished fluidity in mixtures, alterations in rheological properties, heightened equipment abrasion, inadvertent electrical conductivity, potential incompatibility with other additives, and adverse environmental effects. In this context, hybrid fillers have garnered significant attention in the rubber compounds field in recent years. Even more those that contain lignin, since they respond to the need to valorize by-products of renewable biomass.

Lignin is one of the most abundant biopolymers, yet it remains underutilized. Despite its attractive aromatic structure, approximately 98% of the estimated 80 million tonnes of lignin produced annually is burned for energy recovery purposes, which is not in line with circular economy principles. Recent life cycle assessment (LCA) studies highlight that lignin can reduce environmental impacts when added to materials and/or as a filler, compared to its petroleum-based counterparts [5]. The valorization of lignin is a potential alternative in the development of new materials with high added value and is key in the development of the circular economy, especially in biorefineries. It is important to consider these data when making decisions about resource allocation [6].

In recent years, there has been a proliferation of research projects related to lignin, which is studied as a renewable polymer and composite material [7]. Lignin is considered an excellent substitute for raw materials in the preparation of chemicals and polymers [8]. However, one of the main difficulties that persist is the lack of a well-defined structure and its partial flexibility, which is linked to its origin, including fragmentation and extraction procedures [9].

Lignin valorization is a research area that seeks to convert lignin into useful chemicals and materials, including biofuels, plastics, and chemicals for use in industries such as pharmaceuticals, cosmetics, and agriculture [10].

One strategy for using lignin in rubber applications is to blend it with a rubber matrix. However, these blends often have low compatibility and poor dispersibility. To improve lignin dispersion and enhance composite properties, further research has explored the use of high-temperature dynamic heat treatment. Several studies have reported successful outcomes using the latex co-precipitation method, although it involves a complex process. However, this method is effective in improving the distribution of lignin within the rubber matrix, resulting in rubber composites with enhanced mechanical properties. Additionally, modifying the surface using coupling agents and compatibilizers also has a significant impact on the properties of lignin-filled rubber composites [11].

The aim of this chapter is therefore to summarize the most recent research on lignin and rubbers and the progress made in using lignin or modified lignins as reinforcing, flame-retardant, antioxidant and in agricultural and biomedical applications.

8.2 Stabilizers in Rubber Nanocomposites

Natural rubber (NR), obtained from *Hevea brasilensis*, is a poly(cis-1,4-isoprene) elastomer [12], which after the vulcanization process has excellent properties such as resistance to traction, tearing and abrasion [13, 14]. Although these properties are of great interest for different applications, the NR has problems with low resistance to the action of heat, oxygen, ozone and sunlight [15], due to the oxidation of these agents on the double bonds present in repetitive units (see Fig. 8.1). Besides, the dried NR suffers crosslinking density between polymer chains, which results in a decrease in its mechanical properties [16].

A possible solution to delay or inhibit oxidative processes is the use of stabilizers and therefore reduce the degradation of NR [17]. In particular, different synthetic products have been used although in recent times the use of biomass products has been attractive [18–20].

In particular, several authors have studied the addition of lignin to rubber matrices as filler [11, 21–23] and found that it improved the physical and mechanical properties of lignin-filled NR vulcanizes prepared by the conventional method, and also increased the thermos-oxidative value.

In a recent study the group of Crespo et al. [24] studied the behavior of different stabilizers in rubber vulcanization processes in aging processes. The authors evaluated at predetermined times of mechanical properties and content of different additives and their combinations. The authors worked with paraffin wax, N-1,3-dimethyl-butyl-N′-phenyl-*p*-phenylenediamine (6PPD), oligomerized 2,2,4-trimethylquinoline (TMQ), and eucalyptus lignin [25]. It is important to remark that this study allowed them to evaluate the synergistic effect of the presence of different stabilizers, and in particular the performance of lignin as a co-stabilizer. Figure 8.2 shows the chemical structure of the stabilizers.

The authors [24] prepared nine preparations: one without with stabilizer, four with individual additive and the rest with combinations of different stabilizers. They performed different studies to evaluate the performances of the samples, such as elemental analysis, migration, mechanical properties, oxidation induction time, crosslinking, thermo-oxidation, photochemical and the evaluation of ozone aging. From the experimental results, the authors concluded:

1. Significant differences in tear resistance property, compared to tensile strength.
2. The combination of wax and lignin presented a longer oxidation induction time.
3. Thermo-oxidative aging, where a higher temperature is used (100 °C), a greater decrease in mechanical properties and stabilizer consumption was observed.
4. Photo-oxidative aging, the samples with the least reduction in properties were those that used the combination of three stabilizing agents (6PTC and 6PLC).

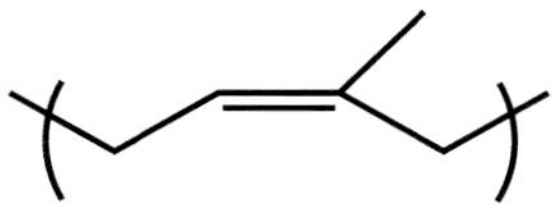

Fig. 8.1 Repetitive units in natural rubber

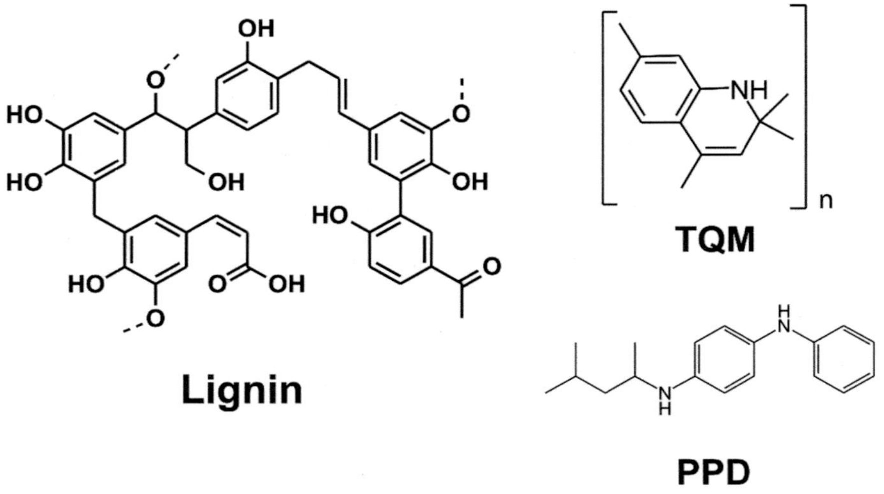

Fig. 8.2 Chemical structure of different stabilizers

5. Accelerated aging with ozone, the samples showed greater retention of properties.

Finally, the authors conclude that the addition of lignin (agricultural waste product) has the potential to be used as a stabilizer, replacing synthetic stabilizers, because its use in vulcanized NR formulations did not affect their performance.

Considering that nanomaterials based on rigid NR with high resistance are of great interest in advanced industrial applications, the addition of products that increase hardness is very important. In a work by Yu et al. [26], the authors replaced a percentage of silica with alkaline lignin and found that the NR vulcanizate presented optimal general mechanical properties, although due to the large particle size and the lack of interfacial interaction with the rubber matrix, more work had to be done in the processing. In this regard, nanofillers from biomass are a promising approach to environmentally friendly chemistry [27–29].

In this context, Annamalai et al. [30] carried out a study of the preparation and addition of nanoscale organosolv lignin (OSL) to NR latex to improve its mechanical properties. SEM studies of dried OSL indicated that the structure occurs layer by layer with particle sizes larger than 2 μm. The solubility of OSL in water resulted in dispersion due to its hydrophobic character due to the presence of the phenolic and carboxylic hydroxyl groups present in the OSL (determined by 2D 13C-1H HSQC NMR), then to solve this problem the authors increased the PH with ammonia solutions, resulting in the corresponding salts. The authors performed a dynamic light scattering (DLS) analysis to evaluate the level of aggregation and dispersion of the alkaline OSL solutions and observed that the dispersions were clusters of 8 and 30 nm.

The nanocomposites showed a substantial improvement in the mechanical properties (tensile strength, toughness and shore hardness), maintaining a high elongation at break, when the NR films were reinforced with 5% by weight of OSL. These properties are attributed substantially to the enhanced nanoscale dispersion of OSL and the associated favorable interfacial adhesion of OSL to the NR matrix. Furthermore, the impact of an industrially applicable leaching process on the toughness and thermo-oxidative stability of rubber films is also analyzed [30]. Different works described [31, 32] the macromolecular structure is successfully dissociated into nanoscale colloidal particles.

Taking into account the problems caused by the particle size of lignin, Zhu et al. [33] reported, based on spectroscopic studies, chemical interactions between lignin and the silylated coupling agent in poly(L-lactic) composite materials. On the other hand, Bahl et al. [34] found that carbon black (CB)/lignin hybrid fillers dramatically enhanced lignin dispersion through π–π stacking interaction at the interface. In the same way, literature [35] describes that polybutadiene-g-polypentafluorostyrene copolymer increased the interaction between lignin and the styrene-butadiene rubber matrix through arene-perfluoroarene interactions.

In this context, Wang et al. [36] performed a comparative study referrer the synergic effect on the carbon black/lignin and silica/lignin hybrid filled natural rubber composites. The authors used a co-precipitation method to prepare mixtures to guarantee good dispersion of lignin in NR matrices and with the possibility of being industrially scalable. The materials were characterized by determining their mechanical and physical properties that include a degree of cross-linking, amount of lignin. Likewise, work was carried out on the study of the reinforcement mechanisms for both the NR composites, such as carbon black/NR and the silica/NR.

Taking into account the studies carried out, the authors propose two interaction mechanisms for the two composites. In the case of CB/lignin, has good tensile strength and elongation at break, especially with the addition of lignin lower than 30 phr, and the π–π stacking interaction between the aromatic rings of lignin with CB is proposed. These interactions improving between both components produce the increase in entanglements with greater amounts of lignin. Figure 8.3 shows the mechanism proposed [37].

NR silica fillers interaction is enhanced due to the silylation reaction of the silane coupling agent, the interactions between polar lignin and polar silica weaken the lignin-silica network, which is beneficial to improve the dispersion of the filler [38]. Figure 8.4 presents the mechanism proposed.

It is important to note that the addition of lignin plays an important role in the reinforcing effect of silica-filled NR composites. The enhanced interaction between silica/lignin and NR and the improved dispersion of the filler contributed significantly to the improved reinforcement. Despite these data, the authors observed severe infill networks when the lignin concentration was too high (up to 50 phr), producing a decrease in tensile strength, however taking into account the results of mechanical properties would be very interesting to optimize this methodology for possible industrial applications.

Incorporating lignin into elastomers can improve their properties by reinforcing the soft fraction with a rigid fraction derived from lignin [11], but dispersing lignin

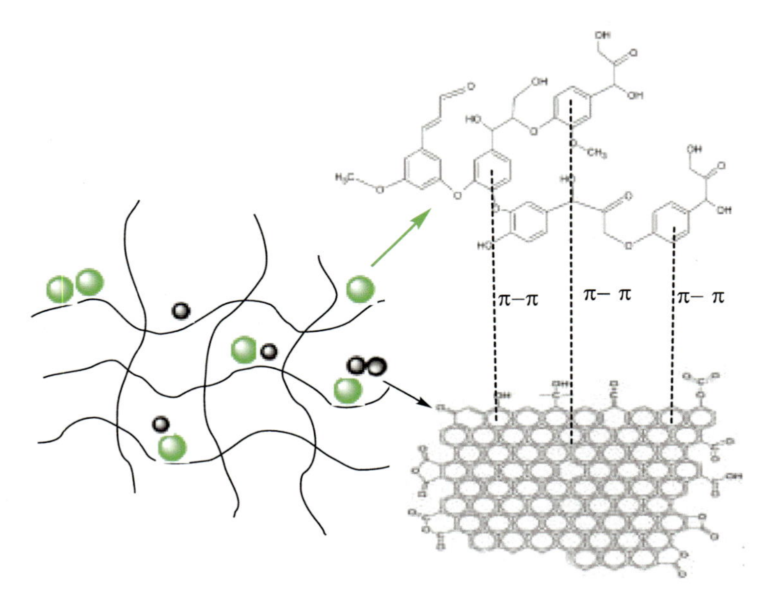

Fig. 8.3 Stacking interaction between the aromatic rings of lignin with CB is proposed

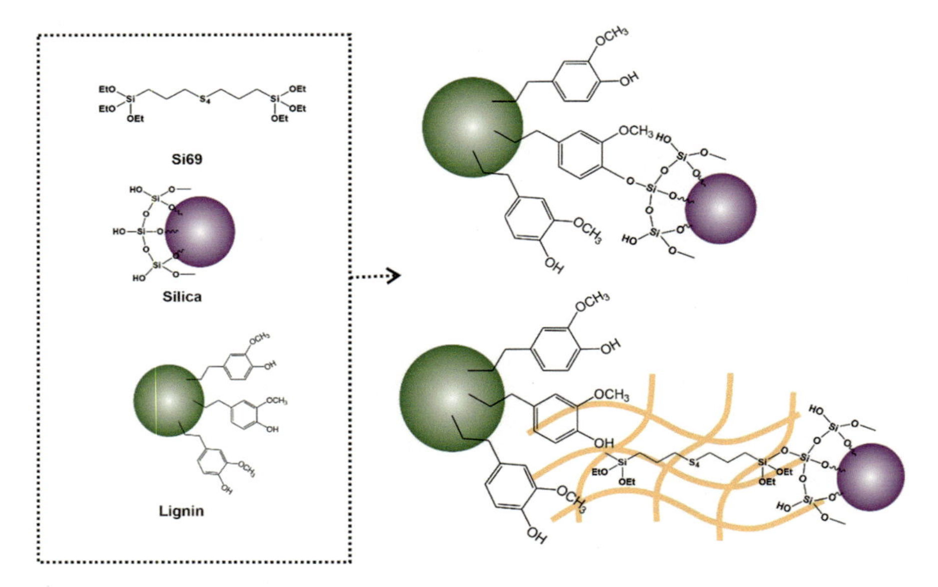

Fig. 8.4 Possible reinforcement mechanism for silane coupling agent-modified silica/lignin-filled NR composites

particles in rubber is often inefficient due to the incompatibility of the materials: rubber is highly hydrophobic, while lignin has a high number of hydroxyl groups.

Chemical modifications of lignin are of attractive interest since they offer the possibility of derivatizing these hydroxyl groups to reduce their hydrophilicity and improve their dispersion and compatibility with these elastomers [39].

To achieve this goal, Shorey et al. [40] silylated kraft lignin (KL) using 3-(triethoxysilyl) propyl isocyanate (TEPI) as the silylating agent and 2-ethyl tin hexanoate (II) as the catalyst (Fig. 8.5).

The results of the wettability test in solvents of opposite polarity, such as water and toluene, indicated that the modified lignin formed a stable layer at the toluene-water interface or was slightly dispersed in toluene, depending on the degree of modification. No dispersion was observed in the aqueous phase (Fig. 8.6).

The resulting silylated lignin particles were ground and then incorporated into the elastomeric matrices. The composite sheets of NR-lignin were produced using a solution casting process [41]. The modification of lignin polarity improved its dispersibility in the NR matrix compared to the unmodified version. This was demonstrated through compatibility analysis, water absorption study, and SEM morphology analysis. Additionally, the modified lignin-based composite samples exhibited superior tensile properties compared to NR, with a 44.4% increase in tensile strength, improving the reinforcing capacity of lignin in natural rubber.

In this study, a dual-phase carbon–silica nanohybrid LDPCS derived from black liquor lignin (BLL) and sodium silicate was fabricated through a simple co-gelation/

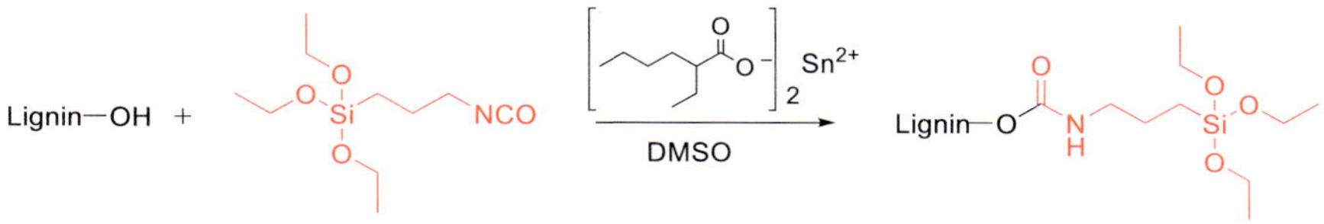

Lignin—OH + (EtO)$_3$Si–(CH$_2$)$_3$–NCO → [Sn^{2+} 2-ethylhexanoate$_2$], DMSO → Lignin—O–C(=O)–NH–(CH$_2$)$_3$–Si(OEt)$_3$

Fig. 8.5 Proposed reaction mechanism

Fig. 8.6 Wettability test of the unmodified and modified samples in 1:1 toluene water mixture. Reproduced with permission from Elsevier

self-assembly and carbonization strategy. By incorporating 40 phr LDPCS into SBR by conventional dry mixing method, the rubber compound could achieve a tensile strength of 17.92 ± 0.87 MPa and elongation at a break of 708 ± 10%, both of which are superior to those of CB-filled SBR compounds. Besides, DMA and TGA indicated that LDPCS-filled SBR rubber compounds had a lower rolling resistance and comparable thermal stability with the CB-filled SBR compound (Scheme 8.1). Overall, this study provided the possibility for the high-value application of lignin as reinforcing or functional nanofillers in rubber and other polymers [42] (see Fig. 8.7).

On the other hand, He et al. [43] developed hybrid nanoparticles composed of lignin-based biochar-silica (LB-S) such as an innovative filler comparable to commercial carbon black. Hybrid nanoparticles were synthesized by combining a two-step acid precipitation process with carbonization, using lignin black liquor extracted from xylose residue along with sodium silicate as starting materials.

LB-S offers improvements in the mechanical properties of the rubber, generated by the excellent compatibility and dispersibility of the hybrid in the rubber matrix. Although the reinforcement performance of the synthesized is comparable to that of commercial carbon, its reinforcement mechanisms differ due to disparities in particle structure and surface activity.

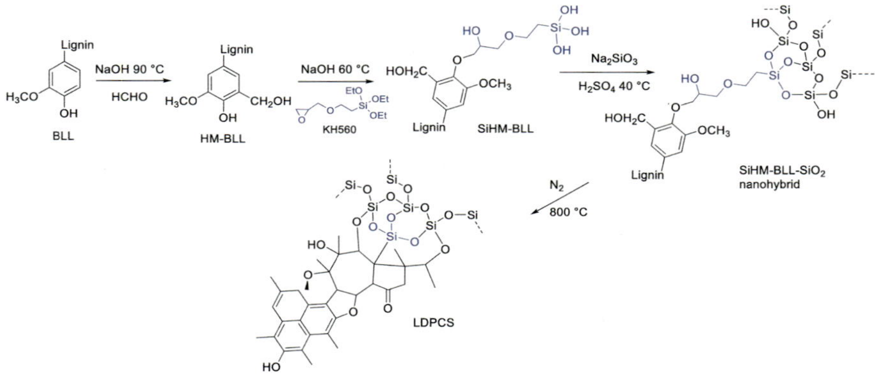

Scheme 8.1 Synthetic scheme to prepare LDPS

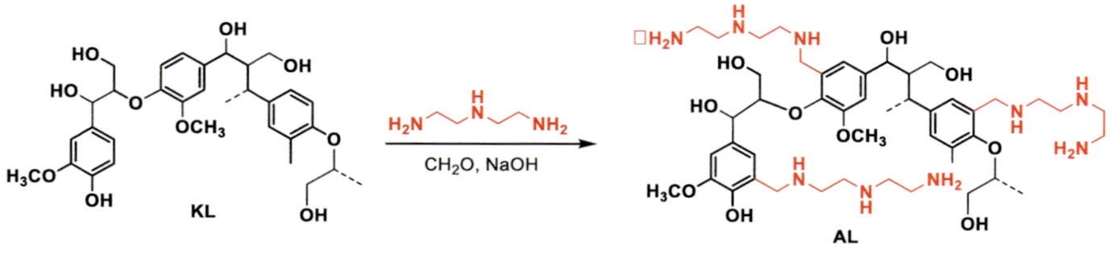

Fig. 8.7 Process used to obtain AL via the amination reaction

8.3 Flame Retardant Applications

Flame retardants are essential for rubber due to their irreplaceable role in complex industrial systems [44]. They reduce the release of heat, smoke and toxic gases without affecting the properties of the material. Lignin could be used as a flame retardant due to its phenolic structure and thermal properties, but the difficulty in dispersing it with many polymer matrices means that the final product often fails to meet industrial requirements. Modification of lignin with chemicals containing nitrogen and/or phosphorus or metallic elements, and reduction of lignin to the nanoscale to prevent aggregation, improves the behaviour of lignin as a flame retardant [45].

Styrene butadiene rubber (SBR) is highly flammable [46], so effective flame retardants must be added to enable it to be used, for example, in conveyor belts in the coal mining industry and in decorative materials such as carpets.

Lignin grafted with Si, P and N (Lig-K-DOPO), synthesized by an Atherton-Tod reaction (Scheme 8.2), showed excellent flame retardancy when added as a filler to SBR composites filled with silica [47].

Sheets with a thickness of 2 mm were prepared containing SBR, silica, hexaphenoxy-cyclophosphazene/expandable graphite, Zn, stearic acid, antioxidant, accelerators, sulphur, anti-burn and the silane coupling agent Si69, as well as lignin or Lig-K-DOPO particles.

The samples containing Lig-K-DOPO increased the curing rate of the sulphur-cured SBR composites, probably due to the presence of the amine group, and also exhibited superior dispersion compared to the samples containing Lig.

In addition, the presence of Lig-K-DOPO reduced the heat release rate and smoke production rate of the SBR vulcanizates and increased the time corresponding to the maximum heat release rate, indicating that Lig-K-DOPO is an excellent flame retardant.

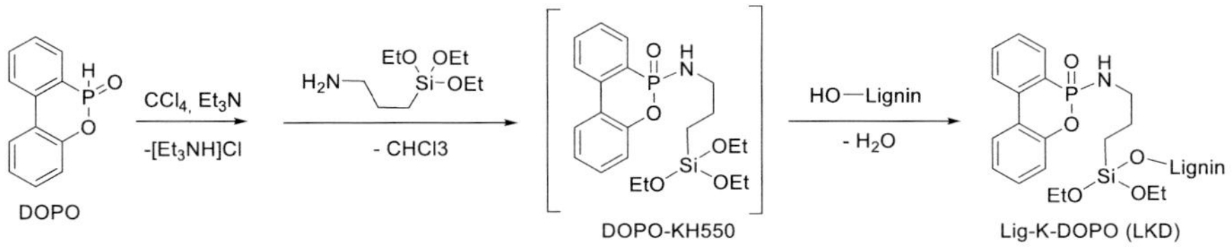

Scheme 8.2 Synthetic route of Lig-K-DOPO

8.4 Antioxidants Applications

Polyphenolic compounds from various fruits and vegetables have emerged as potent natural antioxidants [48]. Lignin is an abundant source of polyphenolic compounds and therefore represents an interesting option in this regard as an ecological and cost-effective alternative to conventional antioxidants.

Lignin used in polymeric materials can improve the anti-aging effect, mainly due to the presence of phenolic groups in lignin, which can convert alkyl ($R^·$) or peroxy ($ROO^·$) radicals generated by external factors such as heat and oxygen into stable phenoxy radicals [49]. However, because lignin contains one or two –OH per monomer, it turns out to be a polar polymer, which generally affects its interaction with non-polar polymer matrices, presenting poor miscibility that will eventually affect the thermal stability and mechanical properties of the material.

The covalent chemical bonds in the rubber backbone are susceptible to ozone and oxygen radicals. The ozonation reaction breaks the double bond, causing cracks that are fatal and reduce the service life of the rubber [50]. To scavenge active radicals, antioxidants containing polar groups such as amines, phenolics, phosphites or thioesters are usually added [51]. 6PPD is commonly used as an antioxidant in automobile tires. However, it transforms into 6PPD-quinone, a highly toxic chemical, when released into the environment [52].

Nam's group proposed synthesizing a new green antioxidant by grafting amine groups onto the kraft lignin (KL) surface [53]. The process of functionalizing lignin with amino groups was carried out by adding KL to an aqueous solution of NaOH, diethylenetriamine, and formaldehyde (Fig. 8.7).

The success of the reaction was confirmed through the use of EDS, XPS, and IR techniques. EDS elemental mapping images of AL showed an increase in the atomic percentage of nitrogen after the reaction, while the carbon content decreased. XPS analysis confirmed that the chemical composition on the surface of AL particles after the reaction had a higher N/C ratio (0.18) compared to that of KL (0.01) and a significant increase in the area ratio of C/O. It was indicated that alkylamine groups, which contain carbon chains, had been incorporated into AL. This resulted in higher carbon content and lower oxygen content. The infrared spectrum of AL exhibited a broader band in the 3500–3300 cm^{-1} range, which is attributed to the NH stretching vibrations of the primary and secondary amines that were grafted through the amination reaction. Additionally, the peaks at 2928 and 2840 cm^{-1}, which correspond to the asymmetric and symmetric C–H stretching vibrations of the methyl and methylene groups, respectively, also showed a significant increase after modification.

The new materials were produced by blending natural rubber and butadiene rubber in an internal mixing kneading machine. Oil, zinc oxide, stearic acid and several types of antioxidants (6PPD, KL and AL) were added and processed at 80 °C and 40 rpm. The resulting rubber compounds were mixed with sulphur, CTP and TBBS using a two-roll mill at room temperature. Finally, the composite samples were compression molded.

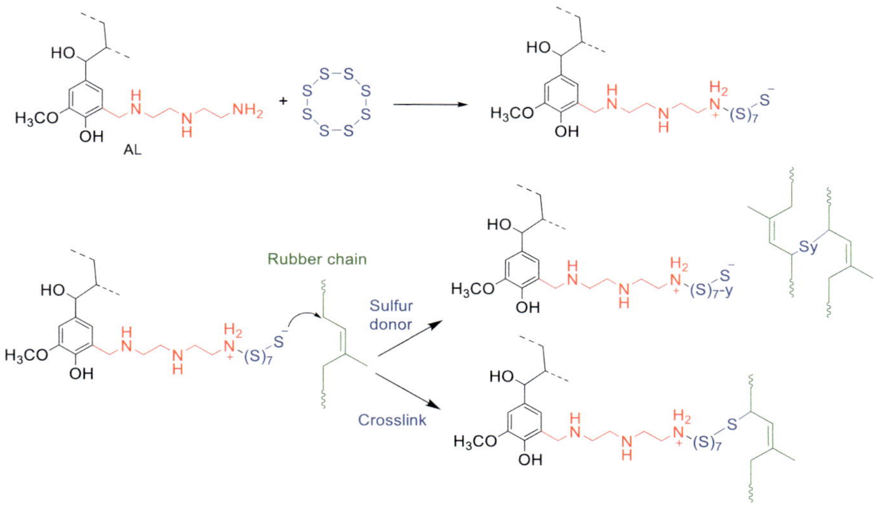

Fig. 8.8 Expected mechanisms of the heterolytic ring-opening reaction between AL and sulphur and the subsequent rubber cross-linking reaction

The addition of AL improved the rheological properties of the rubber compounds compared to 6PPD. This improvement is likely due to the amine groups on the AL particles, which acted as nucleophilic activators. These groups not only accelerated the vulcanization reaction of the rubber but also increased the degree of cross-linking, as shown in Fig. 8.8.

This led to an improvement in fatigue resistance through physical and chemical bonding with the rubber.

The material that incorporated the AL particles into the rubber also exhibited superior resistance to thermal and ozone aging, comparable to that of 6PPD. The reduction in the uncured area of the matrix and the radical scavenging properties of the AL particles are responsible for the exceptional resistance of this material to ozone.

Based on the above, the lignin-based antioxidant has the potential to reduce pollution and pave the way for next-generation rubber.

Unlike other types of precursor, lignin is a low-cost precursor for the large-scale production of carbon dots (CDs). However, due to the complex structure of lignin, the production of CDs with excellent optical properties from lignin has some drawbacks [54].

Recently, Zhao et al. synthesized carbon dots using crude lignin (CLCD) through hydrothermal treatment with H_2O_2 and triethylenetetramine [55]. The resulting product exhibited a strong antioxidant capacity. CLCDs were prepared by dispersing lignin in water, followed by the addition of triethylenetetramine and hydrogen peroxide. The mixture was then transferred to a stainless steel reactor with Teflon lining and heated at 180 °C for 12 h. During this period, lignin is oxidized and decomposed by hydrogen peroxide, producing low molecular weight carboxylic acids [56]. These acids then undergo complex reactions with triethylenetetramine. The prepared mixture was filtered and subsequently evaporated to obtain a concentrated aqueous solution of CLCD.

The CLCDs were characterized by FTIR and XPS, which revealed the presence of N–H and O–H bands, as well as signals attributed to C–N and C=O, indicating the highly hydrophilic nature of the CLCDs. Thus, the compatibility between polar CLCDs with different polar groups and non-polar NR is poor, which would lead to poor dispersion of CLCDs in the rubber matrix and therefore low antioxidant efficiency. Vinyl pyridine-styrene-butadiene terpolymer (VPR) was then added to natural rubber as a compatibilizer [57]. Rubbers were prepared from NR or NR/VPR latex and prevulcanized by adding various additives to the latex to obtain a uniform mixture at room temperature. The desired amount of CLCD and other rubber additives was then added by vigorous stirring with a magnetic stirrer. It was poured into a PTFE mold to obtain films and, after drying, subjected to a high-temperature treatment (120 °C for 20 min) in a vacuum oven.

It was observed that the addition of CLCD to the NR matrix significantly degrades the mechanical performance of the composites. Additionally, the ductility of the composites could be reduced and their modulus increased. However, the inclusion of VPR significantly improved these properties probably due to its effective dispersion of CLCDs. CLCDs with VPR also have a positive antioxidant effect on the rubber compound. After undergoing thermo-oxidative aging for several days, the antioxidant effect of CLCDs can be comparable to that of the commercial antioxidant 2-tert-butyl-6-(3-tert-butyl-2-hydroxy-5-methylbenzyl)-4-methylphenyl acrylate.

Therefore, it is possible to develop new antioxidants for lignin-based elastomers that are low cost, effective and beneficial for a sustainable world.

To improve the mechanical properties and anti-aging of styrene-butadiene rubber (SBR) compounds, Zhao and colleagues modified lignin by grafting the antioxidant p-aminodiphenylamine (RT) onto the surface through a silane bond [58]. Scheme 8.3, shows the chemical modification Lig-g-RT is obtained by preheating RT and KH560 in toluene at 85 °C and then adding an alkaline aqueous solution of lignin which is allowed to react at 100 °C for 48 h with stirring to produce the epoxy derivative. The FTIR spectra showed that the alkoxy group of KH560 had been successfully grafted onto lignin to form the Lig-g-RT. The TG and DTA curves showed that the temperature of maximum weight loss of Lig-g-RT was 403 °C, which is 51.9 °C higher than that of lignin. More importantly, Lig-g-RT showed better thermal stability than lignin.

In order to achieve a good dispersion of lignin in the SBR matrix, the SBR/lignin and SBR/Lig-g-RT composites were prepared by the latex co-precipitation method, as shown in Fig. 8.9. The resulting solid was then premixed with silica and other rubber additives, except sulphur and accelerators. Finally, the vulcanizing ingredients were added and the sheets were press-cured at 160 °C. SEM images showed that SBR vulcanized with Lig-g-RT has relatively better filler dispersion than

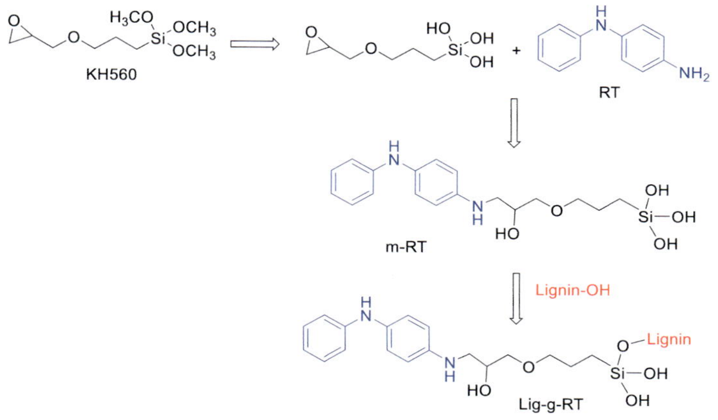

Scheme 8.3 **a** Possible reaction mechanism among lignin, RT and KH560; **b** Preparation of SBR/Lig-g-RT composites

that of unmodified lignin, which is beneficial for maintaining the physico-mechanical properties of SBR vulcanisates.

When the effect of lignin/Lig-g-RT on the thermo-oxidative ageing properties of vulcanized SBR was evaluated, it was found that composites filled with lignin or Lig-g-RT showed better retention of tensile strength, elongation at break and aging coefficient compared to SBR without filler.

It is noteworthy that the SBR composite with a loading of 10 phr of Lig-g-RT has a better aging coefficient than a composite filled with unmodified lignin with the same loading, indicating that Lig-g-RT has a better resistance to thermo-oxidative degradation than lignin.

A possible proposed antioxidant mechanism of the SBR/Lig-g-RT vulcanizate is shown in Fig. 8.9, where the generated alkyl peroxide radicals can not only be trapped by the phenolic structure of lignin, but also the p-aminodiphenylamine structure in Lig-g-RT could trap these radicals generated in the thermo-oxidative ageing process, resulting in an improvement of the antioxidant properties of SBR vulcanizates.

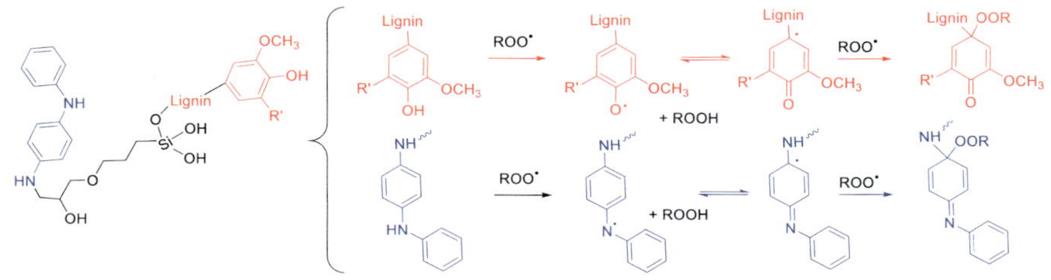

Fig. 8.9 Possible antioxidant mechanism of the SBR/Lig-g-RT vulcanizate

Therefore, the synergistic antioxidant capacity between the *p*-phenylenediamine groups of RT and the hindered phenol groups on the surface of Lig-g-RT may be the reason for the improvement.

8.5 Agriculture Applications

The demand for agricultural products has increased globally due to the growth in food consumption, resulting in a rise in the use of chemical fertilizers [59]. Urea, the most popular nitrogen fertilizer in agriculture, has had problematic in its application. Its low effectiveness when directly applied to soils was generated due to volatilization, interaction with organic compounds in the soil, and leaching into water systems. Moreover, the economic feasibility of mineral fertilizers has significantly decreased, particularly due to the sharp increase in fertilizer prices, which has led to higher food prices.

In this context, different research materials were developed new coating materials, including sulfur-based and polymer-based coatings, and were used as an alternative in the production of controlled-release fertilizers (CRFs). However, synthetic polymer-based coatings can be problematic due to their resistance to degradation in soil and potential environmental hazards. In this sense, researchers have increasingly focused on environmentally friendly materials. However, many of these biomaterials exhibit high hydrophilic properties, which can lead to the rapid release of fertilizers within a short timeframe, typically less than 30 days.

Biomaterials-based coating for slow-release fertilizer was prepared from a composite of lignin. Boonying et al. developed a coating composite of lignin and natural rubber modified with natural rubber-graft-polyacrylamide. The composite material released a total of N substances over a period of 112 days. The nitrogen release in Li/NR-g10 coated urea is limited due to the restricted hydrophilic channels within the coating. This limitation is a result of NR-g-PAM's role as a compatibilizer for natural rubber (NR) and lignin. NR-g-PAM improves the integrity of Li/NR-g10, reducing interfacial defects and voids between lignin and NR [60].

Previous research has demonstrated the beneficial effect of lignin in achieving uniform dispersion of hybrid nanofillers in the rubber matrix. For example, Boonying et al. (2023b) developed a pre-vulcanized natural rubber latex/lignin green biocomposite coating for controlled-release urea fertilizers. The research shows that the layer containing lignin improved the mechanical strength and had a significant effect on urea release, lasting up to 224 days. In contrast, uncoated urea only lasted 7 days [61].

8.6 Printing Applications

The emergence of additive manufacturing (AM), also referred to as 3D printing, has motivated academic and industrial researchers to merge elastomeric properties with design freedom and the potential for easy mass customization. This has facilitated the production of personalized items such as hearing aid inserts, dental retainers, footwear, and headphones. In 2015, a major footwear manufacturer reported the first 3D-printed shoe sole using selective laser sintering of thermoplastic polyurethanes [62].

By relying on polymer deposition only at the desired 3D pixels (voxels) in a 3D design, rather than relying on standard subtractive manufacturing methods, AM significantly reduces material waste, enables lightweight by printing low-density, high-strength lattice geometries and provides the ability for mass customization, where each part remains tailored to the needs of the individual. Two major tire companies have recently released their vision of the future tire, inspired by the potential of 3D printing. They propose 3D printing airless and lightweight tires using recycled or bio-based materials. The design would provide improved shock-absorbing properties and enhance tire safety due to improved wet grip and puncture resistance. Additionally, one of these companies envisions a customized tread pattern.

The customer can have the tread pattern reprinted at a service station to better suit demanding weather conditions, such as ice or snow. Both concepts highlight the significance of additive manufacturing and materials design in achieving a more sustainable future.

There are distinct types of additive manufacturing (AM) and between most employed in the 3D printing of elastomers we can find:

(1) Vat photopolymerization (VPP) involves directed energy deposition upon the surface of a photocurable mixture of monomers, oligomers and/or polymers, termed a photopolymer, inside a container.
(2) Fused filament fabrication (FFF), otherwise known as fused deposition modeling (FDM) and a member of the material extrusion AM class, involves the directed deposition of a thermoplastic polymer filament in a layer-by-layer fashion.
(3) Direct ink writing (DIW) materials extrusion process, which enables printing of viscoelastic materials at ambient temperatures. It is also known by the name robocasting.

Lewis et al. provided two outstanding reviews describing the concept of DIW and discussed suitable materials [63]. Figure 8.10 illustrates a common DIW set-up, including a syringe with an ink deposition nozzle and a computer-aided positioning stage. The apparatus continuously extrudes material out of the nozzle, generating 3D architectures layer-by-layer. Suitable materials for DIW should possess specific rheological properties to be processable [64].

One of the characteristics is that the material should be shearing thinning to enable extrusion out of the printing nozzle. It should also possess a shear yield

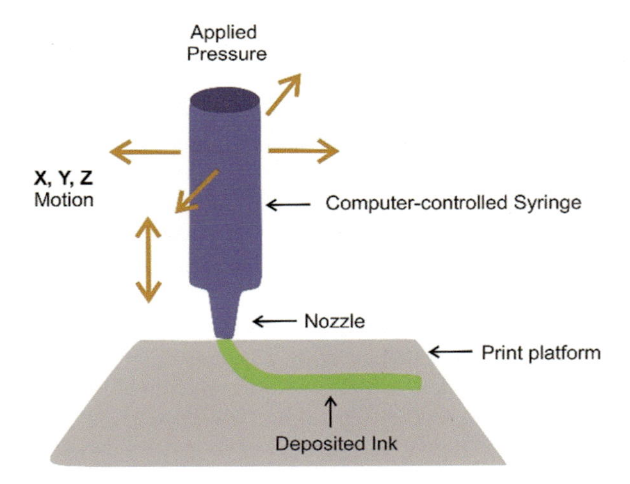

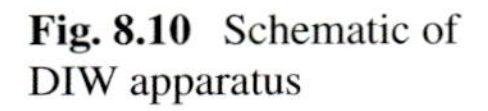
Fig. 8.10 Schematic of DIW apparatus

stress. To induce flow, a shear stress above the yield stress of the resin is applied. Subsequently, the shear stress is released, and the resin recovers its rigidity when placed on a substrate. For elastomers, the curing of the precursors often requires extended time, and the resulting networks possess low storage moduli. The latter might cause deformation or collapse of the printed structures under their own body weight. One method to overcome this challenge is the use of freeform embedded 3D printing. This AM technique utilizes a support bath, which holds the printed structure in place during the print. However, removal of the support medium from printed structures, e.g. hollow features remains challenging, because the rapid solidification after placement on the substrate, requires careful resin design. In this case, the use of rigid and resistant fillers plays an important role.

Polymer resins are commonly blended with fillers, e.g. silica particles or nano-clay to afford the described rheological properties. The fillers induce shear thinning flow behavior and at optimal resin/filler compositions, afford a material that possesses a shear yield stress.

Examples of thermoplastic substances containing lignin, like polyhydroxy butyrate (PHBs), polylactic acid (PLA), polyethylene (PE) and acrylonitrile butadiene styrene (ABS), are extensively used for synthesis and responsible for enhancing antimicrobial antioxidant activity, biodegradable properties and UV radiation stabilization when creating composites or bio-composites through 3D printing. In addition to its ability to reduce the number of free radicals or reactive oxygen species present, lignin could offer some of these antioxidant properties. Moreover, in addition to lignin to PLA composites thermal and hydrophobic properties may be improved [65].

Whole lignocellulosic materials (not deconstructed) usually do not melt, such as lignin. It is a fraction that degrades at elevated temperatures and forms a rigid char, resulting in increased resistance to flow or deformation. Some isolated lignin fractions, particularly those with linear sinapyl alcohol units from hardwood (HW) plants [66, 67] and oligomeric fractions of extracted softwood (SW) lignins [21, 68, 69], however, exhibit good melt stability.

Nguyen et al. [70] reported earlier that combining lignin with the nitrile group-containing rubbers produced a new class of melt-processable thermoplastic elastomers. These exhibited unique yield stress and high toughness primarily due to morphology control of the dispersed lignin phase [21, 66].

In another work, this group synthesized ABL thermoplastic elastomeric compositions based on Acrylonitrile-Butadiene Rubber (NBR41) melt mixed with HW lignin. The low viscosity and the low thermal transition temperature of HW lignin were preferred to the characteristics of SW lignin. ABL is a renewable analog of ABS, in which the styrene component of ABS is replaced with lignin in ABL. ABS has a glassy plastic matrix where the rubbery phase is dispersed; in contrast, ABL has a rubbery matrix where the lignin phase is dispersed as discrete domains or as a co-continuous phase [21].

The incorporation of a stiffer component (lignin) leads to higher stiffness of nitrile rubber-lignin composites. Lignin does not form a freestanding film due to its very low molecular weight and inherent brittleness; hence, NBR41 is used as a binding segment between lignin molecules. Varying the concentrations of these components results in variable properties of the ABL composites, with increasing lignin content leading to increased brittleness. However, the reduction in the viscosity of ABL melt compared to neat rubber is unique. Good shear-thinning behavior and low viscosity are critical to enhance flow characteristics during the 3D printing process.

The dominance of b-O-4′ linkages, aliphatic segments, aliphatic ether groups, and oxygenated aromatic carbons in sinapyl alcohol-rich HW lignin offers good thermal processing characteristics [68]. Kraft SW lignin has significant amounts of stiff segments consisting of radically coupled biphenyl and biphenyl ethers based on G and H units, and this leads to it having a high softening point and higher viscosity. They successfully defined a viscosity window, from ca. 70 to 500 Pa s, for good printability via melt extrusion-based 3D printing of lignin-based materials. The characteristics of different HW lignin compositions analogous to the most common polymer 3D-printing materials, ABS and nylon, were compared to evaluate their printability. ABS was modified by substituting lignin with styrene by mixing lignin with nitrile-butadiene rubber to form ABL, and nylon was modified by simply blending it with lignin.

While ABL revealed properties that were not conducive for 3D printing, lignin/nylon blends demonstrated promising properties for 3D printability. This allowed lignin to be used as a renewable feedstock to develop green materials for 3D-printing applications. The rheological properties and their correlated chemical structures helped us design and prepare renewable polymeric compositions for 3D printing.

Other commonly used 3DP technologies like fused deposition modeling (FDM), include selective laser sintering (SLS), thermal inkjet (TIJ), polyjet and stereolithography apparatus (SLA) [71].

Cellulose nanocrystals (CNC) possess inherent stiffness and strength and a high aspect ratio, making them an ideal material to enhance the mechanical properties of elastomers and meet end-user requirements. As a biobased nanoscale filler material,

CNC also offers distinct advantages for achieving material sustainability in the final product, as evidenced by corporate industrial roadmaps and government regulatory initiatives. The inhomogeneous dispersion of CNC in an elastomeric matrix can have detrimental effects on the structural integrity of the nanocomposite material and may even cause 3D printing defects. It is important to ensure a homogeneous dispersion of CNC to avoid these issues. The inhomogeneous dispersion of CNC in an elastomeric matrix can have detrimental effects on the structural integrity of the nanocomposite material and may even cause 3D printing defects.

In 2017, Feng et al. [72] used lignin-coated CNC (L-CNC) to reinforce a methacrylate-based resin (MA) photopolymerized via SLA and post-cured using UV and heat treatment at 120 °C for 40 min [25]. They show a small increase of 1 and 3% in tensile strength at 0.1 and 0.5 wt% L-CNC, respectively, with a corresponding slight increase in the elongation at break of samples loaded with 0.1 wt% L-CNC.

On the other hand, a novel genre of thermoplastic elastomers has been developed by Tran et al. [21] incorporating nanoscale dispersed lignin into nitrile rubber. Temperature-controlled miscibility between the lignin and rubber during high-shear melt phase synthesis enables tuning of material morphology and performance. The sustainable material has an unprecedented yield strength (15–45 MPa), is strain-hardened under large deformation, and has excellent recyclability. The multiphase polymers created from an equal mass blend of a melt-stable lignin fraction and nitrile rubber with optimal acrylonitrile content using the method described here exhibit 5–100 nm lignin lamellae with a high modulus rubbery interphase. Molded or printed elastomeric products made from the lignin-nitrile material provide an additional revenue stream for pulp mills and biorefineries.

Lignin content has been limited to about 30 wt%, often requiring chemical modification, solvent fractionation of lignin, or prohibitively expensive additives. Each of these factors is a deterrent to industrial adoption of lignin-based polymers, limiting the potential of this renewable resource. Lignin-derived thermoplastics and elastomers with both versatile performance and commercialization potential have been elusive for several decades. Bova et al. [66] report on high-performance multiphase thermoplastics prepared from a blend of 41 wt% unmodified industrial lignin and low-cost additives in a general-purpose acrylonitrile-butadiene rubber (NBR) matrix. Hardwood soda lignin (HSL) and softwood kraft lignin (SKL) were blended under high shear conditions with NBR, carbon black (CB), polyethylene oxide (PEO), boric acid (BA) and dicumyl peroxide (DCP). This association with SKL lignin in the appropriate proportions gave a thermoplastic with a tensile strength and elongation at break of 25.2 MPa and 9%, respectively; it displayed an unusual tensile yield similar to that of ABS, a commodity thermoplastic. The similar HSL lignin compositions are tough materials with tensile strengths of 7.3–16.7 MPa and elongation at a break of 80–140%. The distinct ductility and yield behavior were analyzed based on the morphology of the compositions and the interfacial structure resulting from the chemical nature of each lignin studied. The roles of CB as reinforcement in the rubbery phase, DCP and BA as crosslinkers to create multiphase networks, and PEO to promote lignin adhesion and compatibility in commercial grade NBR.

Abdelwahab et al. [73] worked with 61% renewable resource content exhibiting superior toughness using polylactide (PLA) and organosolv lignin (OL). Although they did not work with the material in 3D printing, the material is very promising for printing because they used reactive extrusion of PLA with OL in the presence of poly (vinyl acetate), PVAc and glycidyl methacrylate (GMA) and proved that improved interfacial adhesion resulting in a novel biobased elastomer. A constrained mixture design was adopted to evaluate the effect of OL, PVAc and GMA on the performance of the composites. Optimized composites displayed superior toughness (elongation at break of ~340% and notched impact strength of ~900 J/m) with lowered tensile strength and modulus. FTIR analysis suggested that interfacial adhesion between GMA and PLA occurred during melt processing. One of the most significant findings of this study is the generation of single glass transition temperature of the complex structured composites. This new material offers a sustainable novel strategy for producing high-performance PLA materials with value-added uses of lignin to replace petroleum-based thermoplastic elastomers.

Murillo et al. [74] proposes a biotechnological strategy consisting of the partial degradation of alkali lignin using *Bacillus ligniniphilus* laccase (Lacc) L1 for its use as a nucleating agent in a polylactic acid/thermoplastic polyurethane (PLA/TPU) blend. The results revealed that the addition of enzymatically modified lignin (EL) increased the modulus of elasticity to a maximum of 2.5 times that of the control and provided a maximum biodegradation rate of 15% after 6 months under the soil burial method. In addition, the print quality provided satisfactory smooth surfaces, geometries, and tunable addition of a woody color. These results open a new door for the use of laccase as a tool to enhance the properties of lignin and its use as a scaffold to produce more environmentally sustainable filaments with improved mechanical properties for 3D printing.

Interesting is the case presented by Zhou et al. [75]. In this study, thermoplastic polyurethane (TPU) was incorporated to modulate the rheological properties of lignin for 3D printing. The 3D printed lignin/TPU composite exhibited a smooth surface, non-plastic, warm wood feel, and natural color at 50 wt% lignin loading. To better enhance the mechanical properties of the composite, carbon fiber (CF) was added to the lignin/TPU composite. The obtained CF/Lignin/TPU composites had 1.7 times higher tensile strength and 2.4 times higher elongation at break than the Lignin/TPU composites. Moreover, the smooth surface of the filament and the dense interlayer bonds of the printed specimens are also accomplished. This work brings new insights to realize the high-value utilization of lignin and extends the practical implementation of lignin in 3D printing.

However, in the fields of electronics and biomedical industry, manufacturing devices for application is still a challenge, specifically in the case of shape memory polymer (SMP) materials with complicated shapes in response to remote stimuli precisely and instantly. Wang et al. [76] achieved ultraviolet light-responsive SMP composites by direct ink writing (DIW) 3D printing of shape memory polyurethane (SMPU) with the introduction of laboratory-prepared lignin nanotube (LNT) as a UV absorber. They prepared a lignin nanotube with uniform morphology by dissolution and then controllable molecular assembly way during dialysis in deionized water.

The 3D-printed SMP composite of PU/LNT was found to achieve a shape recovery rate of 99% under UV irradiation (370 nm) for 15 min. In addition, the tensile strength of the 3D printed SMP composite of PU/LNT is increased to 16.4 MPa, which is 131% higher than that of the original PU (7.11 MPa). It is worth noting that the glass transition temperature (T_g) of the 3D printed SMP composite of PU/LNT is continuously changed from 47 to 33 °C by increasing the LNT content to 9%, which provides the composites with much wider application feasibility. The research on photo-responsive 3D printed SMP composites with accurate shape recovery rates provides new candidates to improve the development of biomaterials and intelligent robots.

In conclusion, the low cost of lignin, its biobased nature, good thermal stability, and excellent antioxidant properties make reinforced elastomer composites a promising option for the 3D printing industry. This combination has been demonstrated to be a good replacement, resulting in a novel biobased elastomer. Significant exploration has been conducted on natural polymers derived from wood and their functionalized variations in the realm of 3D-printed products. This advanced method is used for creating organ models and producing personalized scaffolds, particularly in biomedical applications. Research has shown that lignin-based composites can have mechanical performance equivalent to, or even superior to, that of conventional petroleum-based thermoplastics [77].

8.7 Conclusions

Lignin, the second-largest biomass resource in nature, has excellent environmental advantages and versatile properties, including UV-blocking capability, thermostability, and high stiffness. Elastomers are a significant class of polymers with applications in various industries, such as healthcare, aerospace, automotive, and apparel. However, the additive manufacturing (AM) of elastomers remains challenging due to their inherent physical, thermal, and mechanical properties. The combination of both enhances and proposes new materials (lignin-based composites) with tunable properties. The 3D-printed composites presented in the last chapter, demonstrated that these materials have adjustable properties, providing a new way to fabricate smart materials for robotics and smart biomaterials in the future. Therefore, achieving a high lignin content as a raw material for preparing composite filaments in the case for example of FDM-based 3D printing remains a challenge. This would improve bioenergy utilization and reduce the need for petroleum-based products.

References

1. Low DYS, Supramaniam J, Leong WD, Soottitantawat A, Charinpanitkul T, Tanthapanichakoon W et al (2023) Self-healing synthetic rubber composites: review of recent progress and future directions towards sustainability. Mater Today Sustain 24:100545
2. Strohmeier L, Schrittesser B, Schlögl S (2022) Approaches toward in situ reinforcement of organic rubbers: strategy and recent progress. Polym Rev [Internet]. 62:142–174. https://doi.org/10.1080/15583724.2021.1897998
3. Wemyss AM, Bowen C, Plesse C, Vancaeyzeele C, Nguyen GTM, Vidal F et al (2020) Dynamic crosslinked rubbers for a green future: a material perspective. Mater Sci Eng R Rep 141:100561
4. Tadiello L, Guerra S, Giannini L (2022) Sepiolite-based anisotropic nanoparticles: a new player in the rubber reinforcement technology for tire application. Appl Sci 12
5. Garlapati VK, Chandel AK, Kumar SPJ, Sharma S, Sevda S, Ingle AP et al (2020) Circular economy aspects of lignin: towards a lignocellulose biorefinery. Renew Sustain Energy Rev 130:109977
6. Yang Y, Wang Y, Zhu M, Zhao J, Cai D, Cao H (2023) Valorization of lignin for renewable non-isocyanate polyurethanes: a state-of-the-art review. Mater Today Sustain 22:100367
7. Lawoko M, Berglund L, Johansson M (2021) Lignin as a renewable substrate for polymers: from molecular understanding and isolation to targeted applications. ACS Sustain Chem Eng 9:5481–5485
8. Zevallos Torres LA, Lorenci Woiciechowski A, de Andrade Tanobe VO, Karp SG, Guimarães Lorenci LC, Faulds C et al (2020) Lignin as a potential source of high-added value compounds: a review. J Clean Prod 263:121499
9. Collins MN, Nechifor M, Tanasă F, Zănoagă M, McLoughlin A, Stróżyk MA et al (2019) Valorization of lignin in polymer and composite systems for advanced engineering applications—a review. Int J Biol Macromol 131:828–849
10. Tardy BL, Lizundia E, Guizani C, Hakkarainen M, Sipponen MH (2023) Prospects for the integration of lignin materials into the circular economy. Mater Today 65:122–132
11. Mohamad Aini NA, Othman N, Hussin MH, Sahakaro K, Hayeemasae N (2020) Lignin as alternative reinforcing filler in the rubber industry: a review. Front Mater 6
12. Guerra NB, Sant'Ana Pegorin G, Boratto MH, de Barros NR, de Oliveira Graeff CF, Herculano RD (2021) Biomedical applications of natural rubber latex from the rubber tree Hevea brasiliensis. Mater Sci Eng C 126:112126
13. Wei YC, Zhu D, Xie WY, Xia JH, He MF, Liao S (2022) In-situ observation of spatial organization of natural rubber latex particles and exploring the relationship between particle size and mechanical properties of natural rubber. Ind Crops Prod 180:114737
14. Zhao X, Hu H, Zhang D, Zhang Z, Peng S, Sun Y (2019) Curing behaviors, mechanical properties, dynamic mechanical analysis and morphologies of natural rubber vulcanizates containing reclaimed rubber. E-Polymers 19:482–488
15. Manaila E, Stelescu MD, Craciun G (2018) Degradation studies realized on natural rubber and plasticized potato starch based eco-composites obtained by peroxide cross-linking. Int J Mol Sci 19:2862
16. Men X, Wang F, Chen GQ, Zhang HB, Xian M (2019) Biosynthesis of natural rubber: current state and perspectives. Int J Mol Sci 20
17. Carpenedo GA, Demori R, Carli LN, Giovanela M, de Paoli MA, Crespo JS (2020) Evaluation of stabilizing additives content in the mechanical properties of elastomeric compositions subject to environmental and accelerated aging. Mater Res 23:e20201039
18. Barana D, Ali SD, Salanti A, Orlandi M, Castellani L, Hanel T et al (2016) Influence of lignin features on thermal stability and mechanical properties of natural rubber compounds. ACS Sustain Chem Eng 4:5258–5267
19. Rahimi A, Azarpira A, Kim H, Ralph J, Stahl SS (2013) Chemoselective metal-free aerobic alcohol oxidation in lignin. J Am Chem Soc 135:6415–6418

20. Visakh PM, Thomas S, Oksman K, Mathew AP (2012) Crosslinked natural rubber nanocomposites reinforced with cellulose whiskers isolated from bamboo waste: processing and mechanical/thermal properties. Compos Part A Appl Sci Manuf 43:735–741
21. Tran CD, Chen J, Keum JK, Naskar AK (2016) A new class of renewable thermoplastics with extraordinary performance from nanostructured lignin-elastomers. Adv Funct Mater 26:2677–2685
22. Košíková B, Gregorová A, Osvald A, Krajčovičová J (2007) Role of lignin filler in stabilization of natural rubber-based composites. J Appl Polym Sci 103:1226–1231
23. Makhalema M, Hlangothi P, Motloung SV, Koao LF, Motaung TE (2021) Influence of kraft lignin on the properties of rubber composites. Wood Res. 66:285–296
24. Carpenedo GA, Guerra NB, Giovanela M, De Paoli MA, Da Silva Crespo J (2022) Evaluation of lignin as stabilizer in vulcanized natural rubber formulations. Polimeros 32
25. Wang B, Sun YC, Sun RC (2019) Fractionational and structural characterization of lignin and its modification as biosorbents for efficient removal of chromium from wastewater: a review. J Leather Sci Eng 1:1–25
26. Yu P, He H, Jia Y, Tian S, Chen J, Jia D et al (2016) A comprehensive study on lignin as a green alternative of silica in natural rubber composites. Polym Test 54:176–185
27. Hosseinmardi A, Annamalai PK, Martine B, Pennells J, Martin DJ, Amiralian N (2018) Facile tuning of the surface energy of cellulose nanofibers for nanocomposite reinforcement. ACS Omega 3:15933–15942
28. Hosseinmardi A, Annamalai PK, Wang L, Martin D, Amiralian N (2017) Reinforcement of natural rubber latex using lignocellulosic nanofibers isolated from spinifex grass. Nanoscale 9:9510–9519
29. Riyajan SA (2015) Robust and biodegradable polymer of cassava starch and modified natural rubber. Carbohyd Polym 134:267–277
30. Hosseinmardi A, Amiralian N, Hayati AN, Martin DJ, Annamalai PK (2021) Toughening of natural rubber nanocomposites by the incorporation of nanoscale lignin combined with an industrially relevant leaching process. Ind Crops Prod 159:113063
31. Norgren M, Edlund H, Wågberg L (2002) Aggregation of lignin derivatives under alkaline conditions. Kinetics and aggregate structure. Langmuir 18:2859–2865
32. Vainio U, Maximova N, Hortling B, Laine J, Stenius P, Simola LK et al (2004) Morphology of dry lignins and size and shape of dissolved kraft lignin particles by X-ray scattering. Langmuir 20:9736–9744
33. Zhu J, Xue L, Wei W, Mu C, Jiang M, Zhou Z (2015) Modification of lignin with silane coupling agent to improve the interface of poly(L-lactic) acid/lignin composites. BioResources 10:4315–4325
34. Bahl K, Miyoshi T, Jana SC (2014) Hybrid fillers of lignin and carbon black for lowering of viscoelastic loss in rubber compounds. Polymer (Guildf). 55:3825–3835
35. Bahl K, Swanson N, Pugh C, Jana SC (2014) Polybutadiene-g-polypentafluorostyrene as a coupling agent for lignin-filled rubber compounds. Polymer (Guildf). 55:6754–6763
36. Liu R, Li J, Lu T, Han X, Yan Z, Zhao S et al (2022) Comparative study on the synergistic reinforcement of lignin between carbon black/lignin and silica/lignin hybrid filled natural rubber composites. Ind Crop Prod 187:115378
37. Datta J, Parcheta P, Surówka J (2017) Softwood-lignin/natural rubber composites containing novel plasticizing agent: preparation and characterization. Ind Crops Prod 95:675–685
38. Zhang X, Zhao Z, Ran G, Liu Y, Liu S, Zhou B et al (2013) Synthesis of lignin-modified silica nanoparticles from black liquor of rice straw pulping. Powder Technol 246:664–668
39. Komisarz K, Majka TM, Pielichowski K (2023) Chemical and physical modification of lignin for green polymeric composite materials. Materials (Basel) 16
40. Shorey R, Gupta A, Mekonnen TH (2021) Hydrophobic modification of lignin for rubber composites. Ind Crops Prod 174:114189
41. Adibi A, Kim J, Mok J, Lenges C, Simon L, Mekonnen TH (2021) Enzymatic polymerization designed alpha-1,3 glucan particle morphology as reinforcing fillers of dipped and casted rubber films. Carbohydr Polym 267:118234

42. Jiang C, Shen H, Bi X, Wang Z, Yao M, Wu Y et al (2022) A green dual-phase carbon-silica nanohybrid derived from black liquor lignin for reinforcing styrene-butadiene rubber. Compos Sci Technol 230:109775
43. He Z, Li Y, Liu C, Yang J, Qian M, Zhu Y et al (2022) Turning lignin into treasure: an innovative filler comparable to commercial carbon black for the green development of the rubber industry. Int J Biol Macromol 218:891–899
44. Lai L, Liu J, Lv Z, Gao T, Luo Y (2023) Recent advances for flame retardant rubber composites: mini-review. Adv Ind Eng Polym Res 6:156–164
45. Solihat NN, Hidayat AF, Taib MNAM, Hussin MH, Lee SH, Ghani MAA et al (2022) Recent developments in flame-retardant lignin-based biocomposite: manufacturing, and characterization. J Polym Environ 30:4517–4537
46. Rybiński P, Janowska G, Jóåwiak M, Jóåwiak M (2013) Thermal stability and flammability of styrene-butadiene rubber (SBR) composites: effect of attapulgite, silica, carbon nanofiber, and the synergism of their action on the properties SBR composites. J Therm Anal Calorim 113:43–52
47. Li J, Yan Z, Liu M, Han X, Lu T, Liu R et al (2023) Triple silicon, phosphorous, and nitrogen-grafted lignin-based flame retardant and its vulcanization promotion for styrene butadiene rubber. ACS Omega 8:21549–21558
48. Arias A, Feijoo G, Moreira MT (2022) Exploring the potential of antioxidants from fruits and vegetables and strategies for their recovery. Innov Food Sci Emerg Technol 77:102974
49. Li K, Zhong W, Li P, Ren J, Jiang K, Wu W (2023) Recent advances in lignin antioxidant: antioxidant mechanism, evaluation methods, influence factors and various applications. Int J Biol Macromol 251:125992
50. Kamaruddin S, Muhr AH (2018) Investigation of ozone cracking on natural rubber. J Rubber Res 21:73–93
51. Zhao W, He J, Yu P, Jiang X, Zhang L (2023) Recent progress in the rubber antioxidants: a review. Polym Degrad Stab 207:110223
52. Xu J, Hao Y, Yang Z, Li W, Xie W, Huang Y et al (2022) Rubber antioxidants and their transformation products: environmental occurrence and potential impact. Int J Environ Res Public Health 19
53. Chung JY, Hwang U, Kim J, Kim NY, Nam J, Jung J et al (2023) Amine-functionalized lignin as an eco-friendly antioxidant for rubber compounds. ACS Sustain Chem Eng 11:2303–2313
54. Zhang T, Li H, Zhou J, Wang X, Xiao L, Zhang F et al (2022) Review of carbon dots from lignin: preparing, tuning, and applying. Pap Biomater 7:51–62
55. Zhao W, Dufresne A, Li A, An H, Shen C, Yu P et al (2023) Use of lignin-based crude carbon dots as effective antioxidant for natural rubber. Int J Biol Macromol 253:126594
56. Chio C, Sain M, Qin W (2019) Lignin utilization: a review of lignin depolymerization from various aspects. Renew Sustain Energy Rev 107:232–249
57. Xu T, Li K, Wang Y, Ban J, Wu D, Shi B et al (2019) Enhancing interfacial and mechanical properties of NR/montmorillonite composites using butadiene-styrene-vinylpyridine rubber as compatilizer. Mater Chem Phys 231:357–365
58. Zhao S, Li J, Yan Z, Lu T, Liu R, Han X et al (2021) Preparation of lignin-based filling antioxidant and its application in styrene-butadiene rubber. J Appl Polym Sci 138:1–10
59. Skorupka M, Nosalewicz A (2021) Ammonia volatilization from fertilizer urea—a new challenge for agriculture and industry in view of growing global demand for food and energy crops. Agriculture 11
60. Boonying P, Boonpavanitchakul K, Amnuaypanich S, Kangwansupamonkon W (2023) Natural rubber-lignin composites modified with natural rubber-graft-polyacrylamide as an effective coating for slow-release fertilizers. Ind Crops Prod 191:116018
61. Boonying P, Boonpavanitchakul K, Kangwansupamonkon W (2023) Green bio-composite coating film from lignin/pre-vulcanized natural rubber latex for controlled-release urea fertilizer. J Polym Environ 31:1642–1655
62. Saunders S (2018) New balance uses ultrasonic equipment and sieves to ensure high powder throughput for 3D printed midsoles, p 199087. [Internet]. https://3dprint.com/199087/new-balance-powder-case-study/. Accessed May 2018

63. Lewis JA (2006) Direct ink writing of 3D functional materials. Adv Funct Mater 16
64. Herzberger J, Sirrine JM, Williams CB, Long TE (2019) polymer design for 3D printing elastomers: recent advances in structure, properties, and printing. Prog Polym Sci [Internet] 97:101144. https://doi.org/10.1016/j.progpolymsci.2019.101144
65. Shafiq A, Ahmad Bhatti I, Amjed N, Zeshan M, Zaheer A, Kamal A et al (2024) Lignin derived polyurethanes: current advances and future prospects in synthesis and applications. Eur Polym J 112899
66. Bova T, Tran CD, Balakshin MY, Chen J, Capanema EA, Naskar AK (2016) An approach towards tailoring interfacial structures and properties of multiphase renewable thermoplastics from lignin-nitrile rubber. Green Chem 18:5423–5437
67. Sun Q, Khunsupat R, Akato K, Tao J, Labbé N, Gallego NC et al (2016) A study of poplar organosolv lignin after melt rheology treatment as carbon fiber precursors. Green Chem 18:5015–5024
68. Nguyen NA, Meek KM, Bowland CC, Barnes SH, Naskar AK (2018) An acrylonitrile-butadiene-lignin renewable skin with programmable and switchable electrical conductivity for stress/strain-sensing applications. Macromolecules 51:115–127
69. Akato K, Tran CD, Chen J, Naskar AK (2015) Poly(ethylene oxide)-assisted macromolecular self-assembly of lignin in ABS matrix for sustainable composite applications. ACS Sustain Chem Eng 3:3070–3076
70. Nguyen NA, Barnes SH, Bowland CC, Meek KM, Littrell KC, Keum JK et al (2018) A path for lignin valorization via additive manufacturing of high-performance sustainable composites with enhanced 3D printability. Sci Adv 4:1–15
71. Palaganas NB, Palaganas JO, Doroteo SHZ, Millare JC (2023) Covalently functionalized cellulose nanocrystal-reinforced photocurable thermosetting elastomer for 3D printing application. Addit Manuf [Internet] 61:103295. https://linkinghub.elsevier.com/retrieve/pii/S2214860422006844
72. Feng X, Yang Z, Chmely S, Wang Q, Wang S, Xie Y (2017) Lignin-coated cellulose nanocrystal filled methacrylate composites prepared via 3D stereolithography printing: mechanical reinforcement and thermal stabilization. Carbohydr Polym [Internet] 169:272–281. https://doi.org/10.1016/j.carbpol.2017.04.001
73. Abdelwahab MA, Jacob S, Misra M, Mohanty AK (2021) Super-tough sustainable biobased composites from polylactide bioplastic and lignin for bio-elastomer application. Polymer (Guildf) [Internet]. 212:123153. https://doi.org/10.1016/j.polymer.2020.123153
74. Murillo-Morales G, Sethupathy S, Zhang M, Xu L, Ghaznavi A, Xu J et al (2023) Characterization and 3D printing of a biodegradable polylactic acid/thermoplastic polyurethane blend with laccase-modified lignin as a nucleating agent. Int J Biol Macromol [Internet] 236:123881. https://doi.org/10.1016/j.ijbiomac.2023.123881
75. Zhou X, Ren Z, Sun H, Bi H, Gu T, Xu M (2022) 3D printing with high content of lignin enabled by introducing polyurethane. Int J Biol Macromol [Internet] 221:1209–1217. https://doi.org/10.1016/j.ijbiomac.2022.09.076
76. Wang F, Jiang M, Pan Y, Lu Y, Xu W, Zhou Y (2023) 3D printing photo-induced lignin nanotubes/polyurethane shape memory composite. Polym Test [Internet] 119:107934. https://doi.org/10.1016/j.polymertesting.2023.107934
77. Nguyen NA, Bowland CC, Naskar AK (2018) A general method to improve 3D-printability and inter-layer adhesion in lignin-based composites. Appl Mater Today 12:138–152

Chapter 9
Advancements in Green Nanocomposites: A Comprehensive Review on Cellulose-Based Materials in Biocomposites and Bionanocomposites

Jayvirsinh Atodariya, Manav Agrawal, Ansh Singh, and Neha Patni

Abstract With an emphasis on cellulose-based components in biocomposites and bionanocomposites, this thorough overview examines the most recent developments in green nanocomposites. As the need for environmentally friendly materials and concerns about environmental sustainability grow, cellulose-based nanocomposites are emerging as viable substitutes in a number of industries. This work offers a methodical review of the synthesis techniques, characteristics, and uses of cellulose-based nanocomposites, emphasizing their potential in the packaging, automotive, building, and biomedical industries. Incorporating cellulose nanofibers, and nanoparticles into biopolymer matrices is covered in detail in this paper, along with the principles underlying their improved mechanical, thermal, and barrier properties. In addition, it looks at how processing methods and reinforcing agents affect the functionality of cellulose-based nanocomposites. The evaluation also discusses the difficulties and potential outcomes related to the commercialization and broad use of these sustainable materials, highlighting the need for more study in areas like scalability, cost-effectiveness, and end-of-life issues. This review is an invaluable tool who are interested in creating and applying green nanocomposites for a more sustainable and environmentally friendly future.

Keywords Green nanocomposites · Cellulose-based materials · Biocomposites and bionanocomposites · Cellulose-based nanocomposites · Environmentally friendly materials · Packaging · Automotive · Building · Biomedical industries

J. Atodariya · M. Agrawal · A. Singh · N. Patni (✉)
Chemical Engineering Department, Institute of Technology, Nirma University, Ahmedabad, Gujarat, India
e-mail: neha.patni@nirmauni.ac.in

Visakh P. M. *Rubber Based Bionanocomposites*, Advanced Structured Materials 210,
https://doi.org/10.1007/978-981-10-2978-3_9

9.1 Introduction

In the past few decades, there has been a significant rise in the demand for developing sustainable materials that can overcome environmental concerns too without compromising their effectiveness. Cellulose is one the most important derivative found in plant fibers, which is then useful for any chemical product synthesis of biocomposite or bio nanocomposite. It is a plant derivative. These materials are of high value and can be taken into use for multiple commercial applications such as automotive, aerospace, packaging, construction, medical, and more [1]. This chapter is focused on the PVC and cellulose-based biocomposite and bionanocomposite materials, their synthesis, properties and varied applications [2].

9.1.1 Cellulose: Dynamic Bio Polymer

Cellulose is the most common and fundamental structural element in plant fiber (Fig. 9.1). Cellulose contains a linear change of D–anhydro glucopyranose which is then combined with b–1–4 glycosidic [3]. This structural combination results in the formation of the microfibrils. This structure has a compatibility of both crystalline and amorphous structures. The main reason behind the adaptability of this kind of structural behavior is because of its fibrous, tough, and hydrophilic nature [4]. Cellulose contains an active hydroxyl group that is readily functional to react with organic, inorganic, polymers, and nanomaterials. The environmental acceptance, economic feasibility, and ease of extraction make cellulose the ideal material for various applications, including the production of films with the addition of high tensile strength, thermal conductivity and antibacterial capabilities [5].

There are ways to modify the crystal of the cellulose by the formation of hydrogen bonding and its solubility, which is then extended to its derivatization. Cellulose has the flexibility of linking with other materials to enhance structural activities such as lignin nanoparticles which enhance antimicrobial activity. In addition to this, the new technological advancement of nano cellulose has opened up many new opportunities for its application in the field of manufacturing [6].

9.1.2 Renewable Focused Cellulose

Cellulose is one of the most abundant renewable polymers available on the earth, which is a crucial part of this chapter. The microfibrils and intermolecular forces and bonds, more precisely the hydrogen bond of the cellulose are naturally organized, and hence the stability is enhanced for this material. It makes it useful in various industries. The review work is focused on the biosynthesis of cellulose, which is more focused on the linear structure and the role of the hydroxyl group in the hydrogen

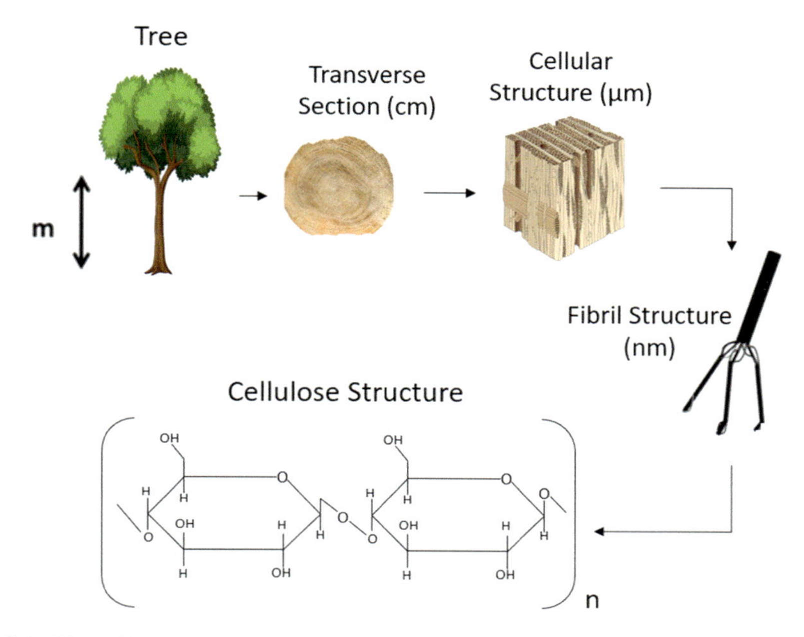

Fig. 9.1 Hierarchical structure of cellulose

bond, which is then governed by the physical characterization of the cellulose [7]. In the further section of the chapter, the special emphasis is on the synthesis, properties, and applications of PVS/cellulose-based biocomposite and bio nanocomposite. The chapter is all about the various aspects, starting from the types of cellulose nanoparticles to the synthesis, with a precise focus on sustainable and innovative solutions for the material challenges.

9.1.3 Nanocomposites: A Nanoscale Approach

Nanocomposite materials are formed with nanoscale dimensions and have unique properties which make them attractive for their significant use in any industry. Most precisely the Polymer Matrix Nanocomposites (PMNC) is the most common one with its effective barrier, optical characteristics, and mechanical properties [8]. These all properties make them individual elements for diverse applications. When it comes to the industrial point of view the nanocomposite material has played an important role in the automotive and packaging industry with their lightweight and cost-effective solutions. The past decade has shown significant development in the field of nanocomposite which includes advanced nanomaterials, functionality changes, biodegradable nanocomposites and fillers along with enhanced coating technology. The future trends with the nanocomposite are related to smart fibers, multifunctionality, biomedical use, energy section and sustainable nanocomposite development. The detailed advancement and future trends are mentioned below [9].

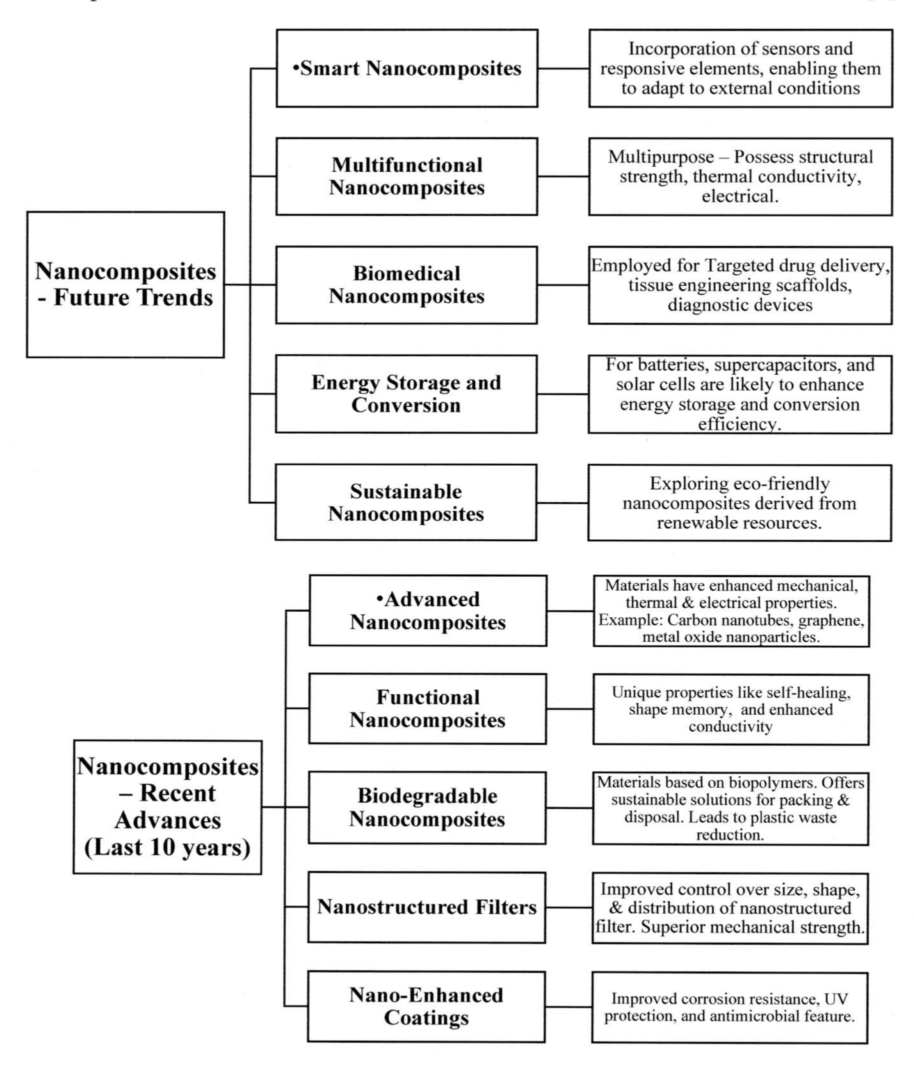

9.1.4 Bionanocomposite: Integration of Biology & Nanotechnology

The term bionanocomposite states that the material is made up of biological components and more diligently with polymers and inorganic solids. Compared to traditional or conventional nanocomposite material the bionanocomposite material has some unique characteristics such as biodegradability, biocompatibility and unique functionality. Some of the research work in the field of nanobiocomposites, green composites and biohybrids has gained attention for the convergence of nanotechnology and biology [10]. When nanocomposites have evolved into a combination of scientific discoveries, technologies and innovations the growing awareness of sustainability has shown the major advancements in the field of bionanocomposites as well. The future trends suggest a multipurpose focus on the functionality, smart innovative technologies and multipurpose applications such as in health care, energy, and the environmental sector. The detailed analysis is as follows [11].

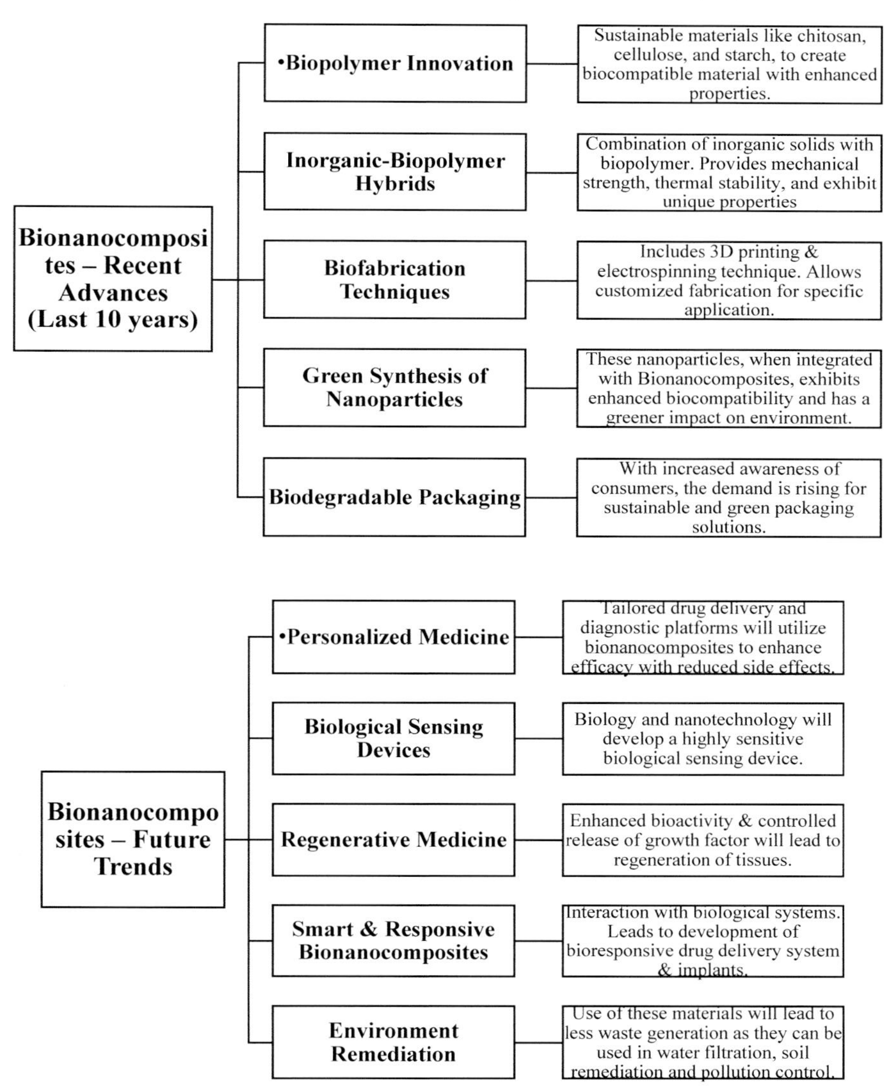

Biocomposites containing cellulose, a plentiful polysaccharide presents in the cell walls of plants, have become a major subject of intensive research and development in recent years due to their exceptional characteristics and environmentally benign properties. This study explores the complex field of cellulose-based biocomposites, offering a thorough examination of their composition, advanced manufacturing techniques, and wide range of varied uses [12].

9.2 Cellulose: Revealing the Structure and Properties of Nature's Architectural Wonder

The complex structure and distinct characteristics of cellulose itself are the fundamental components of cellulose-based biocomposites. Cellulose, which is the main component of plant cell walls, possesses a molecular structure that highlights its unique qualities [13]. Cellulose is composed of linear chains of glucose units, which exhibit both crystalline and amorphous regions. These regions collectively contribute to cellulose's exceptional tensile strength, natural capacity to degrade, and remarkable stiffness. The molecular composition of cellulose is not simply a random arrangement of glucose units. Cellulose's mechanical and thermal abilities are defined by the harmonious interaction between its crystalline and amorphous domains, with each domain playing a vital role. Cellulose possesses formidable tensile strength due to its crystalline sections, rendering it a strong candidate for applications that demand structural integrity [14]. Conversely, the shapeless regions of cellulose enhance its ability to undergo biological decomposition, a trait that is environmentally conscientious and in line with the worldwide effort for sustainable materials. Upon further examination of cellulose's molecular structure, the arrangement of hydrogen bonds becomes a crucial determinant of its mechanical and thermal characteristics. The hydrogen connections in cellulose molecules form a robust network, which gives the biocomposite its durability and structural stability [15]. Gaining insight into this intricate chemical process enables researchers and producers to precisely adjust the characteristics of cellulose-based biocomposites for particular uses, spanning from packaging materials to sophisticated structural elements. The hydrogen bond arrangement of cellulose contributes to its thermal qualities, enhancing its appeal. Cellulose-based biocomposites exhibit exceptional resistance to heat degradation, making them highly suitable for applications that require strong temperature resistance. This provides opportunities in various industries, such as automotive manufacture, construction, and aerospace, where materials need to endure different heat conditions while maintaining their structural integrity [16].

9.2.1 Methods of Processing for Biocomposites: Crafting Nature's Green Marvels

The process of creating cellulose-based biocomposites involves a dynamic interaction between renewable resources found in nature and advanced processing processes. This section explores the fascinating processes that convert cellulose into adaptable and environmentally friendly wonders, ranging from the extraction of cellulose to the complex techniques used for composite synthesis [17].

9.2.1.1 Extracting Cellulose: Harnessing the Abundance of Nature

The journey commences with cellulose, obtained from sustainable resources such as timber, cotton, and diverse botanical fibers. These natural reservoirs provide a renewable source for extracting cellulose, which serves as the basis for creating biocomposites. The extraction process, which encompasses both conventional techniques and cutting-edge technologies, guarantees the integrity and excellence of cellulose, thus laying the foundation for the environmentally conscious path that lies ahead [18].

9.2.1.2 Unveiling Isolation Techniques in Fiber Fantasia

Delving further into the process of isolating cellulose, strategies are employed to free cellulose fibers from their plant-based origins. These techniques involve a combination of chemical and mechanical processes that work together harmoniously. Chemical procedures, such as acid hydrolysis, and mechanical ones, such as grinding or steam explosion, release the fibrous nature of cellulose. This process of extracting cellulose not only improves its quality, but also opens up several opportunities for customizing the features of biocomposites.

9.2.1.3 Alchemy Processing: Improving Compatibility

After cellulose is included in the biocomposite, processing procedures are employed to facilitate a seamless collaboration between cellulose and the other components of the composite. The essence of the art rests in enhancing compatibility, and the techniques utilized resemble a complex alchemical formula. The objective is to combine cellulose with polymers, natural fibers, or nanoparticles in order to create a synergistic effect that improves the overall performance of the biocomposite [19].

9.2.1.4 Extrusion, Injection, and Compression: The Intricate Dance of Manufacturing

Imagine cellulose-based biocomposites as an elaborate dance performance taking place on the factory floor. Extrusion, injection molding, and compression molding are the primary methods used to create cellulose-based biocomposites. Extrusion, similar to a synchronized dance, guarantees the smooth integration of components, while injection molding introduces accuracy to the process, resulting in elaborate designs and structures. Compression molding utilizes high pressure to solidify the performance of the biocomposite, resulting in superior dispersion and reinforcement that significantly enhance its quality. This section elucidates the techniques involved in processing cellulose-based biocomposites and also encourages readers to visualize the various stages of this environmentally benign phenomenon. As we observe the shift from extracting cellulose to the final process of production, we appreciate the combination of natural resources and human creativity, leading to the creation of environmentally friendly wonders that provide a promising and thrilling future. Participate in this captivating investigation of cellulose-based biocomposites, where scientific knowledge and natural elements combine in a remarkable demonstration of inventive thinking and environmental awareness (Fig. 9.2).

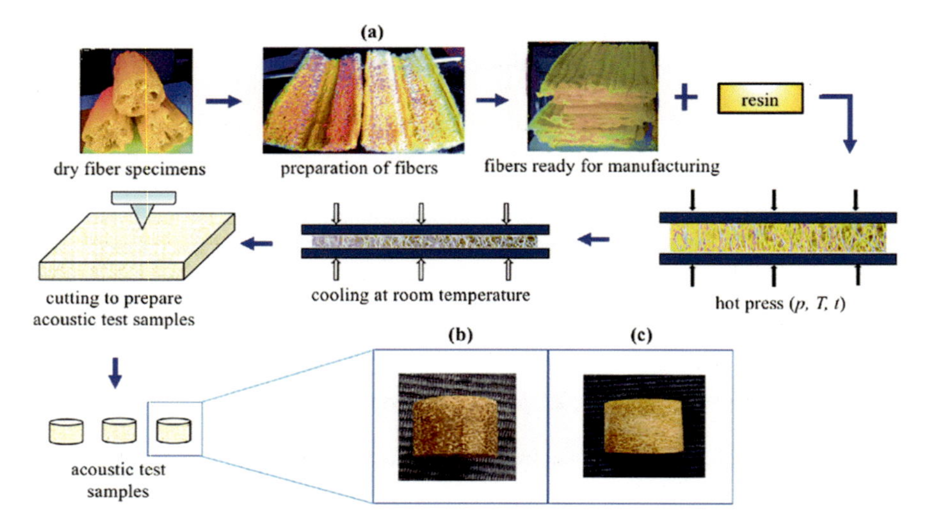

Fig. 9.2 Manufacturing process for biocomposite [20]

9.3 Application of Cellulose-Based Biocomposites

The realm of cellulose-based biocomposites presents limitless opportunities, providing a wide range of applications in various industries. Within this part, we undertake an enthralling exploration of the diverse terrain of these verdant marvels, highlighting their profound influence on a wide range of industries, including packaging and biomedical advancements. Cellulose-based biocomposites are at the forefront of the packaging industry, leading a shift towards materials that are both lightweight and environmentally benign. Envision packaging options that not only safeguard products but also actively contribute to a sustainable future. When combined with other polymers, cellulose creates packaging materials that redefine industry norms and promote environmental consciousness [1]. Cellulose, when introduced into the automotive industry, revolutionizes it by increasing the ratio of strength to weight in components. This results in a decrease in the total weight of cars, leading to a significant improvement in fuel efficiency. Cellulose-based biocomposites are being used to create lighter and more fuel-efficient automobiles, which are helping to make the automotive industry more environmentally friendly and sustainable. Cellulose-based biocomposites excel in the building industry, offering notable fire resistance and thermal insulation properties. Envision structures reinforced with materials that provide both fire resistance and the ability to enhance energy efficiency [21]. Cellulose, when used in its biocomposite form, becomes a fundamental element in building, allowing for the creation of structures that align with both safety and sustainability. Cellulose-based biocomposites are increasingly being utilized in biomedical applications as the field of material science continues to advance. Their exceptional capacity to flawlessly interact with live organisms provides opportunities for revolutionary medical devices and implants. The convergence of eco-friendly materials with healthcare developments ushers in a new era of biocompatible innovations, offering a healthier and more sustainable future [22].

9.4 Challenges of Cellulose-Based Biocomposites

Cellulose-based biocomposites exhibit remarkable adaptability, but they also present certain problems that require the focus of researchers and industry leaders. The challenges of establishing proper dispersion of cellulose inside polymer matrices, optimizing interfacial adhesion, and addressing compatibility difficulties require further investigation. An important obstacle is the attainment of optimal dispersion of cellulose inside polymer matrices. The pursuit of advanced processing techniques becomes crucial, as researchers traverse the complexities of harmoniously combining cellulose with other components. The successful exploration of cellulose-based biocomposites requires a joint endeavor between material scientists and engineers to fully exploit their capabilities [23]. Explore the domain of nanocellulose, a minuscule entity that possesses the potential to improve the properties of cellulose-based biocomposites. Subsequent investigations should prioritize the incorporation of nanocellulose into the mixture, thereby revealing novel aspects of robustness, adaptability, and environmental friendliness. The partnership between material scientists and nanotechnology experts serves as the driving force for surpassing current constraints and advancing cellulose-based biocomposites into unexplored domains [24]. The exploration of new horizons for cellulose-based biocomposites beckons researchers, extending the journey beyond present uses. The combined contributions of material scientists, engineers, and environmental experts play a crucial role in expanding knowledge and finding innovative uses that transform the field of sustainable, high-performance materials.

9.5 Introduction to Nanocellulose

Explore the fascinating realm of nanocellulose, a remarkable substance obtained from cellulose, where cellulose nanocrystals (CNC) and cellulose nanofibrils (CNF) play a prominent role. Imagine CNC as the mighty Hercules of the nanoworld, with extraordinary mechanical strength and impressive aspect ratios that surpass conventional boundaries [25]. However, CNF stands out as a very adaptable and versatile material, with a large surface area that offers numerous possibilities. Embark on the CNC story, where strength is not merely a characteristic, but rather a defining attribute. Cellulose nanocrystals have exceptional mechanical strength, establishing unprecedented benchmarks in the field of nanocomposites. Envision materials enhanced with the power of Computer Numerical Control (CNC), offering extraordinary strength and durability, and enabling applications that require exceptional resilience. The CNF chronicles highlight the prominence of flexibility as cellulose nanofibrils reveal their multifaceted nature. CNF exhibits exceptional agility, surpassing conventional standards, and assumes the role of a master in promoting flexibility. It empowers materials to effortlessly bend, twist, and conform to intricate designs. This attribute not only amplifies the versatility of nanocomposites but also stimulates innovation in domains where flexibility is crucial for achieving success [26]. In addition to their particular excellence, both CNC and CNF possess three specific characteristics that position them as highly promising materials for bionanocomposites. Their eco-friendly signature lies in their biodegradability, ensuring a graceful departure from the material world after their intended use is complete. Nanocellulose, with its ability to regenerate and sustainably meet growing world needs, can be likened to a phoenix that emerges from destruction. The low toxicity of nanocellulose ensures that innovation does not result in any harmful consequences. Within the complex realm of material science, nanocellulose possesses characteristics that not only serve as constituents of bionanocomposites, but also act as advocates for a sustainable future [27].

9.6 Enhancing Polymer Matrices in Bionanocomposites

As the next act begins, the focus turns to the significance of nanocellulose in bionanocomposites. Imagine polymer matrices being turned into incredibly strong fortresses, thanks to the reinforcing power of CNC (carbon nanotube composites) and the flexible support of CNF (cellulose nanofibers). This symbiotic interaction results in bionanocomposites that defy expectations, coupling strength with flexibility in an exquisite equilibrium [28] (Fig. 9.3).

To summarize, the investigation of cellulose-based bionanocomposites is not merely a scientific endeavor; it is an epic voyage into a realm where nanocellulose, with its prominent CNC and CNF components, fundamentally transforms the materials domain. Participate in this fusion of power and adaptability, where the qualities of being able to decompose naturally, being able to be renewed, and having minimal harmful effects combine seamlessly with originality, creating a clear representation of a sustainable future guided by bionanocomposites made from cellulose.

9.6.1 Production of Cellulose-Based Bionanocomposites

Step into the captivating world where cellulose and nanocomposites harmoniously combine, resulting in a symphony of ground-breaking advancements. The manufacturing of cellulose-based bionanocomposites involves the complex task of incorporating nanocellulose into polymer matrices, a process that extends beyond simple blending. Envision a realm where the amalgamation of nanocellulose and polymer

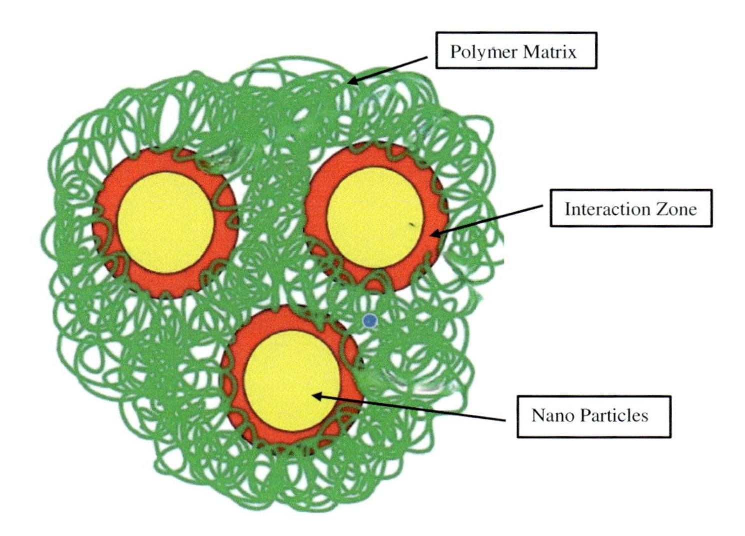

Fig. 9.3 Structure of bio nano composite

transcends into an artistic endeavor, molding the attributes of the bionanocomposite in a manner that redefines the realm of potential [29]. Within this intricate fabric of assimilation, the methods utilized resemble captivating chapters in a mesmerizing narrative. Solution blending involves the combination of nanocellulose and polymer in a fluidic manner [30]. Melt compounding refers to the fusion of these materials in a molten state. In-situ polymerization takes place within the matrix being created. Every approach employed contributes to the uniform distribution and strong bonding between nanocellulose and polymer, resulting in a distinct narrative where each detail holds significance [31]. The selected synthesis technique plays a crucial role in determining the final properties of the cellulose-based bionanocomposite, thereby molding its overall characteristics. The pursuit of optimal dispersion and preservation of the particular properties of nanocellulose is a deliberate choice. The production process evolves into a refined craftsmanship, a laborious voyage that converts cellulose and nanocomposites into a harmonious combination, prepared to leave a lasting impression on the world [32]. As we continue to explore the use of cellulose-based bionanocomposites, the focus turns to nanocellulose and its significant impact on several industries [33]. Imagine nanocellulose as the central figure in a Renaissance, fundamentally transforming the fields of electronics, medicine, and packaging through its distinct characteristics. Nanocellulose plays a prominent role in the field of electronics, serving as the foundation for pliable and lightweight conductive materials. Envision wearable gadgets that effortlessly adapt to the body, utilizing the mechanical properties of nanocellulose for power. This is more than just a progression in technology; it signifies a complete transformation in our interaction with electronic devices, opening up possibilities for a new era characterized by adaptability and conductivity [33].

Nanocellulose, derived from cellulose, demonstrates its potential in medicine through applications such as drug delivery systems, tissue engineering, and bioimaging [34]. The beneficial characteristics of biocompatibility and sustained release, which are inherent to nanocellulose, serve as the fundamental components of ground-breaking medical advancements. Envision customized drug delivery systems and engineered tissues that harness the potential of nanocellulose, offering revolutionary advancements in health care [35]. Cellulose-based bionanocomposites in the packaging industry redefine conventional norms by providing enhanced barrier properties against gases and moisture. What is the outcome? Extended durability of packaged goods, minimizing wastage and guaranteeing that products are delivered to consumers in their prime state. It is not solely about the way items are packaged, but rather the coming together of sustainability and efficiency to fundamentally transform how we manage and use products [36].

9.6.2 Overcoming Difficulties, Planning Sustainable Futures

As the symphony of cellulose-based bionanocomposites reaches its peak, problems arise as a contrasting element, requiring careful consideration and creative solutions. The challenges to overcome include achieving homogeneous dispersion of nanocellulose, controlling its alignment within the polymer matrix, and mitigating potential aggregation issues [37]. The synthesis process presents itself as a perplexing puzzle that requires a diligent pursuit of improvement. Researchers explore the complexities of synthesis procedures, aiming to discover new dimensions that overcome the problems presented. This is not merely a scientific undertaking; it is an expedition that aims to expand the limits of what can be accomplished using cellulose-based bionanocomposites [38]. After overcoming the obstacles, there are unknown regions that are ready to be discovered. The utilization of cellulose-based bionanocomposites extends to the fields of nanomedicine and electronics, providing new opportunities for creative advancements. The synergy of scientists, engineers, and dreamers plays a pivotal role in understanding the possible applications in various sectors, leading to a future when cellulose-based materials become essential for revolutionary breakthroughs [39]. The narrative of cellulose-based bionanocomposites does not conclude with their synthesis and usage; it encompasses the ecological ramifications that occur during their entire lifecycle. Researchers and environmental experts collaborate to understand the impact caused by these materials, guaranteeing that the progress made in innovation is in line with sustainability objectives. It is important to prioritize responsible progress that aligns with the well-being of the planet, rather than focusing solely on progress itself [40].

9.7 Introduction to PVCs

Dive deeper into the world of Polyvinyl Chloride (PVC), an extraordinary polymer that transcends conventional boundaries with its unparalleled adaptability and durability. PVC, or polyvinyl chloride, stands as a synthetic plastic marvel, uniquely capable of undergoing a transformative process—melted and reshaped to meet diverse needs. Its chemical resistance, electrical insulation properties, and resilience to harsh environmental conditions make PVC an indispensable material in various sectors, from construction to healthcare and packaging [41]. The evolution of PVC is a testament to its versatility. Initially employed in construction sites, where its robustness and enduring nature contribute significantly to pipelines, fittings, and overall building structures, PVC has expanded its footprint across industries. In health care, PVCs reliability take center stage as it is seamlessly integrated into critical applications such as medical tubing, blood bags, and pharmaceutical packaging, ensuring the secure and high-quality storage of essential medicinal resources. Meanwhile, the packaging industry embraces PVC for its barrier qualities, extending the longevity of products and minimizing environmental impact through reduced waste

[42]. PVC's influence extends even further with its increasing role as an effective electrical insulator. Picture a scenario where PVC acts as a protective shield against electrical currents, safeguarding both infrastructure and electronic equipment. Within the realm of wiring, cables, and electrical components, PVC serves as an underappreciated yet vital asset, facilitating the smooth transmission of electricity while prioritizing safety. Its contribution to the electrical industry not only underscores its practical utility but also exemplifies its versatility, transcending boundaries and finding applications in unexpected sectors. As we delve into the myriad facets of PVC, it becomes evident that its adaptability and durability are not confined to specific industries but rather manifest in a broad spectrum of applications. From construction and healthcare to packaging and electrical systems, PVC stands tall as a polymer that consistently delivers on performance, showcasing its ability to evolve and adapt to the ever-changing demands of diverse sectors [43].

9.7.1 Techniques for PVC/Cellulose Biocomposites Synthesis

Polyvinyl Chloride (PVC) and cellulose biocomposites are created using a variety of techniques designed to bring these two different materials together harmoniously. These synthesis methods are essential in establishing the ultimate characteristics and uses of the biocomposites [44]. We explore the three main synthesis techniques in this section: injection molding, extrusion, and blending.

9.7.1.1 Mixing

A basic method for combining PVC with cellulose is blending, which involves mechanical mixing procedures. This process is easy to use, economical, and scalable for large-scale manufacturing. Nonetheless, there are still issues with guaranteeing that cellulose is evenly distributed throughout the PVC matrix and preventing any possible deterioration during processing [45].

9.7.1.2 The Process of Extrusion

To achieve a continuous profile, extrusion involves melting the PVC and cellulose combination and pushing it through a die. This process can be used to create a variety of forms, including films, sheets, and profiles, because it gives exact control over the final product's dimensions and shape [46]. However, the high processing temperatures could affect the characteristics of cellulose, making it difficult to achieve uniform distribution, particularly in intricate geometries.

9.7.1.3 Using Injection Molding

In order to create complex three-dimensional structures, the molten PVC/cellulose blend is injected into a mold during the injection molding process. Due to its high production rates and adaptability in terms of product design, this approach is preferred. However, controlling possible deterioration during high-temperature processing and attaining homogeneous cellulose dispersion within the matrix present difficulties [47] (Table 9.1).

9.7.2 Methods for Characterizing PVC/Cellulose Composites

In order to fully understand the complex characteristics and structural subtleties of PVC and cellulose biocomposites, characterization procedures are essential. Researchers can learn a great deal about the composition, morphology, and mechanical behavior of the biocomposite materials by using these techniques, which range from spectroscopy to microscopy and mechanical testing [48]. Three important characterization techniques—spectroscopy, microscopy and imaging, and mechanical testing—are discussed in this section.

Table 9.1 Synthesis methods of PVC-based biocomposites

Synthesis method	Description	Advantage	Challenges	References
Blending	Using mechanical mixing techniques, cellulose and PVC are combined to create a homogenous slurry	Easy to use, economical, and scalable for manufacturing	Cellulose dispersion uniformity and possible processing degradation	Perera et al. [45]
Extrusion	To form a continuous profile, the PVC and cellulose combination is melted and pressed through a die	Exact control over form, appropriate for a variety of types	Elevated processing temperatures can impact the characteristics of cellulose and result in an uneven distribution	Biswal et al. [46]
Injection modeling	Into a mold is poured a molten PVC/cellulose combination to create complex three-dimensional forms	High rates of production and design flexibility	Uniform dispersion of cellulose and possible breakdown at high processing temperatures	Ojok et al. [47]

9.7.2.1 Using Spectroscopy

PVC/cellulose biocomposites' molecular structures and chemical bonding are examined using spectroscopic methods like nuclear magnetic resonance (NMR) and Fourier-transform infrared spectroscopy (FTIR). These methods provide insightful information on the materials' chemical makeup and intermolecular interactions [49]. However, there are restrictions for capturing nanoscale features, and their application necessitates a certain amount of skill.

9.7.2.2 Imaging and Microscopy

A closer look at the structure and surface morphology of PVC/cellulose biocomposites can be obtained by microscopic and imaging techniques such as atomic force microscopy (AFM) and scanning electron microscopy (SEM). These techniques provide the capacity to view micro and nanostructural characteristics, which facilitates the evaluation of dispersion in the materials. The impact of sample preparation on outcomes and the restriction to surface observations are challenges in this field [50].

9.7.2.3 Testing Mechanically

Tensile, flexural, and impact tests are examples of mechanical testing that are essential for assessing the mechanical characteristics of PVC/cellulose biocomposites. These tests provide vital information about the performance of the material by quantifying important characteristics like strength, flexibility, and toughness [51]. Sample variability and the impact of processing conditions on test results are challenges in this field (Table 9.2).

9.7.3 The Environmental Impact and Biodegradability of PVC/cellulose Biocomposites

The evaluation of biodegradability is a crucial component in comprehending the environmental impact of PVC/cellulose biocomposites. This procedure entails a thorough investigation into these biocomposites' capacity for biological breakdown aided by microbes. The benefits of these evaluations stem from their ability to analyze the materials' environmental friendliness and possibilities for impact reduction [52]. Examining the biodegradation properties gives scientists and industry practitioners important information about how to dispose of PVC/cellulose biocomposites sustainably, which may help to lessen the long-term environmental effects of conventional plastics. However, this approach is made more difficult by the difficulties that come

Table 9.2 Characterization techniques

Characterization technique	Description	Advantages	Challenges	References
Spectroscopy	PVC/cellulose biocomposites' molecular structures and chemical bonding are examined using methods such as FTIR and NMR	Provide information about intermolecular interactions and chemical makeup	Needs skill and might miss nanoscale details	Ghozali et al. [49]
Imaging & microscopy	At different scales, surface morphology and structure are investigated using atomic force microscopy (AFM) and scanning electron microscopy (SEM)	Evaluates dispersion and displays micro and nanostructural characteristics	Effects of sample preparation on outcomes are restricted to surface observations	Awad et al. [50]
Testing mechanically	Tensile, flexural, and impact tests are used to assess the PVC/cellulose biocomposites' mechanical characteristics	Measures toughness, flexibility, and strength	Sample fluctuation and the impact of processing parameters	Zuwanna et al. [51]

with doing biodegradability assessments. The kind of cellulose utilized, the particular processing conditions used during the production process, and the many end-use settings these biocomposites may be exposed to are some of the elements that greatly affect how effective biodegradation is [53]. A sophisticated grasp of the ways in which these variables interact and affect the biodegradation process is necessary to navigate this unpredictability. The dynamic nature of biocomposite materials presents challenges for researchers, who must strike a balance between the pursuit of biodegradability and other important material qualities. Life Cycle Assessment (LCA) is required for a comprehensive analysis of the environmental impact of PVC/cellulose biocomposites, going beyond biodegradability evaluations. From the extraction of raw materials to the disposal of the biocomposites at the end of their useful lives, this analytical framework offers a thorough understanding of the environmental sustainability of the materials [54]. The benefits of using life cycle assessment (LCA) stem from its capacity to take into account the wider environmental consequences linked to every phase of the biocomposite's existence, offering a more comprehensive knowledge than isolated evaluations.

However, there are certain difficulties in putting LCA into practice. Precise data gathering is a major challenge, especially when it comes to the environmental effects of manufacturing processes, end-of-life situations, and raw material extraction. Careful thought must be given to establishing suitable system boundaries, guaranteeing data veracity, and interpreting intricate environmental relationships. These obstacles must be overcome by LCA researchers and practitioners in order to draw insightful conclusions about the overall environmental sustainability of PVC/

cellulose biocomposites. To put it simply, the study of biodegradability and its effects on the environment goes beyond the lab and involves a dynamic interaction between industrial processes, environmental responsibility, and scientific examination. Untangling the intricacies of PVC/cellulose biocomposites' biodegradability and life cycle evaluation is crucial to promoting a more ecologically conscious future as the need for sustainable materials grows [55].

9.7.4 The Combination of Polyvinyl Chloride (PVC) and Cellulose: A Sustainable Partnership in Biocomposites

As we explore the various aspects of PVC's adaptability, an intriguing development arises when PVC forms a collaboration with cellulose, resulting in a relationship that enhances the sustainability of the polymer. The integration of biocomposites into the field expands possibilities for collaborative results that redefine the standards for PVC-based products. Experience the emergence of the biocomposite revolution, as PVC and cellulose combine to create a sustainable tapestry [39]. Cellulose, obtained from sustainable sources, introduces environmental sustainability into PVC, converting it into a material that not only endures over time but also aligns with the worldwide movement toward eco-friendly solutions. The combination of PVC with cellulose forms a biocomposite that synergistically combines the advantages of both materials, providing a route towards environmentally friendly and more sustainable alternatives. The combination of PVC and cellulose in biocomposites involves more than just the coming together of components; it entails a complex interplay of processing procedures [40]. Envision the meticulousness of shaping sustainability through the techniques of solution mixing, melt compounding, and in situ polymerization, which serve as the skilled craftsmen creating the biocomposite masterpiece. These approaches, which are intended to guarantee uniform dispersion and strong adhesion, serve as the means to mold the future of PVC-based products, combining flexibility with environmental awareness [41]. The combination of PVC and cellulose in biocomposites not only ensures sustainability but also enables a wide range of applications in many industries. Imagine a world where PVC, along with the fundamental qualities of cellulose, becomes a powerful catalyst in various domains. The combination of PVC and cellulose in construction is a significant advantage, offering environmentally friendly options for pipes, fittings, and structural elements [42]. The long-lasting nature of PVC combines effectively with the renewable properties of cellulose, offering structures that not only endure over time but also contribute to a more environmentally friendly construction industry. A revolution is occurring in the packaging sector as PVC and cellulose combine to form biocomposites that have enhanced barrier qualities. Packaging materials that effectively control gases and moisture help to prolong the shelf life of products, hence minimizing waste and promoting a more environmentally conscious packaging strategy. PVC, once seen as a standard option, now represents an advancement in environmentally conscious packaging options [43].

9.7.5 Challenges and Potential of PVC/Cellulose Biocomposites

While we admire the successful collaboration between PVC and cellulose in biocomposites, it is important to acknowledge the obstacles encountered along the way. This section delves into the challenges encountered and the optimistic opportunities that await, as the pursuit of sustainable materials ventures into unfamiliar domains. An obstacle encountered in the manufacturing of PVC/cellulose biocomposites is the attainment of homogeneous dispersion. The interaction between PVC and cellulose, although harmonic, requires meticulousness to guarantee that each component fulfills its duty flawlessly [56]. Researchers are undertaking a search for novel processing procedures in order to find the optimal rhythm that ensures excellent dispersion and maintains the distinct properties of each material. The impediments transform into stepping stones when the synthesis process assumes a prominent position [57].

Researchers, engineers, and material scientists work together to uncover the mysteries of cutting-edge synthesis methods. The objective is not only to surmount existing obstacles but also to establish the groundwork for future progress in the field of PVC/cellulose biocomposites. The interaction between these elements undergoes a dynamic process of development, where each synthesis approach contributes to the enhancement of the biocomposite storyline [58]. Amidst the difficulties, opportunities arise on the horizon, offering potential for novel uses and progress. The application of PVC/cellulose biocomposites extends beyond the fields of construction, healthcare, and packaging, and also explores uncharted areas. The synergy of these materials serves as a catalyst for innovation in the fields of electronics, automobiles, and beyond. Researchers are investigating the untapped potential of PVC/cellulose biocomposites, which are emerging as pioneers in crafting a sustainable future [59].

9.8 Applications and Future Prospects of PVC/Cellulose Biocomposites

The combination of PVC and cellulose not only enhances their performance, but also enables their utilization in several sectors, hence creating opportunities for environmentally-friendly solutions. The PVC biocomposite canvas serves as a platform for innovative architects, who imagine a future in which PVC and cellulose revolutionize several industries. PVC/cellulose biocomposites are becoming increasingly prominent in the field of construction, as they offer a sustainable solution for building structures. The durability of PVC aligns with the renewable nature of cellulose, offering enduring and environmentally friendly construction materials [60]. Envision a cityscape where buildings serve as symbols of both exceptional engineering and ecological conscientiousness. PVC/cellulose composites are becoming increasingly prominent in the automotive industry, leading to a significant shift towards environmentally friendly practices. These biocomposites, which are both lightweight and sustainable, play a crucial role in enhancing fuel efficiency and minimizing environmental harm. The utilization of PVC and cellulose in cars serves as a representation of sustainability, so facilitating a future in which transportation harmoniously integrates with environmental awareness [61].

The consumer products industry, which was formerly associated with environmental issues, is undergoing a transformation with the introduction of PVC/cellulose biocomposites, which provide sustainable alternatives. Envision common commodities, ranging from domestic articles to technology devices, imbued with the environmentally friendly properties of cellulose. The consumer goods industry is evolving to include sustainability as an intrinsic characteristic, rather than just an optional element. The development of PVC/cellulose biocomposites presents various hurdles, nevertheless, each difficulty serves as a catalyst for progress toward a sustainable future [62]. As researchers and innovators plan for the future, they view difficulties as opportunities and see limitless possibilities on the horizon. The difficulty lies in achieving optimal synergy between PVC and cellulose. Researchers undertake a pursuit for compatibility, aiming to find the optimal equilibrium that allows both materials to function together, utilizing their advantages without any concessions. This endeavor is not solely focused on surmounting obstacles; rather, it revolves around establishing a mutually beneficial connection that shapes the destiny of PVC-based materials [63]. The challenge of intricate production processes presents itself as an enigma that demands a solution. Researchers give priority to the creation of production techniques that are both cost-effective and capable of being expanded for use in industrial manufacturing. The objective is not only to produce sustainable materials within the laboratory, but also to release them into the global market, thereby generating a significant influence on various businesses and customers. The pursuit of sustainable waste management becomes a crucial factor as PVC/cellulose biocomposites gain prominence. Scientists explore the complexities of end-of-life considerations, guaranteeing that the advantages of these sophisticated biocomposites go beyond their useful duration [64]. The objective is not merely to generate things, but to establish a lasting heritage of conscientious invention. Ultimately, the development of PVC/cellulose biocomposites represents a significant evolution, as PVC, a resilient polymer, incorporates the environmentally friendly properties of cellulose to produce materials that revolutionize various industries. Participate in this story where obstacles are viewed as chances for growth, and the future of PVC-based products is not only environmentally friendly but also filled with dynamic advancements. The canvas is continuously developing, with each brushstroke of cellulose-infused PVC portraying a vision of a future that is more environmentally friendly, durable, and sustainable.

9.9 PVC/Cellulose Biocomposites Market Trends

Gaining knowledge of PVC/cellulose biocomposites' market trends is essential for assessing their prospective applications, growth, and customer preferences. Currently, a wide range of industries, including consumer products, automotive, construction, and packaging, are using these biocomposites. PVC/cellulose biocomposites are becoming more and more popular among automakers as component suppliers because of their exceptional strength and sustainability [65]. Similarly, these biocomposites are used in the building industry to create materials like panels and profiles, which helps the sector in its search for environmentally friendly substitutes. Investigating expanding industries reveals encouraging prospects for PVC/cellulose biocomposites as market dynamics change. Novel fields like 3D printing offer a frontier where these biocomposites' adaptability and sustainability can be used for complex, bespoke manufacturing. Another emerging market is the medical device industry, where the customized characteristics and biocompatibility of biocomposites make them more appealing for a range of applications, including implants and prosthetics. In addition, the incorporation of PVC/cellulose biocomposites into electronic components demonstrates their versatility and potential to raise the bar for innovation and technology [66]. Customer preferences are a major influence on market trends, and the adoption of PVC/cellulose biocomposites is fueled by the need for eco-friendly and sustainable materials. Consumer behavior has changed as a result of the conversation surrounding environmental consciousness, prompting industries to give preference to products made using sustainable methods [67]. PVC/cellulose biocomposites are well-positioned to satisfy the changing demands of environmentally conscious consumers due to their lower environmental impact and potential for biodegradation.

One of the main factors advancing market trends is innovation. Developments and partnerships in the field of PVC/cellulose biocomposites have spurred revolutionary shifts. These innovations frequently entail cutting-edge material compositions, creative manufacturing techniques, and business alliances that support market growth [68]. Research institutes, businesses, and environmental groups working together highlight how dynamic the area is and open up new possibilities for increased market acceptability and innovative uses. In summary, the PVC/cellulose biocomposites market trends show how rapidly evolving industries, customer preferences, and new developments interact with existing applications. PVC/cellulose biocomposites have a potential future ahead of them, as companies continue to look for sustainable alternatives. They will be able to make a significant contribution to a variety of industries while also being in line with the larger social shift towards environmental responsibility.

9.10 Strategies for Improving PVC/Cellulose Biocomposites

The optimization of PVC/cellulose biocomposites' mechanical characteristics, processability, and overall efficacy can be achieved through the selective integration of reinforcements and additives.

9.10.1 Supplements

Adding additives, including compatibilizers and plasticizers, appears to be a key tactic for improving the properties of PVC/cellulose mixtures. Plasticizers boost the biocomposites' flexibility, which makes them more malleable and simpler to handle [69]. Compatibilizers, on the other hand, tackle the difficulty of mixing two dissimilar materials, improving their compatibility and encouraging a more homogenous mixture. The benefits of adding additives go beyond processability improvements and include improvements in mechanical qualities.

9.10.2 Supports

A clever tactic to modify the characteristics of PVC/cellulose biocomposites is the use of reinforcing elements, like fibers or nanoparticles. By enhancing particular properties like strength and stiffness, these reinforcements help to improve the biocomposite material's overall mechanical performance [70]. Because of their minuscule size, nanoparticles have special chances for nanoscale reinforcement, which increases strength and durability. To fully utilize their potential, it is necessary to achieve a uniform dispersion of these reinforcements throughout the matrix; nevertheless, this presents obstacles. Alongside the pursuit of enhanced mechanical performance are potential effects on biodegradability that need to be taken into account [71]. In this enhancement technique, maintaining environmentally friendly features while maximizing the efficiency of reinforcement becomes central to the strategy [72]. To conclude, the procedures employed for enhancement have a significant impact on the characteristics and uses of PVC/cellulose biocomposites. The dynamic character of material engineering is highlighted by the thoughtful addition of reinforcements and additives, which provide a way to maximize the advantages of these sustainable composite materials while addressing obstacles. It takes a careful and nuanced approach to achieve a harmonious blend that maximizes both mechanical performance and environmental impact, highlighting the significance of research and development in the field of PVC/cellulose biocomposites.

9.11 Conclusion

The review chapter explores the relationship between Polyvinyl Chloride (PVC) and cellulose, focusing on their role in creating biocomposites, which are materials that have the potential to contribute to sustainable innovation. The amalgamation of these elements beyond mere collaboration; epitomizes a dedication to sustainability, reshaping the domain of polymer-based products. By incorporating cellulose into PVC matrices, a new era of mechanical resilience and thermal endurance is achieved. Additionally, the integration of nanocellulose at the nanoscale enhances PVC matrices with exceptional strength and barrier characteristics. Each section of this review chapter contributes to the development of eco-conscious materials. The research explores the potential applications and prospects of PVC/cellulose biocomposites, revealing a future where sustainability is embraced in industries such as building, automotive, and consumer goods. The compatibility, manufacturing complexity, and end-of-life considerations present significant hurdles that serve as opportunities for a more sustainable future. Researchers, engineers, and innovators are actively working on finding ways to efficiently and affordably create and apply new synthesis methods and applications. This review chapter provides a comprehensive assessment, charting the progress of PVC/cellulose biocomposites toward a future characterized by the seamless integration of polymer innovation and ecological responsibility. This promises a more dynamic and sustainable future for both industry and consumers.

References

1. Al Mahmud MZ (2023) Exploring the versatile applications of biocomposites in the medical field. Bioprinting 36:e00319. https://doi.org/10.1016/j.bprint.2023.e00319
2. Aaliya B et al (2024) Fabrication and characterization of talipot starch-based biocomposite film using mucilages from different plant sources: a comparative study. Food Chem 438:138011. https://doi.org/10.1016/j.foodchem.2023.138011
3. Thanapornsin W et al (2024) Preparation and characterization of biocomposite film made of activated carbon derived from microalgal biomass: an experimental design approach for basic yellow 1 removal. S Afr J Chem Eng 47:178–196. https://doi.org/10.1016/j.sajce.2023.11.009
4. Salama A (2023) Novel cellulose derivative containing aminophenylacetic acid as sustainable adsorbent for removal of cationic and anionic dyes. Int J Biol Macromol 253:126687. https://doi.org/10.1016/j.ijbiomac.2023.126687
5. Zhu Y et al (2024) Regenerated cellulose hydrogel with excellent mechanical properties for flexible sensors. Ind Crop Prod 210:118026. https://doi.org/10.1016/j.indcrop.2024.118026
6. Prakash S et al (2024) Unlocking the potential of cotton stalk as a renewable source of cellulose: a review on advancements and emerging applications. Int J Biol Macromol 129456. https://doi.org/10.1016/j.ijbiomac.2024.129456
7. Wang C et al (2023) A sustainable strategy to transform cotton waste into renewable cellulose fiber self-reinforcing composite paper. J Clean Prod 429:139567. https://doi.org/10.1016/j.jclepro.2023.139567

8. Vishnupriya C et al (2024) Evaluation of heavy metal removal and antibiofilm efficiency of biologically synthesized chitosan- silver nano-bio composite by a soil actinobacterium Glutamicibacter uratoxydans VRAK 24. Int J Biol Macromol 255:128032. https://doi.org/10.1016/j.ijbiomac.2023.128032
9. Naeimi A et al (2022) First and efficient bio-nano composite, SnO_2/Calcite based on cypress leaves and eggshell wastes, for cytotoxic effects on HepG2 liver cancer cell lines and its antioxidant and antimicrobial activity. J Mol Struct 1259:132690. https://doi.org/10.1016/j.molstruc.2022.132690
10. Kumar CMP et al (2022) Natural nano-fillers materials for the bio-composites: a review. J Indian Chem Soc 99(10):100715. https://doi.org/10.1016/j.jics.2022.100715
11. Kasbaji M et al (2023) Future trends in dye removal by metal oxides and their nano/composites: a comprehensive review. Inorg Chem Commun 158:111546. https://doi.org/10.1016/j.inoche.2023.111546
12. Shu R et al (2020) Recent researches of the bio-inspired nano-carbon reinforced metal matrix composites. Compos Part A: Appl Sci Manuf 131:105816. https://doi.org/10.1016/j.compositesa.2020.105816
13. Verma J, Petru M, Goel S (2024) Cellulose based materials to accelerate the transition towards sustainability. Ind Crop Prod 210:118078. https://doi.org/10.1016/j.indcrop.2024.118078
14. Dacrory S et al (2023) Evaluation of biocompatible amino acid-functionalized cellulose composites: characterizations, molecular modeling, anticoagulant activity, and cytocompatibility. Bioact Carbohydr Diet Fibre 30:100372. https://doi.org/10.1016/j.bcdf.2023.100372
15. Brahma R, Ray S (2024) Optimization of extraction conditions for cellulose from jackfruit peel using RSM, its characterization and comparative studies to commercial cellulose. Meas: Food 13:100130. https://doi.org/10.1016/j.meafoo.2023.100130
16. Li X et al (2024) Promising cellulose–based functional gels for advanced biomedical applications: a review. Int J Biol Macromol 260:129600. https://doi.org/10.1016/j.ijbiomac.2024.129600
17. Das O et al (2023) Functionalised biochar in biocomposites: the effect of fire retardants, bioplastics and processing methods. Compos Part C: Open Access 11:100368. https://doi.org/10.1016/j.jcomc.2023.100368
18. Foret S et al (2023) Thermoplastic starch biocomposites reinforced with hemp shives obtained via extrusion. Ind Crop Prod 206:117707. https://doi.org/10.1016/j.indcrop.2023.117707
19. Srivastava V, Singh S, Das D (2024) Development and characterization of peppermint essential oil/rice husk fibre/corn starch active biocomposite film and its performance on bread preservation. Ind Crop Prod 208:117765. https://doi.org/10.1016/j.indcrop.2023.117765
20. Hasan K et al (2021) Jute and luffa fiber-reinforced biocomposites: effects of sample thickness and fiber/resin ratio on sound absorption and transmission loss performance. J Nat Fibers. https://doi.org/10.1080/15440478.2021.1907832
21. Joseph A et al (2024) Crashworthiness of biocomposites in automotive applications. In: Biocomposites for industrial applications. Elsevier, pp 169–194. https://doi.org/10.1016/B978-0-323-91866-4.00003-2
22. Parthasarathy V et al (2024) Physicochemical and biological properties of the biocomposite in the dental applications. In: Biocomposites for industrial applications. Elsevier, pp 113–124. https://doi.org/10.1016/B978-0-323-91866-4.00015-9
23. Okwuwa CC et al (2023) Cellulose dissolution for edible biocomposites in deep eutectic solvents: a review. J Clean Prod 427:139166. https://doi.org/10.1016/j.jclepro.2023.139166
24. Ruz-Cruz MA et al (2022) Thermal and mechanical properties of PLA-based multiscale cellulosic biocomposites. J Mater Res Technol 18:485–495. https://doi.org/10.1016/j.jmrt.2022.02.072
25. Lv Q et al (2024) Nanocellulose-based nanogenerators for sensor applications: a review. Int J Biol Macromol 259:129268. https://doi.org/10.1016/j.ijbiomac.2024.129268
26. Shen T, Dong H, Wang P (2024) Research progress on nanocellulose and its composite materials as orthopedic implant biomaterials. Alex Eng J 87:575–590. https://doi.org/10.1016/j.aej.2024.01.003

27. Jaouahar M et al (2024) Preparation and characterization of sulfated nanocellulose: from hydrogels to highly transparent films. Int J Biol Macromol 260:129464. https://doi.org/10.1016/j.ijbiomac.2024.129464
28. Palanisamy S et al (2024) Unlocking sustainable solutions: nanocellulose innovations for enhancing the shelf life of fruits and vegetables–a comprehensive review. Int J Biol Macromol 129592. https://doi.org/10.1016/j.ijbiomac.2024.129592
29. Agbakoba VC et al (2023) Preparation of cellulose nanocrystal (CNCs) reinforced polylactic acid (PLA) bionanocomposites filaments using biobased additives for 3D printing applications. Nanoscale Adv 5(17), 4447–4463. https://doi.org/10.1039/D3NA00281K
30. Das AK et al (2023) Cellulose-based bionanocomposites in energy storage applications-a review. Heliyon 9(1):e13028. https://doi.org/10.1016/j.heliyon.2023.e13028
31. Jeevanandam J et al (2024) Cellulose-based bionanocomposites: synthesis, properties, and applications. In: Advances in Bionanocomposites. Elsevier, pp 191–210. https://doi.org/10.1016/B978-0-323-91764-3.00011-5
32. Mahmoud ME, El-Sharkawy RM, Ibrahim GAA (2022) A novel bionanocomposite from doped lipase enzyme into magnetic graphene oxide-immobilized-cellulose for efficient removal of methylene blue and malachite green dyes. J Mol Liq 368 120676. https://doi.org/10.1016/j.molliq.2022.120676
33. Leite LSF et al (2020) Scaled-up production of gelatin-cellulose nanocrystal bionanocomposite films by continuous casting. Carbohydr Polym 238:116198. https://doi.org/10.1016/j.carbpol.2020.116198
34. Chaturvedi S et al (2023) Bionanocomposites reinforced with cellulose fibers and agro-industrial wastes. In: Cellulose fibre reinforced composites. Elsevier, pp 317–342. https://doi.org/10.1016/B978-0-323-90125-3.00017-3
35. Zubair M, Wu J, Ullah A (2022) Bionanocomposites from spent hen proteins reinforced with polyhedral oligomeric silsesquioxane (POSS)/cellulose nanocrystals (CNCs). Biocatal Agric Biotechnol 43:102434. https://doi.org/10.1016/j.bcab.2022.102434
36. Khairnar Y et al (2021) Cellulose bionanocomposites for sustainable planet and people: a global snapshot of preparation, properties, and applications. Carbohydr Polym Technol Appl 2:100065. https://doi.org/10.1016/j.carpta.2021.100065
37. Ali A et al (2023) Polysaccharides and proteins based bionanocomposites as smart packaging materials: from fabrication to food packaging applications a review. Int J Biol Macromol 252:126534. https://doi.org/10.1016/j.ijbiomac.2023.126534
38. Palechor-Trochez JJ et al (2021) A review of trends in the development of bionanocomposites from lignocellulosic and polyacids biomolecules as packing material making alternative: a bibliometric analysis. Int J Biol Macromol 192:832–868. https://doi.org/10.1016/j.ijbiomac.2021.10.003
39. Jayakumar A et al (2022) Recent innovations in bionanocomposites-based food packaging films–a comprehensive review. Food Packag Shelf Life 33:100877. https://doi.org/10.1016/j.fpsl.2022.100877
40. Ghosal K et al (2024) Polyvinyl alcohol-based bionanocomposites: synthesis, properties, and applications. In: Advances in bionanocomposites. Elsevier, pp 117–132. https://doi.org/10.1016/B978-0-323-91764-3.00010-3
41. Mansouri I et al (2024) Optoelectric characteristics of PVC/SrxZnO nanocomposite films. Opt Mater 147:114693. https://doi.org/10.1016/j.optmat.2023.114693
42. Wu Y, He J, Xiong L (2024) Price discovery in Chinese PVC futures and spot markets: impacts of COVID-19 and benchmark analysis. Heliyon 10(2):e24138. https://doi.org/10.1016/j.heliyon.2024.e24138
43. Shapouri L et al (2024) Preparation, characterization, and fouling analysis of PVC/ND-PEG ultrafiltration membranes for whey separation. Diam Relat Mater 142:110776. https://doi.org/10.1016/j.diamond.2023.110776
44. Khodayari J et al (2024) Synthesis of eco-friendly carboxymethyl cellulose/metal–organic framework biocomposite and its photocatalytic activity. J Photochem Photobiol A: Chem, 446:115097. https://doi.org/10.1016/j.jphotochem.2023.115097

45. Perera UP, Foo ML, Chew IML (2023) Synthesis and characterization of lignin nanoparticles isolated from oil palm empty fruit bunch and application in biocomposites. Sustain Chem Clim Action 2:100011. https://doi.org/10.1016/j.scca.2022.100011
46. Biswal T, Rekha Sahoo D, Acharya SK (2023) Synthesis and study of mechanical properties of polypropylene (PP)/banana nano-filler biocomposites. Mater Today: Proc 74:726–729. https://doi.org/10.1016/j.matpr.2022.10.299
47. Ojok W et al (2022) Synthesis and characterization of hematite biocomposite using cassava starch template for aqueous phase removal of fluoride. Carbohydr Polym Technol Appl 4:100241. https://doi.org/10.1016/j.carpta.2022.100241
48. Zhang W et al (2024) Characterization of tool-ply friction behavior for treated jute/PLA biocomposite prepregs in thermoforming. Compos Part A: Appl Sci Manuf 107875. https://doi.org/10.1016/j.compositesa.2023.107875
49. Ghozali M et al (2024) Preparation and characterization of Arenga pinnata thermoplastic starch/bacterial cellulose nanofiber biocomposites via in-situ twin screw extrusion. Int J Biol Macromol 261:129792. https://doi.org/10.1016/j.ijbiomac.2024.129792
50. Awad S et al (2023) Evaluation of characterisation efficiency of natural fibre-reinforced polylactic acid biocomposites for 3D printing applications. Sustain Mater Technol 36:e00620. https://doi.org/10.1016/j.susmat.2023.e00620
51. Zuwanna I et al (2023) Preparation and characterization of silica from rice husk ash as a reinforcing agent in whey protein isolate biocomposites film. South African J Chem Eng 44:337–343. https://doi.org/10.1016/j.sajce.2023.03.005
52. Safaripour M et al (2021) Environmental impact tradeoff considerations for wheat bran-based biocomposite. Sci Total Environ 781:146588. https://doi.org/10.1016/j.scitotenv.2021.146588
53. Abdellaoui H (2024) Durability evaluation of biocomposites by aging: hydrothermal and environmental factors. In: Biocomposites-bio-based fibers and polymers from renewable resources. Elsevier, 183–206. https://doi.org/10.1016/B978-0-323-97282-6.00017-7
54. Zheng G et al (2023) Recent advances in functional utilisation of environmentally friendly and recyclable high-performance green biocomposites: a review. Chin Chem Lett 108817. https://doi.org/10.1016/j.cclet.2023.108817
55. Bushra A, Subhani A, Islam N (2023) A comprehensive review on biological and environmental applications of chitosan-hydroxyapatite biocomposites. Compos Part C: Open Access 12:100402. https://doi.org/10.1016/j.jcomc.2023.100402
56. Chen Y et al (2024) High-performance self-bonding bio-composites from wood fibers. Ind Crop Prod 209:117944. https://doi.org/10.1016/j.indcrop.2023.117944
57. Song G et al (2024) Application of Hashin–Shtrikman bounds homogenization model for frequency analysis of imperfect FG bio-composite plates. J Mech Behav Biomed Mater 151:106321. https://doi.org/10.1016/j.jmbbm.2023.106321
58. Saada K et al (2024) Exploring tensile properties of bio composites reinforced date palm fibers using experimental and modelling approaches. Mater Chem Phys 314:128810. https://doi.org/10.1016/j.matchemphys.2023.128810
59. Sharma A et al (2023) Bio-inspired nacre and helicoidal composites: from structure to mechanical applications. Thin-Walled Struct 192:111146. https://doi.org/10.1016/j.tws.2023.111146
60. Xu G et al (2023) Evaluation of properties of bio-composite with interpretable machine learning approaches: optimization and hyper tuning. J Mater Res Technol 25:1421–1446. https://doi.org/10.1016/j.jmrt.2023.06.007
61. Sathish Kumar, RK et al (2023) Advancements in bio-polymeric composite materials for active and intelligent food packaging: a comprehensive review. Mater Today: Proc. https://doi.org/10.1016/j.matpr.2023.08.271
62. Bhayana, M et al (2023) A review on optimized FDM 3D printed Wood/PLA bio composite material characteristics. Mater Today: Proc [Preprint]. https://doi.org/10.1016/j.matpr.2023.03.029
63. Lawan I et al (2023) Development of cashew apple bagasse based bio-composites for high-performance applications with the concept of zero waste production. J Clean Prod 427:139270. https://doi.org/10.1016/j.jclepro.2023.139270

64. Zhu Z et al (2023) A formaldehyde-free bio-composite sheet used as adhesive with excellent water-wet bonding performance. Ind Crop Prod 198:116680. https://doi.org/10.1016/j.indcrop.2023.116680
65. Dhal MK et al (2023) Polylactic acid/polycaprolactone/sawdust based biocomposites trays with enhanced compostability. Int J Biol Macromol 253:126977. https://doi.org/10.1016/j.ijbiomac.2023.126977
66. Pesaranhajiabbas E, Misra M, Mohanty AK (2023) Recent progress on biodegradable polylactic acid based blends and their biocomposites: a comprehensive review. Int J Biol Macromol 253:126231. https://doi.org/10.1016/j.ijbiomac.2023.126231
67. Trivedi AK, Gupta MK, Singh H (2023) PLA based biocomposites for sustainable products: a review. Adv Ind Eng Polym Res 6(4):382–395. https://doi.org/10.1016/j.aiepr.2023.02.002
68. Enriquez-Medina I et al (2023) From purposeless residues to biocomposites: a hyphae made connection. Biotechnol Rep 39:e00807. https://doi.org/10.1016/j.btre.2023.e00807
69. Zhao X et al (2022) Impact of biomass ash content on biocomposite properties. Compos Part C: Open Access 9:100319. https://doi.org/10.1016/j.jcomc.2022.100319
70. Alavi SE et al (2023) Biocomposite-based strategies for dental bone regeneration. Oral Surg, Oral Med, Oral Pathol Oral Radiol 136(5):554–568. https://doi.org/10.1016/j.oooo.2023.04.015
71. Hong G et al (2022) Cleaner production strategy tailored versatile biocomposites for antibacterial application and electromagnetic interference shielding. J Clean Prod 366:132835. https://doi.org/10.1016/j.jclepro.2022.132835
72. Li S et al (2024) A new strategy for PEEK-based biocomposites to achieve porous surface for bioactivity and adjustable mechanical properties for orthopedic stress matching. Compos Part A: Appl Sci Manuf 177:107909. https://doi.org/10.1016/j.compositesa.2023.107909

Chapter 10
Taraxacum Koksaghyz Rodin as an Alternative Source of Natural Rubber and Inulin

Marina Arias and Neiker Tecnalia

10.1 Introduction

Polymers are natural or synthetic substances composed of very large molecules, called macromolecules that are multiples of simpler chemical units called monomers. Biopolymers are those produced by living organisms. Many materials in living organisms including proteins, polysaccharides, and nucleic acids, are polymers [1]. Two of the most important biopolymers are rubber (a polymer of hydrocarbon isoprene) and inulin (a mixture of oligo- and polysaccharides).

10.2 Rubber

Natural rubber is one of the most important polymers, for human life, produced by plants. It is used in thousands of products and hundreds of medical devices converting it into an essential raw material. It is contained in latex, an aqueous emulsion present in the laticiferous vessels (ducts) or parenchymal (single) cells of rubber-producing plants [2]. It is a biopolymer consisting of isoprene units (C5H8)n linked together in a 1,4 cis-configuration (Fig. 6.1).

An isoprene monomer in natural rubber (C5H8) has a molecular weight of 68 Da. “n” is approximately 18,000, taking as a reference the average molecular weight in *Hevea brasiliensis* (1,200 kDa) [3]. NR combines high strength with incredible resistance to fatigue. It has moderate resistance to environmental damage by heat, light and ozone which are one of its drawbacks. Natural rubber compounds are exceptional for their flexibility, good electrical insulation, low internal friction, and resistance to most inorganic acids, salts and alkali, even though it has poor resistance

M. Arias (✉) · N. Tecnalia
Campus Agroalimentario de Arkaute. Apto 46. E-01080, Vitoria-Gasteiz (Araba), Spain
e-mail: marinillaarias@gmail.com; marina_arias@hotmail.com

Visakh P. M. *Rubber Based Bionanocomposites*, Advanced Structured Materials 210,
https://doi.org/10.1007/978-981-10-2978-3_10

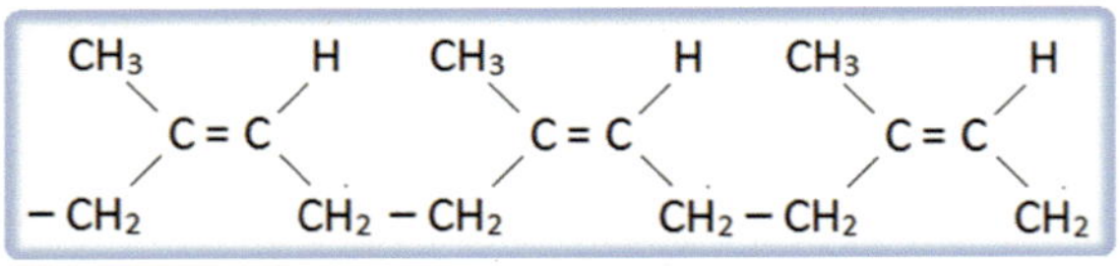

Fig. 10.1 Chemical structure of natural rubber. *Source* quimitube.com

to petroleum products such as oil, gasoline and Naphtha [4]. Natural rubber has a strategic importance, since after seven decades of industrial research it cannot be replaced by synthetic alternatives in most of its applications [5] (Fig. 10.1).

10.2.1 Rubber History

Ancient Mesoamerican peoples were already processing rubber by 1,600 B.C., as mentioned in "Prehistoric Polymers: Rubber Processing in Ancient Mesoamerica" [6], predating in this way the development of the vulcanization process around 3,500 years. As explained in their book, Mesoamericans used several items made of rubber in their daily life and liquid rubber for medicines, paintings and rituals. Maurice Morton recognized Christopher Columbus, in his book "Rubber Technology", as the first European to discover natural rubber. It was in the early 1490 s, when he travelled to Haiti and found natives playing with a ball made of an extract from a tree [7]. F. Juan de Torquemada wrote in 1615 about Indians and Spanish settlers wearing shoes, clothing and hats made by dipping cloth into latex. The use of latex was, little by little, expanded by white conquerors. But its use was still limited, since it became sticky with warm weather and hardened and cracked with cold weather. (Whaley, 1948). When Charles Marie de la Condamine went to South America one century later (1734), trying to solve whether the Earth was an elongated or flattened sphere, he landed just North of Guayaquil. He travelled from the West coast to Esmeraldas, on the East coast, passing over the Andes. Along his trip, he found different uses of rubber, like in torches and candles. But he wasn't a botanist and he couldn't appreciate the difference between the trees bearing rubber that he found on his trip (Hevea or Siphonia and Castilla trees).

A few years later rubber reached Europe. Several pairs of royal boots were sent to Para (Belem), in the North of Brazil, in order to be waterproofed. Para was the first Portuguese settlement. In 1759 its government presented a coat made of fabric covered with rubber!! [9]. Hevea succeeded with respect to the Castilla rubber tree because of the way their latex is transported along the tree and because of Ridley's discovery. In the Hevea tree, there are connected latex tubes forming a network; it allows the latex to flow continuously when it is extracted. In Castilla trees, there is not a connected system of tubes and latex doesn't flow continuously. On the other hand, Ridley discovered a method for harvesting Hevea plantations by continuous tapping, while Castilla trees can die by severe tapping. This discovery rapidly changed civilizations and social relations. Rubber became an essential material in the development of countries [10]. In 1839, the vulcanization process was invented by Charles Goodyear. He combined sulfur and heat hardening rubber and keeping its elasticity. Rubber changes when exposed to variations of temperatures disappeared. It wouldn't crack in the winter and melt in the summer anymore [11]. The pneumatic tier was invented by Dunlop in 1888, and it meant an important push for the international industry of rubber. It was first used for bicycles and in the 1890s it became essential for the emerging car industry. Dunlop's wheel boosted rubber as a raw material of worldwide strategic importance [12].

Rubber was also strategic for the Industrial Revolution. Rubber demand increased constantly from 1850 through 1912 in Industrial Europe and the United States. The search for high-quality rubber trees to be exploited was very important, the reason why businessmen pushed entrepreneurs and traders to penetrate Amazonian rainforests. From 1880 to 1920, Amazonia had the monopoly of rubber [12] and the developing industry growing in Europe and the EEUU and the non-stopping discovery of new applications of rubber, made its price and volume of extraction increase in a geometrical way in the world's markets [12]. Knowing its strategic importance, Brazil prohibited the export of rubber seeds or seedlings. But it was a very tantalizing treasure and H. A. Wickham smuggled 70.000 rubber seeds hidden in banana leaves and brought them to England in 1876. Out of the stolen seeds, 1900 seedlings germinated and survived and they were used to start the rubber plantations in Malaya late in the nineteenth century [13]. It was the beginning of the end of Brazil as the main rubber producer. It took only 12 years for the Eastern plantations to reach the peak of production in the Amazon and become the main world's natural rubber supplier. It was more competitive in price. From these days until the First World War, the rubber collection from wild sources of tropical America declined tremendously [8]. During the war, the borders were closed to commerce, and the supply of rubber was cut off. EEUU recognized, early in 1942, that they didn't have sufficient natural rubber to meet the demands, encouraging the search for alternative rubber sources either natural or synthetic [8]. After the War, the supplies of rubber by the Eastern plantations started again, when commerce was re-opened, and the effort to look for new rubber sources decreased tremendously. The price of Eastern rubber was much more competitive than the one obtained from new rubber sources.

10.2.2 Rubber Localization and Extraction Methods

Rubber is compartmentalized within sub-cellular rubber particles, located in the cytosol of cells, whether these be specialized laticifers (pipe-like anastomized cell systems that produce latex) (Fig. 10.2), as in *H. brasiliensis* [14, 15], *Taraxacum koksaghyz* R, [3] and *Ficus elastica* [15, 16], or generalized cells, such as the bark parenchyma of *P. argentatum* [15, 17] (Fig. 10.3).

Hevea latex is tapped by incision into the bark [18], Russian dandelion must be either pressed or homogenized to extract its latex [3] and guayule's tissue must be

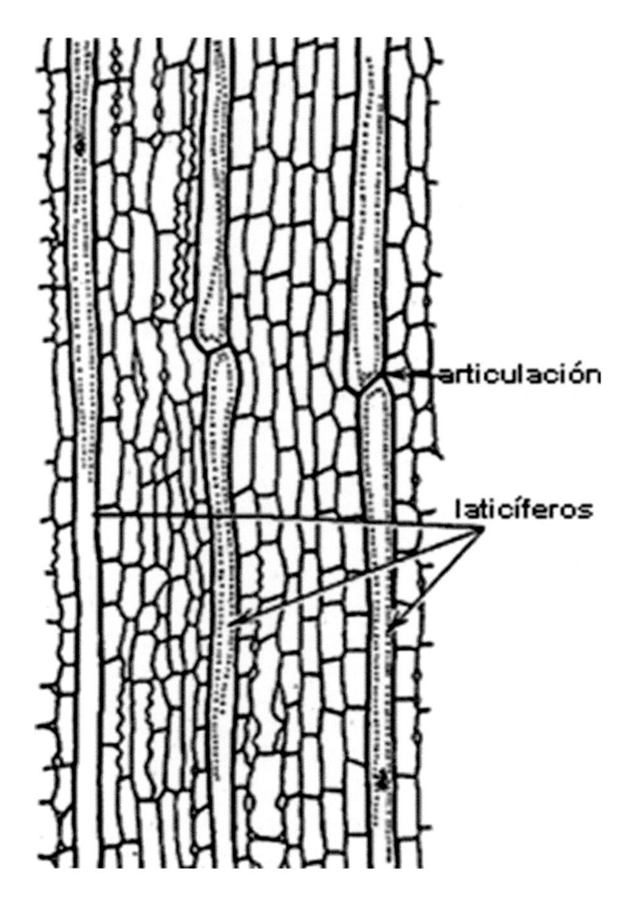

Fig. 10.2 Laticifers. *Source* www.biologia.edu.ar

thoroughly macerated to free the rubber particles [17]. It is one of the few rubber-containing plants that do not contain laticifers [17]. The content of poly (cis-1,4-isoprene) in Hevea latex is approximately 30% and is harvested by a "tapping" method after the bark of the plants is cut diagonally. One tree yields 100 to 200 ml per 3 h. The tapping can be carried out every 2–3 days and this results in yielding up to 2,500 kg of natural rubber per year per ha [19]. Trees are tapped for 25–30 years, starting when the young tree is 5–7 years of age. Guayule is a very good alternative for hypoallergenic latex extraction [20] and it is already being commercialized by Yulex Corporation [21]. Even when it has a good rubber quality, it has some undesired characteristics that make it a difficult species to work with from a breeder's point of view. Since *Taraxacum koksaghyz* Rodin is not a commercial species yet, its rubber production is much lower, but it has the advantage of being an annual species presenting all the conditions to be an ideal rubber-producing crop [3].

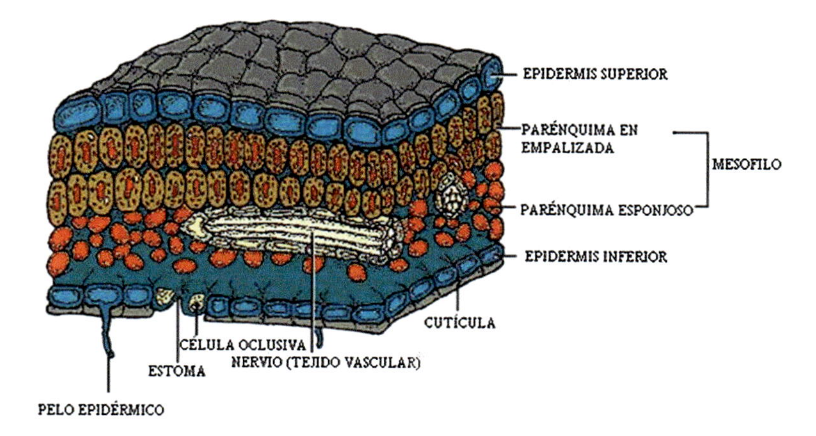

Fig. 10.3 Parenchyma cells. *Source* www.abc.com.py

A patent called "Dandelion processes, compositions and products", with publication number US 20,140,370,579 A1 [22] gives a detailed explanation of how to get profit from all parts of the Russian dandelion, using a simultaneous rubber and inulin extraction process. The rubber extraction process is done by the floatation method. Other methods such as the "toluene extraction method" (used in the present research) obtain good extraction results, but the one that shows the best results is the ASE method (Accelerated Solvent Extraction).

10.2.3 Natural Rubber Production

Since 1914 the global rubber market has been in Asian hands. Up to 91. 2% is produced in Asia. Thailand is the largest producer at the world level (Fig. 6.4), but it occupies the second place in rubber plantation extension. The difference between plantation extension and yield is due to the higher production per plant compared to other countries. Higher productivity was raised in the Asian countries due to the improvement of the production techniques over more than 80 years, the usage of high-production clone material and plant breeding programs [23]. Improvements of botanists through grafting and breeding have increased the productivity by 1000% (Fig. 10.4).

The biggest harvested area of *Hevea brasiliensis* in the world is the Asian continent, followed from a distance by Africa and Africa (Fig. 6.5).

In Europe, *H. brasiliensis* cannot be produced, since it doesn't meet the needs for cropping. The ideal growing conditions are high temperature, altitude of no more than 400 m and high humidity (NMCE, 2013) [25] (Fig. 10.5).

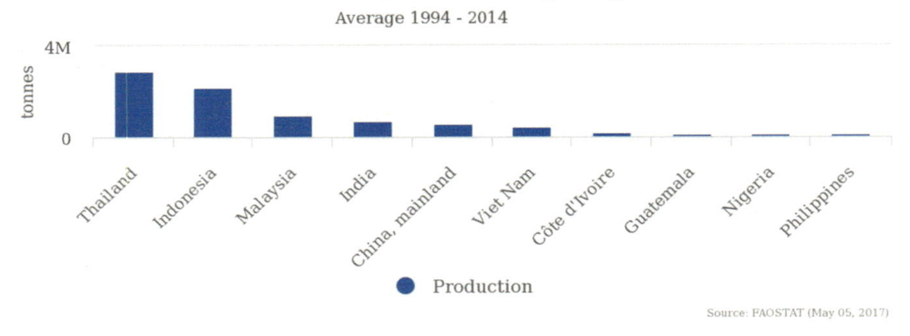

Fig. 10.4 Top 10 producers of Natural Rubber. Average 1994–2014. [24] *Source* FAOSTAT 2017

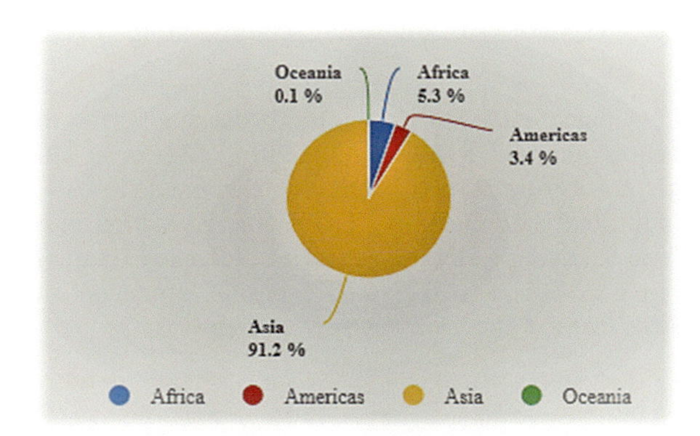

Fig. 10.5 Natural Rubber production by continent. Average 1994–2014. [24] *Source* FAOSTAT 2017

10.2.4 Natural Rubber Consumption

Even though humans have tried to replace it with synthetic rubber, NR consumption cannot be replaced [26]. It is essential in the use of heavy goods vehicles, agricultural vehicles and aircraft and in civil engineering, due to qualities like low heating and the ability to regain its original shape. 65–70% of the global consumption of NR is done by the world's tyre manufacturing industry [27]. 24% of a typical tyre composition is NR (Fig. 10.6), so it has a privileged place in the world's economy.

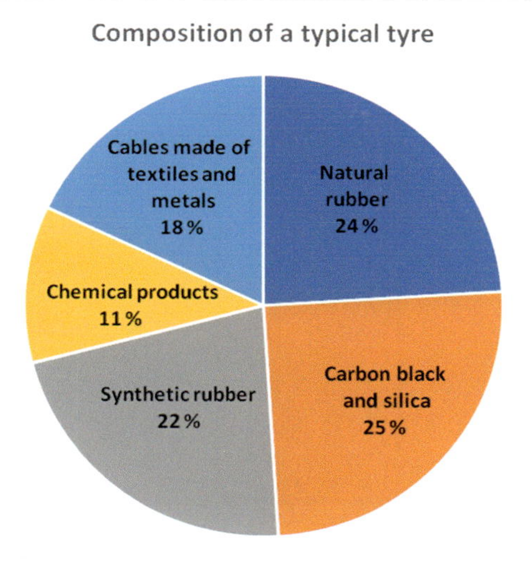

Fig. 10.6 Composition of a typical tyre. *Source* Société Internationale de Plantations dHévéas

10.3 Why Should We Look for an Alternative Natural Rubber Source?

Throughout history, humans have tried to replace natural rubber with synthetic rubber (SR), but it has been demonstrated that they are both needed and that they complement each other (Fig. 10.7). Around 50,000 products over the World use natural rubber in their composition [26]. NR is synthesized in over 7,500 plant species, confined to 300 genera of seven families [28]. Although approximately 2,000 plants synthesize poly (cis-1,4-isoprene), only rubber of two species is commercialized. NR of *Hevea brasiliensis* (99% of the world market) and rubber of *Parthenium argentatum* (1% of the world market) [29]. If *H. brasiliensis* is the most economically profitable rubber source, why should we look for alternative NR sources? *Hevea brasiliensis* production and utilization is threatened by several factors that are placing the World's NR production in a very delicate situation.

10.3.1 Latex Allergy

The exposure to the presence of residual latex proteins and chemical additives in latex products can produce sensitization and development of latex allergy. Latex proteins are potent allergens capable of inducing fatal anaphylaxis [30].

Hevea brasiliensis has very allergenic proteins and the continued contact with them can cause this allergy. The first appearances of rubber allergies were reported

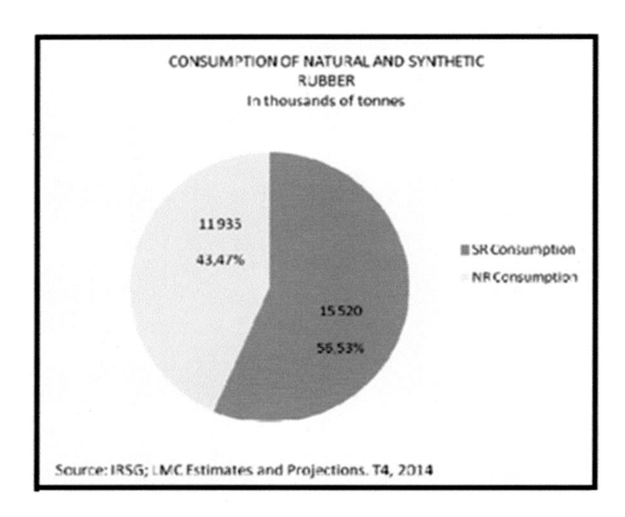

Fig. 10.7 Global distribution of rubber consumption. Legend: SR: synthetic rubber; NR: natural rubber. *Source* IRSG; LMC Estimates and Projections, T4, 2014

in the late 1980s and it's now widespread [31]. Proteins associated with NR are considered as the main cause of latex sensitivity, and the content may vary per source, lot and manufacturing process. Since allergic reactions can be severe, finding a non-allergenic rubber source was needed and K. Cornish described in 1996 the protein profile of *P. argentatum* latex and she proposed it as an alternative source with little or no allergenic properties.

10.3.2 Plant Diseases

Since all Asiatic *H. brasiliensis* individuals come from a small lot of seeds collected in Brazil in the nineteenth century, they present a very narrow genetic variability that means a great danger when threatened by some disease or plague [20]. The South American Leaf Blight disease (SALB) is a very important disease of this species, since it spreads very fast in large-scale plantations, as it happened in South American plantations [8]. Under natural conditions, the trees commonly grow sparsely and it doesn't cause serious damages, since trees growing in between act as non-susceptible species, but under large-scale plantations, it can become lethal [8]. It still remains confined to the tropical Americas [26], but if it arrives in Asia it could mean the extermination of rubber plantations in that continent. Breeding and selection programs are needed to minimize or avoid the disease.

10.3.3 Oil Palm Plantations

Another factor threatening the rubber tree plantations is the very competitive and fast-growing market of palm oil and its side products. Even when both are markets in expansion, Malaysian *Hevea brasiliensis* harvested areas from 1990 to 2011 have started to diminish while oil palm ones for the same period have grown [32]. If the huge and continuous growth of oil palm plantations doesn't stop, either the natural forest or the *Hevea* plantations will have to leave the place for the new crop.

10.3.4 Rubber Monopoly

As has already been mentioned, from the beginning of the twentieth century the global rubber market has been a monopoly in Asian hands. As can be appreciated in Figs. 6.5 and 10.8, Europe has no plantations and production at all and America represents an insignificant contribution to the total production, reason why an alternative NR source suitable for other climates and conditions is pursued.

*Europe hasn't got any *H. brasiliensis* plantation.

10.4 *Taraxacum Koksaghyz* Rodin (TKS) as an Alternative Source of Natural Rubber and Inulin

This species was considered by [3], the ideal rubber-producing crop plant considering that it is an annual crop, fast-growing and producing large amounts of biomass; it is plastic and adaptable to market changes and it is easy to introduce in crop rotation and farming systems. The Russian dandelion (TKS) (Fig. 10.9) was discovered in 1931 in the Tien Shan Mountains of Kazakstan in an expedition that was part of a

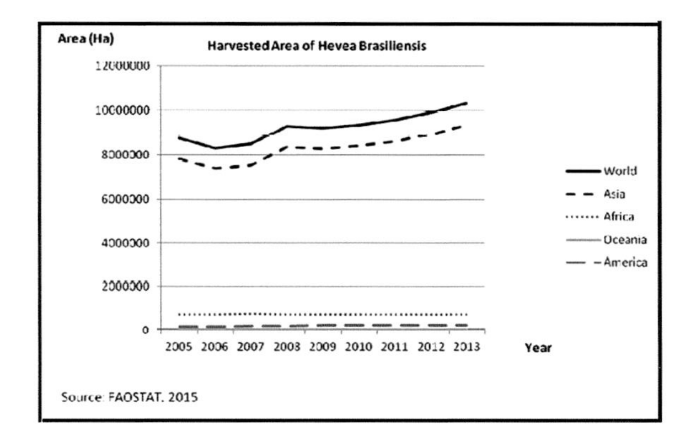

Fig. 10.8 Evolution of *Hevea brasiliensis* harvested area per continent (tones) along the period going from 2005 to 2013. [33] *Source* FAOSTAT, 2015

Fig. 10.9 TKS plant in a pot in the greenhouse

Russian Government program. The aim of the expedition was to discover new or better plants that could supply all critical materials originating from plants [8]. A few years later it became the most promising alternative rubber source in Russia and was planted in a quite considerable extension, yielding 28.35 kg/Ha, with some of the plantings producing several times that much. In the 40 s, as a response to the closure of the rubber commerce in War time, EEUU introduced TKS as part of an emergency program, looking for an alternative rubber source. It was planted in 300 Ha along 42 states. They discovered that the rubber quality was excellent, in some cases, even better than the one coming from *H. brasiliensis* [8]. Together with *P.argentatum* and *Solidago altissima,* it was one of the resources successfully used for rubber production for army vehicle tyres in WWII [15]. After the War rubber from *Hevea brasiliensis* became available again, cheaper and easier to get and the Russian dandelion culture was abandoned in all countries. It only remained in Russia until the 1950s, covering 30% of total natural rubber (NR) Russian consumption with 67,000 Ha of TKS crops. But it finally disappeared from Russia too, maybe because of economic reasons [5].

Due to several risks and conditioning factors, the search for alternative rubber sources has begun again and in 2007 van Beilen et al. reported a yielding of 150–500 kg/Ha per year. The potential of the species as a natural rubber producer was studied and based on the prediction that Polhamus made in 1962 (an increase of 15-25% of dry weight potential rubber production obtained by breeding programs) and the possibility of hybridizing TKS with *T. officinale*, van Beilen et al. predicted a theoretical rubber yield of 1,200–1,800 kg per year and Ha [5]. *Taraxacum koksaghyz* Rodin is a member of the tribe *Cichoriae,* family *Compositae,* that grows erect or in a manner which it is very close to the ground in a disposition of rosettes with a bunch of 25–50 Gy-green colored leaves [8] The optimum flowering and fruiting time is the end of May or beginning of June [34], but in the first year of TKS life only a few plants produce flowers, what radically changes in the second year, producing flowers most of them. From an economical point of view, the first-year plants are more desirable since, apparently, flowering plants have a greater fresh weight than those non-flowering plants, and a smaller % of rubber content [35]. Even when rubber is produced in leaves and roots, they have different qualities. For industrial purposes, only root rubber is desirable, since it has a low percentage of associated resins, resulting in a high-quality product, just the opposite of leaf rubber, the reason why leaves have to be removed before processing [8]. With $2n = 16$ somatic number of chromosomes, it usually presents cross-pollination as the method for reproduction [35]. TKS presents a self-incompatibility system that prevents self-pollination making breeding a quite complicated practice [3], but on the other hand it can be grown by tissue culture and it's relatively easy to transform, which makes it susceptible to becoming a model rubber-producing plant [5].

Apart from rubber, TKS produces a very interesting by-product called inulin. It is present in 25–40% of root dry weight and it can be used in food and non-food applications [3]. Even though TKS is still in the pre-commercial stage, the use of its inulin has been studied for the production of pharmaceutical preparations and dietary supplements [36] and it could be used in the production of the common polyester polyethylene terephthalate (PET) [37]. The quality of TKS inulin is similar to the one obtained from chicory, and it has been tested for bioethanol and biobutanol production as biofuels, with promising results [38–41]. Fermentation of unhydrolyzed inulin media with C. saccharobutylicum P262 produced a fair amount of solvents. The residual root bagasse (after extracting rubber and inulin) and leaves of TKS can also be used for cellulosic biofuel production [42]. In order to obtain an economically profitable species, future work is necessary, including molecular studies, using the tools of genomics, metabolomics, proteomics, and marker-assisted breeding [3].

10.5 Inulin

Inulin, a non-digestible carbohydrate; is a fructan that is not only found in many plants but has also been part of human's daily diet for several centuries [43].

It is a mixture of poly- and oligosaccharides which almost have the same chemical structure GFn (G: glucose, F: fructose, and n: number of fructose units linked to one another). The degree of polymerization (DP) goes between 2 and 60 or higher. The links between the molecules are of a very special type: the β (2–1) form, which makes this molecule indigestible for all higher animals [44]. Many factors such as plant source, climate and growing conditions, harvesting time, and storage conditions, determine inulin's DP (44,45). Even the type of tissue from which inulin is extracted has an influence [45], and the type of use of inulin is determined by the DP, either for food or for pharmaceutical purposes. Inulin with low DP can be added to food as a low-calorie sweetener, whereas inulin with higher DP can be used as a fiber-type prebiotic with several health-promoting effects [46]. In plants, dicot species belonging to Asteraceae are the main sources of inulin-type fructans [47]. Important species are chicory (Chicorium intybus L.), artichoke (Cynara scolymus), Jerusalem artichoke (Helianthus tuberosus L.), dandelion (Taraxacum officinale), dahlia (Dahlia variabilis), yacon (Polymnia sonchifolia) [36, 47, 49] and Russian dandelion (TKS) [35, 48, 50].

10.5.1 Inulin History

Inulin was first isolated in 1804 by the German scientist Rose. It was extracted with boiling water from *Inula helenium* and it was in 1818 when Thomson called it inulin [44]. Together with oligofructose, it is present in more than 36,000 plants [51].

10.5.2 Inulin Uses

10.5.2.1 Human Consumption

It is industrially obtained from chicory roots, being used as a functional food ingredient, considering that it is a non-digestible carbohydrate that has been traditionally present in the common human diet. It offers interesting nutritional properties and important technological benefits (it is taste free, increases the stability of foams and emulsions, etc.). Nowadays it represents a key ingredient for the food industry.

10.5.2.2 Pharmaceutical Uses

As a response to a number of unique chemical and physical properties, it is an element of desire of the pharmaceutical industry. It has many proven and potential applications. For example, soluble inulin is inert and non-toxic. Its glycosidic bonds make it indigestible by humans and higher animals, passing through the human digestive system relatively intact. When it reaches the large intestine, it is digested by bifid bacteria. From the stimulation of healthy intestinal microflora some important by-products are obtained; for example, butyric and propionic acids are relevant in suppressing colon cancer development. Dietary inulin may also reduce the risk of cardiovascular disease and it has also been shown that it increases calcium and magnesium absorption and bone mineralization in young adolescents [52].

10.5.2.3 Biofuel

In the past decade, we are facing the effects of global warming. It is, in a very high degree, attributed to the consumption of fossil fuels. Together with other factors such as the continuously increasing crude oil price and the depletion of crude oil resources have increased the interest in finding alternatives to fossil fuels. Particularly, sustainable biofuels from biomass have brought important attention. Biofuels such as bioethanol and biobutanol can be produced from inulin [53]. The use of inulin as a raw material for bioethanol production is still too expensive, and the process is relatively complex. It is essential to minimize sugar losses and achieve efficient conversion of inulin into fermentescible sugars to decrease the costs of production. Most efforts of research in this area, have been made in the development of new strains of microorganisms able to ferment inulin directly [54]. Biobutanol is a fuel with superior qualities to bioethanol. It has 30% more energy content in comparison being close to gasoline. Due to its low vapor pressure its application in existing gasoline supply channels is facilitated. It is also less hydrophilic, volatile hazardous to handle, and flammable than ethanol and it can be used in unmodified internal combustion engines, blended with gasoline at any concentration (up to 100%) [42]. But, to be able to compete with bioethanol in the biofuel markets it needs low production costs [55]. It was discovered that solventogenic Clostridium species metabolize a wide range of pure carbohydrates, such as starch, sucrose, starch, fructose, cellobiose, arabinose, lactose, glucose, galactose, xylose, glycero, and inulin [56]. High efforts are being made to study new Clostridium strains and systems-level metabolic engineering approaches to develop superior strains [57]. There are still some limitations for biobutanol as a biofuel, even though its results are very interesting. For example, its toxicity for the microorganisms employed, its low heating value compared to gasoline or diesel fuel, the degeneration of microorganisms, a lower octan number, and the potential corrosiveness [42].

10.6 Biofuel Evaluation of Nine Wild Taraxacum Koksaghyz Rodin Populations. Root Biomass, Rubber, and Inulin Contents

10.6.1 Introduction

Russian dandelion has never become an established crop due to the competition with the cheaper rubber price of *Hevea brasiliensis* [58], but some forty years ago, when alternative natural rubber sources started to be actively sought because of several factors threatening the production and utilization of *Hevea brasiliensis* rubber, TKS was again on the top of the list [15]. One of the reasons for its success as a candidate is the high quality of its rubber, comparable to that one extracted from the rubber tree. Its average molecular weight of 2180 kD is even higher compared to only 1310 kD of *Hevea brasiliensis* [3]. In 2008, under the urgency of finding an alternative source for the endangered natural rubber production of *H. brasiliensis*, the European Union supported the international R&D project EUPEARLS (Production and Exploitation of Alternative Rubber and Latex Sources). As part of the Project, the agronomic performance (with respect to root biomass production and rubber and inulin contents) of nine populations of TKS was analyzed under different planting dates.

10.6.2 Materials and Methods

The available seed material in the seed banks around the World was mixed up with its close relative *T. brevicorniculatum*, making essential the collection of new and fresh seeds of true *T. koksaghyz* Rodin [58]. In 2008, Kirschner and his team went on an expedition to Kazakhstan, collecting TKS seeds of several populations. As a population, we understand a collection of individual plants growing together in a specific isolated region. Seeds of nine of those TKS wild populations (Table 10.1), descending from the area near the villages Saryzhaz, Kegenand Komirshi close to the Mongolian border [58], were planted and evaluated in the experimental station of NEIKER in Arkaute (Spain).

TKS seed populations collected in Kazakhstan in 2008 (Kirschner et al. 2012).

The experimental fields were located close to the city of Vitoria-Gasteiz, in the Basque Country (Spain, 42° 51′ 0.29" N –2° 37′ 21.59" W; 517 m elevation) as shown in Fig. 10.10. The tests were performed in the year 2009 in loamy soil and the annual accumulated precipitation was 629 mm, the mean temperature 11.7 °C and the average humidity 85.1%. Planting date (May and June) and population [9] were the main factors being analyzed.

Table 10.1 Origin of TKS populations used in the population trial

Sample code (population)	Original sample code	Location and date of collection
20	–006	1 KM N of NW part of Sarydzhaz, 4 June 2008
21	202	C. 6 KM ENE of Sarydzhaz, 2 June 2008
22	203	C. 5–6 KM ENE of Sarydzhaz, 2 June 2008
27	207	C. 1–2 KM SW of Kegen, 3 June 2008
30	217	Along the road between Saryzhaz and Komirshi, 3 June 2008
35	219	C. 6 KM SW of Komirshi, 3 June 2008
36	236	C. 4 KM NNW of Kegen, 6 June 2008
37	237	2–3 KM E of Zhalauly village, C. 6 KM NW of Kegen, 6 June 2008
38	246	c. 2MK SE of Kegen, 3 June 2008

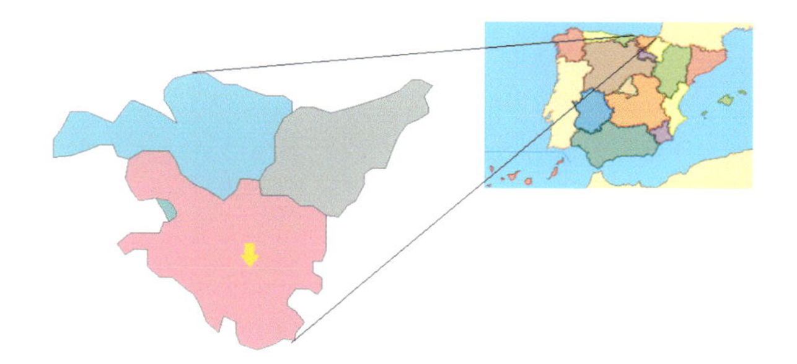

Fig. 10.10 Field trials location

TKS seeds were not planted directly in the field; they were firstly grown in a culture chamber, in wet paper beds, and once the seedlings were 2 cm long, they were transferred to multi pots in the greenhouse, from March to April. When they rose 10 to 15 cm height the seedlings were planted in the field at a depth of 6 to 7 cm, in four rows with five plants per row. The spacing was 24 by 34 cm^2. Trials were set up according to a factorial design with single plant repetitions and elementary plots of 1.2 × 2 m2 (Fig. 6.11). After planting they received 30 min of irrigation.

According to Whaley and Bowen suggestions (1947), May and June were selected as the ideal planting months. The weeds were controlled using a green anti-weed MIPED mesh (2.10 m width) and moles were avoided with an ultrasound mole chaser. Both measures showed very satisfactory results. No extra fertilization was added to the field and water was supplied when needed on the basis of plant withering. Due to the low temperatures, TKS plants stop growing. At the end of October, no further growth was registered and they were harvested. The roots were dragged out and washed manually, and they were kept in a cool chamber until they could be processed, up to 4 days later. Then, morphological data and fresh weight were annotated and the plants were dried for two days at 65 °C. Dry root and aerial part weight were annotated separately and the roots were stored in paper envelopes in cardboard boxes until grinding was needed for rubber and inulin measurements. The roots were milled with a Moulinex AR10031 electric grinder, obtaining a very fine powder. It was stored in 50 ml Falcon tubes at room temperature. Dry rubber content was measured with a gravimetric analysis using the Post et al. [59] method with some modifications. 4 ml of toluene were mixed in a 10 ml Thermo Scientific Teflon FEP tube with an aliquot of 200 mg of root powder. They were incubated at 70 °C for 18 h in a water bath. Then, the toluene supernatant was transferred to a clean tube and vaporized in a thermoblock. 300 μl of toluene was added to the remaining sample for 30 min to re-dissolve it. Rubber was precipitated by incubating with 600 μl of methanol for 30 min. All steps were performed under a gas extraction hood. The tubes were centrifuged at room temperature for 2 min at 16.000 xg. The precipitated rubber could be appreciated as a white layer. It was washed twice with vigorous shaking; first with water for 30 min and later with acetone for another 30 min in an Infors HT Ecotron shaking incubator. Before weighing, the rubber samples were vacuum dried in a HETO Speed Vac, Model DNA mini, for 3 min at 60 °C and cooled down to room temperature. For every combination of population and planting date

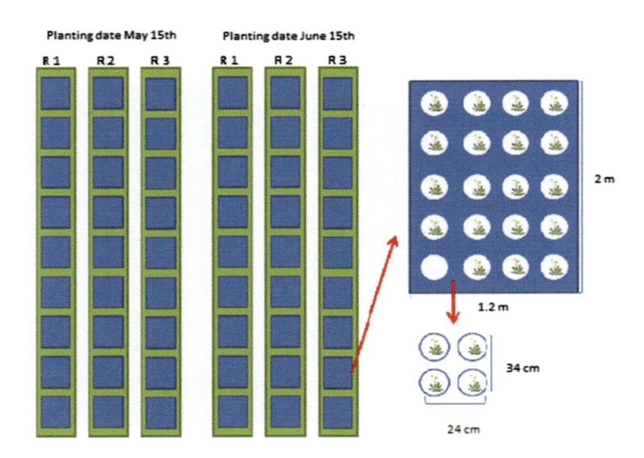

Fig. 10.11 Field trial design

measurements were done in four plants, performing three repetitions per plant. To obtain more consistent data average values were computed (when rubber is dried it is very difficult to grind and the distribution in the flour is sometimes not very homogeneous. This is the reason why three repetitions were performed and average values were computed) (Fig. 10.11).

The UV-method D-glucose/D-fructose of R-Biopharm Kit was used for inulin extraction and quantification. The manufacturer suggested the procedure for inulin hydrolysis. In a 50 ml Falcon tube containing an aliquot of 500 mg of root flour, 15 ml of miliQ boiling water was added. The pH was adjusted to 6.6–8.0 with a Crisson pH-meter Basic 20 + . After adjusting pH, the tube was filled up to 25 ml volume and vigorously shaken. Then it was heated for 15 min in a shaking water bath (Selecta Ultrasonic 320 OR) at 80 °C, 90 u/min. Afterward, the sample was filtered using a Serlab vacuum filtering equipment with Whatman filter paper n° 589/2. The extracted liquid was cooled down to room temperature and filled up with water to 25 ml volume. An aliquot of 12.5 ml was transferred to a 50 ml Falcon tube and 1.25 ml of HCL 25% was added to hydrolyse inulin. They were left in the shaking water bath at 80 °C, 90 u/min for 15 min. Then the sample was neutralized with 5 M KOH and brought to 25 ml volume with miliQ water. For inulin content determination, first, the glucose and fructose contents in the original sample were measured spectrophotometrically. Afterward, the amount of these sugars were measured in the hydrolyzed simple. The amount of inulin present in the sample was obtained by subtracting the corresponding measurements, after converting to inulin contents using a specific formula. The sugar quantification was done using a UV-2401 PC; a Shimadzu UV–Vis Recording Spectrophotometer was used, measuring at 340 nm in a quartz cuvette with a 1 cm light path. For every combination of population and planting date, the measurements were done in three plants. Inulin reduction with time was measured, but no lowering was registered (results not shown).

Once the measurements were done, the statistical analyses were conducted using JMP software version 5.0 (JMP, 2002). The Analysis of variance, based on a factorial design with 4 replicates for rubber content and 3 for inulin content, was employed to analyze the effects of population and planting date. A two-way full factorial analysis of variance was carried out. Tukey's honestly significant difference (Tukey's HSD) was used for comparing treatment means, with an alpha error probability level of 5%. All data were normally distributed and equal variances were assessed with Levene's test of equality of variances [60]. In order to normalize the distribution, a Box-Cox transformation [(RDW- 0.2 - 1) / -0.0384096976766] was used for Root dry weight (RDW) data. Onwards we refer to the transformed data as RDW X.

10.6.3 Results and Discussion

A large degree of variation between single plants of the same population was found for all traits under study, and it was the reason why for rubber content determination, population 20 had to be excluded, since homocedasticity couldn't be adjusted when included because of a very high variance within its values. The big variation could be explained from a reproductive point of view; TKS is a diploid outcrossing species [61], unlike the common dandelion (*T. officinalis*), which is triploid and apomictic [62]. However, visual analysis and repeated measurements of single plants supported the idea of population 20 as a very good candidate for rubber production when planted in June. The analysis of variances revealed different significant main effects for rubber, inulin, and RDW. For rubber content, only a significant main effect of population number was detected, while a highly significant effect of population and a significant influence of the interaction between population and planting date was shown for inulin content. The RDW was only significantly influenced by the interaction effect between population and planting date (Table 10.2a).

The Goodness of Fit test showed that the whole model was able to explain 47.45% of the variance in the response variable with a CV (RMSE) of 51% for rubber content, 74.4% with a CV (RMSE) of 15.7% for inulin content and 46% with a CV (RMSE) of 42.60% for RDW (Table 10.2b). In Table 10.3, the results of Tukey's HSD test can be appreciated, revealing a significantly higher rubber content of population 30 with respect to populations 38,36,21, 27, and 21 and a significantly lower rubber content of population 38 with respect to population 35. The highest inulin contents of level 20-F1 were significantly different from those of both planting dates of populations 30 and 22, and from levels 21-F1, 37-F2, 35-F1, and 36-F1. The largest significant difference was found between populations 20 and 36, both planted in mid-May with a difference of 21.4% in inulin content. The separation of means analyses for RDW revealed as before a large degree of overlapping significances involving many levels. The only significant differences were found between populations 20 and 37, both planted in May and between population 38 planted in June and population 37 planted in May. The rubber content difference between the highest and the lowest population was almost 6.5% for F1 (ranging from 0% of a plant belonging to population 34 and 6.34% of a plant belonging to population 31) and 11.5% for F2 (ranging from 0.03% of a plant belonging to population 38 and 11.50% of a plant belonging to population 20). The range of inulin content for plants planted in May (F1) went from 10.41%,

Table 10.2 Results of analysis of variances for the TKS population trial at two planting dates

Response	Source	DF	Sum of squares	F ratio	Prob.	R^2
(A) Model parameter effect						
Rubber (%)	Population	7	24.324746	4.914	0.0003	0.3765
	Planting date	1	0.019368	0.027	0.8692	0.0001
	Population * planting date	7	6.315005	1.27	0.2823	0.0978
Inulin (%)	Population	8	762.17933	5.815	0.0001	0.3313
	Planting date	1	24.22331	1.479	0.2319	0.0105
	Population * planting date	8	840.70221	6.414	0.0001	0.3654
Root DWX (g)b	Population	8	96.6242221	1.695	0.1209	0.1368
	Planting date	1	23.87673	3.351	0.0727	0.0338
	Population * planting date	8	200.76273	3.522	0.0024	0.2843
(B) Goodness of fit of the model						
Response	Prob.	RSq	RMSE	CV (RMSE) (%)		
Rubber (%)	0.0027	0.47	0.8409	51.0		
Inulin	0.0001	0.74	4.0476	15		
Root DWX (g)	0.0034	4.6	2.6694			

found in a single plant of population 33 to 52.37% of a single plant of population 26. For planting date F2 the range was tighter, going from 16.25% in single plant of population 37 to 35.43% in population 35. For RDW a big difference in values could be found, going from 0.7 g of a single plant of population 37 planted in May to 49.01 g of population 27 also planted in May; for individuals planted in June, it went from 0.4 g of a single plant of population 25 to 42.4 g of population 38.

When deciding which population and planting date we should use, it should be taken into account the purpose of the utilization of the crop, since the decision will be different if we want it for rubber and/or inulin production. From Table 6.4, it can be found the results of the estimation of the potential RDW, rubber, and inulin yield per Ha, based on our trial data and assuming a plant density of 240,000 plants per Ha (Table 10.4).

Estimation data demonstrates that population 20 could be a good candidate for future development of the crop, since when planted in May, apart from being a good rubber producer, it also has the highest potential inulin yield. Population 30 planted in May and population 38 planted in June also showed good rubber and inulin production respectively (because of high rubber concentration and high biomass production) and should also be considered for further studies. We should be aware of these analyses as preliminary results, suggesting how future TKS breeding should

Table 10.3 Results of separation of means using Tukey HSD tests

Rubber % response					Inuline % response						Root DWX (g) response			
Level POP				LSM	Level					LSM	Level			LSM
30	A	B	C	2.96	20-F1	A	B	C	D	36.30	20-F1	A	B	9.96
35	A	B	C	2.22	37-F1	A	B	C	D	36.05	38-F2	A	B	9.87
37	A	B	C	1.68	38-F2	A	B	C	D	33.92	22-F1	A	B	8.51
22		B	C	1.43	38-F1	A	B	C	D	33.91	21-F1	A	B	8.50
27		B	C	1.42	35-F2	A	B	C	D	30.05	36-F1	A	B	7.98
21		B	C	1.40	21-F2	A	B	C	D	29.95	27-F1	A	B	7.23
36		B		1.25	27-F2		B	C		2.40	35-F2	A	B	7.02
38				0.82	20-F2		B	C		26.91	30-F1	A	B	6.68
					36-F2		B	C		26.89	27-F2	A	B	6.46
					27-F1			C		26.14	35-F1	A	B	6.36
					30-F2			C		23.58	30-F2	A	B	5.63
					22-F1					23.34	21-F2	A		4.94
					22-F2					22.79	36-F2	A		4.86
					30-F1					22.13	22-F2			4.71
					21-F1					21.58	38-F1			4.17
					37-F2					20.64	37-F2			3.93
					35-F1					15.32	20-F2			3.80
					36-F1					14.89	37-F1			2.20
Alpha = 0.05; Q = 3.17					Alpha = 0.05; Q = 3.75						Alpha = 0.05; Q = 3.66			

Levels connected by the same letter are not significantly different
Separation of means based on a factorial design with four replicates for rubber content and 3 for inulin content

be done. It should be based on several rounds of selection of single plants with good characteristics (due to the high intra-population variation) and subsequent crossings between them. For single plant selection, our results should be taken into account. The improved resulting genotypes should be selected. Referring to the optimum planting date, spring sowings were chosen based on the experience of the Russian plantations documented by Altukhov, where he states that TKS spring sowings produced higher yields than fall sowings (Whaley and Bowen, 1947). This was also supported by these authors who wrote about the best results for root yield when sowing was done early with good weather conditions. According to our results, even when they were not statistically significant we appreciated that, in general, the aerial part of the plant and the root size were better developed when TKS was planted in May.

October was chosen as the harvesting date based on Mynbaev text (1940) [63], which experienced an increase in the percentage of rubber content until late fall, in the first year of growth, followed by a weak decrease. Choosing this harvesting date, we also wanted to avoid heavy rains that would have supposed a water excess. Wild TKS root rubber content in our studies ranged from 0.4 to 5.6% which is far away from the results exposed by Buranov in 2010 [38] (EU-PEARLS conference), up to 24% rubber content. Of course, he was showing results of improved material, which shows the big improving potential of the species. With respect to rubber yield, basing our results in the means of four plants, we got similar yields as Whaley stated

Table 10.4 Resume of N, mean, and SD of Rubber (% in RDW), Inulin (% in RDW), RDW (g), and potential rubber, inulin, and RDW weight yield in the TKS population trial

Level	N SD Rubber (% in RDW)			N SD Inulin (% in RDW)			N SD (% in RDW) (g)			Expected yield /Ha[a] RDW rubber inulin (kg) (Kg) (kg)		
20F1	4	1.33	2.34	3	36.30	0.64	4	13.1	8.70	3160.8	42.04	1147.37
20F2	4	5.56	4.29	3	26.19	4.96	4	2.19	0.44	525.6	29.22	141.43
21F1	4	1.12	0.20	3	21.58	4.70	4	7.41	3.50	1778.4	19.92	383.78
22F2	4	1.69	0.70	3	29.95	1.57	4	3.46	1.71	830.4	14.03	248.71
22F2	4	1.19	0.78	3	23.34	8.53	4	3.46	9.41	2380.8	28.33	555.68
27F1	4	1.68	1.06	3	22.80	3.21	4	9.92	1.53	708.0	11.89	161.42
27F2	4	1.13	0.71	3	26.14	1.62	4	2.95	4.14	1459.2	16.49	381.43
30F1	4	1.72	0.88	4	28.40	4.85	4	6.08	0.67	969.6	16.68	275.37
30F2	4	3.33	1.75	3	22.13	0.66	4	4.04	4.82	1404.0	46.75	310.71
35F1	4	2.58	0.67	4	23.58	2.73	4	5.85	1.15	835.2	23.80	196.94
35F2	4	2.60	0.70	1	15.32	–	4	3.48	1.25	984.0	25.58	150.76
36F1	4	1.85	0.52	4	30.05	5.63	4	4.10	2.29	1207.2	23.33	362.76
36F2	4	1.01	0.69	4	14.89	5.07	4	5.03	3.13	1562.4	15.78	232.64
37F1	4	1.49	1.01	4	26.89	2.45	4	6.51	0.96	686.4	10.23	184.57
37F2	4	1.45	0.44	1	36.06	–	4	2.86	1.40	458.4	6.65	165.30
38F1	4	1.91	1.07				4	1.91	1.86	684.0	13.06	141.18
38F2	4	1.23	0.59				4	2.85	1.00	600.0	7.38	203.46

[a]240.000 Plants/Ha were assumed for calculations

Legend: F1: Planting date May; F2: Planting date June; RDW: Root Dry Weight. Expected Yield/Ha is based on means of rubber for 4 plants, inulin from 1 to 4 plants, and RDW for 4 plants

in 1948 (28.4 kg/Ha), with some of the studied populations at specific planting dates. It is still far away from the reported 150 to 500 kg/Ha of van Beilen and Poirier (2007a) [3], but if we calculate the rubber yield with single plant (we must remain that we were dealing with wild material), we could state 71.76 kg/Ha, taking as a reference population 20 planted in June (11,5% or rubber in RDW and 2.6 g of root production), what is more than twice our first estimation.

The use of a different extraction method such as Accelerated Solvent Extraction (ASE) [64] could also improve the rubber yield, since we could verify that a remaining rubber amount was retained in TKS root flour with the Post et al. method [59]. For inulin contents, something similar could be reported. In our studies, taking as a base means of 3 or 4 plants, the range went from 15% up to 36%. 42% inulin content was reported in 1946 by Gorham [48], and van Beilen and Poirier [5] reported a range of 25 to 40%. If we had taken single plant values as a base, we could have reported 52% inulin content in population 26 planted in May. The fact that single plant estimations improve so much rubber and inulin yield compared to the mean of several plants estimation, points again in the same direction of the need for a within-population variation study as a possible source of improvement through breeding. Another way of improving rubber and inulin yield and root biomass production could be increasing the plant density per m^2. Taking as a reference Munt et al.[65] studies of the related species *Taraxacum brevicorniculatum*, planting TKS in rows 2–5 cm apart will produce fully developed plants, with a single unbranched root of 40–50 cm long. This variation will produce more roots per m^2 and will facilitate harvesting works.

Nowadays TKS is considered a weed, but with the help of breeding studies, it could be converted into an economically profitable species from a rubber and inulin production point of view.

10.7 Conclusions

Based on our analyses of nine wild populations of TKS cultivated at two planting dates population effects resulted significantly for rubber and inulin contents, but no planting date effects were found. The interaction between population and planting dates showed significant effects on inulin content and RDW. A clearly significant separation between mean values was hindered by a large variation, for all traits, between populations and between individuals of the same population. Populations 20, 30, and 38 were selected as promising candidate genotypes for inulin and / or rubber production. Breeding studies are necessary for converting TKS into an interesting crop from an economic point of view.

References

1. Britannica online Encyclopedia (2015). http://global.britannica.com/topic/synteny
2. Puskas JE, Gautriaud E, Deffieux A, Kennedy JP (2006) Natural rubber biosynthesis- a living carbocationic polymerization. Prog Polym Sci 31:533–548
3. van Beilen J, Poirier Y (2007) Guayule and Russian Dandelion as alternative sources of natural rubber. Crit Rev Biotechnol 27(4):217–231
4. Vijayaram TR (2009) A technical review of natural rubber. Int J Desa Manuf Technol 3(1):25–37
5. van Beilen J, Poirier Y (2007) Establishment of new crops for the production of natural rubber. Trends Biotechnol 25(11):522–529
6. Hosler D, Burkett SL, Tarkanian MJ (1999) Prehistoric polymers: rubber processing in ancient Mesoamerica. Science 284(5422):1988–1991. https://doi.org/10.1126/science.284.5422.1988
7. Ciesielski A (2000) An introduction to rubber technology. Rapra Technology Limited, 183 pp (2000). ISBN: 1–85957–150–6
8. Whaley WG (1948) Rubber-the primary source for american production. Econ Bot 2(2):198–216, https://doi.org/10.1007/BF02859004, Online ISSN: 0013-0001
9. Wolf H, Wolf R (1936) Rubber. A story of glory and green. Covici, Friede Publishers, New York
10. Cook OF (1941) Naming the cultivated rubber tree Siphonia ridleyana. J Wash Acad Sci 31:46–65
11. Vunovic X (2009) Tapping the amazon for victory: brazil's "battle for rubber" of world war II. Doctoral dissertation. Dec, 2, Faculty of the Graduate School of Arts and Sciences of Georgetown University, Washington DC
12. Ullán de la Rosa FJ (2004) La era del caucho en el Amazonas (1870–1920): modelos de explotación y relaciones sociales de producción. An Mus Am 12:183–204
13. Carraher CE (2003) Giant molecules essential materials for everyday living and problem solving. Wiley Inc, 483 p. Online ISBN: 9780471457190

14. D'Auzac J et al (1989) Physiology of rubber tree latex. In: D'Auzac J, Jacob J-L, Chrestin H (eds) CRC Press, Boca Raton, p 470
15. Mooibroek H, Cornish K (2000) Alternative sources of natural rubber. Appl Microbiol Biotechnol 53:355–365
16. Heinrich G (1970) Elektronenmikroskopische Untersuchung der MilchroÈ ren von Ficus elastica. Protoplasma 70:317–323
17. Backhaus RA, Walsh S (1983) The ontogeny of rubber formation in guayule. Parthenium argentatum Gray. Bot Gaz 144(3):391–400
18. Verheye W (2010) Growth and Production of Rubber. In: Verheye W (ed) Land use, land cover and soil sciences. Encyclopedia of life support systems (EOLSS), UNESCO-EOLSS Publishers, Oxford. http://www.eolss.net
19. Rose K, Steinbüchel A (2005) Biodegradation of natural rubber and related compounds: recent insights into a hardly understood catabolic capability of microorganisms. Appl Environ Microbiol 71:2803–2812
20. Venkatachalam P, Geetha N, Sangeetha P, Thulaseedharan A (2013) Natural rubber producing plants: an overview. Afr J Biotechnol 12(12):1297–1310
21. Yulex (2015). http://www.yulex.com/home/
22. Wade J, Swiger D (2014) Patent Application Publication "Dandelion processes, compositions and products." Pub. Nº US 2011(0275142):A1
23. Pulido-Sierra SI, Silva-Calvao ME, Ferreira-Neto JA, Rojo-Alboreca A (2012) Análisis del caucho natural en el Mundo. Spanish J Rural Dev 3(Special 3):57–64
24. FAOSTAT (2017). http://www.fao.org/faostat/en/#data
25. NMCE (National Multy Commodity Exchange), 2013. Report 2012–2013
26. Prabhakaran Nair KP (2010) The agronomy and economy of important tree crops of the developing world, 237–273, ISBN 978–0–12–384677–8. quimitube.com, 2017
27. Bruins J (2003) World agriculture: towards 2015/2030. A FAO prespective. Bruins J (ed) Earthscan Publications Ltd. London, 432 p
28. Cornish K, Siler DJ, Groesjean O, Goodman N (1993) Fundamental similarities in rubber particle architecture and function in three evolutionarily divergent plant species. J. nat. Rubb. Res 8(4):275–285
29. Tanaka Y, Sakdapipanich JT (2005) Chemical structure and occurrence of natural polyisoprene. Biopol. Online, https://doi.org/10.1002/3527600035.bpol2001
30. Sell AM, Laguila-Visentainer JE (2013) Natural rubber latex allergy. In: Pereira C (ed) Allergic diseases- highlights in the clinic, mechanisms and treatments, pp. 289–310
31. Cornish K (1996) Hypoallergenic natural rubber products from Parthenium argentatum (Gray) and other non Hevea brasiliensis species. United States Patent. Patent number: 5.580.942; Dec 3
32. FAOSTAT (2013). http://www.fao.org/faostat/en/#home
33. FAOSTAT (2015). http://www.fao.org/faostat/en/#home
34. van Dijk PJ, Kirschner J, St˘epánek J, Baitulin IO, Cerny T (2010) Taraxacum koksaghyz Rodin definitely is not an example of over collecting in the past: a reply to S. Volis et al. (2009). J Appl Bot Food Qual 83:217–219
35. Krotkov G (1948) Changes in the carbohydrate metabolism of Taraxacum kok-saghyz Rod. During the first and second years of growth. Plant Physiol 25(1):169–180
36. Schütz K, Muks E, Carle R, Schieber A (2006) Separation and quantification of inulin in selected artichoke (Cynara scolymus L.) cultivars and dandelion (Taraxacum officinale WEB. Ex
37. van Beilen J, Poirier Y (2008) Production of renewable polymers from crop plants. Plant J 54:684–701
38. Buranov AU (2010) Russian dandelion seeds and extraction processes in october14-15. In: 2010 meeting: the future of natural rubber. Montpellier, France
39. OARDC (2014). http://www.oardc.ohio-state.edu/images/E_Rubber.pdf
40. OMAFRA (2014). http://www.omafra.gov.on.ca/CropOp/en/indus_misc/chemical/russd.html

41. AGMRC (2014). http://www.agmrc.org/commodities__products/specialty_crops/russian-dandelion/
42. Bharathidasan AK (2013) Production of biobutanol from inulin-rich biomass and industrial food processing wastes. The Ohio State University, Thesis, p 309
43. Franck A, De Leenheer L (2005). Inulin, Biopolymers online 6:439-448. https://doi.org/10.1002/3527600035.bpol6014
44. Coussement P (1999) Inulin and oligofructose as dietary fiber: analytical, nutrition and legal aspects. In: Prosky L, Dreher M (eds) Cho SS. Complex Carbohydrates in Foods, New York, pp 203–212
45. De Leenheer L, Hoebregs H (1994) Progress in the elucidation of the composition of chicory inulin. Starch 46:193–196
46. van den Ende W, Michiels A, Van Wonterghem D, Vergauwen R, Van Laere A (2000) Cloning, developmental, and tissue-specific expression of Sucrose: Sucrose 1Fructosyl Transferase from Taraxacum officinale. Fructan localization in roots. Plant Phys 123:71–19
47. Flamm G, Glinsmann W, Kritchevsky D, Prosky L, Roberfroid M (2001) Inulin and oligofructose as dietary fiber: a review of the evidence. Crit Rev Food Sci Nutr 41:353–362
48. van Laere A, Van den Ende W (2002) Inulin metabolism in dicots: chicory as a model system. Plant Cell Environ 25:803–813
49. Gorham PR (1946) Investigations on rubber-bearing plants: II carbohydrates in the roots of Taraxacum koksaghyz Rod. Can J Res 24C:47–53
50. Wilson RG, Kachman SD, Martin AR (2001) Seasonal changes in glucose, fructose, sucrose, and fructans in the roots of dandelion. Weed Sci 49:150–155
51. Mikhlin DM, Akhumbaeva BO (1955) Fructosans of kok-saghyz roots. Biokhimii͡a 21(2):186–190
52. Carpita NC, Kanabus J, Hously TL (1989) Linkage structure of fructans and fructan oligomers from Triticum aestivum and Festuca arundinacea leaves. J Plant Physiol 134:162–168
53. Barclay T, Ginic-Markovic M, Cooper P, Petrovsky N (2010) Inulin- a versatile polysaccharide with multiple pharmaceutical and food chemical uses. J. Excipients Food Chem 1(3):27–49
54. Tashiro Y, Sonomoto K (2010) Advances in butanol product ion from Clostridia. In: Méndez-Vilas A (ed) Current research, technology and education topics in applied microbiology and microbial biotechnology, pp 1383–1394
55. Neagu C, Bahrim G (2012) Fuel ethanol bioproduction from inulin rich feedstock. Innov Rom Food Biotechnol 11:1–8
56. Green EM (2011) Fermentative production of butanol- the industrial perspective. Curr Opin Biotechnol 22:1–7
57. Ezeji TC, Liu S, Qureshi N (2014) Mixed sugar fermentation by Clostridia and metabolic engineering for butanol production. In: Qureshi N, Hodge D and Vertes A (eds.) Biorefineries: integrated biochemical processes for liquid biofuels, Elsevier BV, pp 191–204
58. Jang Y, Lee J, Malaviya A, Seung DY, Cho JH, Lee SY (2011) Butanol production from renewable biomass: Rediscovery of metabolic pathways and metabolic engineering. Biotechnol J 7:186–198
59. Kirschner J, Stepanek J, Cerny T, De Heer P, van Dijk P (2012) Available ex situ germplasm of potential rubber crop Taraxacum koksaghyz belongs to a poor rubber producer, T. brevicorniculatum (Compositae-Crepidinae). Genet Resour Crop Evol, Published online 19 May 2012, 10.1007/s 10722–012–9848–0
60. Post J, Van Deenen N, Fricke J, Kowalski N, Wurbs D, Schaller H, Eisenreich W (2012) Laticifer-specific cis-prenyltransferase silencing affects the rubber, triterpene, and inulin content of Taraxacum brevicorniculatum. Plant Physiol 158(3):1406–1417. https://doi.org/10.1104/pp.111.187880
61. Levene H (1960) Robust tests for equality of variances. In: Olkin I (ed) Contributions to probability and statistics. University Press, Palo Alto, California, Stanford, pp 278–292
62. Warmke HE (1943) Macrosporogenesis, fertilization, and early embryology of Taraxacum kok-saghyz. Bull Torrey Bot Club 70(2):164–173

63. van Dijk P, de Jong H, Vijverberg K, Biere A (2009) An apomixis-gene's view on dandelions. In: Schön I, Martens K, Van Dijk P (eds), Lost sex. The evolutionary biology of parthenogenesis. Springer, Dordrecht. pp 474–493. https://doi.org/10.1007/978-90-481-2770-2_22
64. Mynbaev K (1940) Vozrastnaia izmenchivost' kok-sagyza. [Changes with the age of kok-saghyz], Vestnik Sel' sko-khoz. Nauki. Tekhn. kul' tur 5:58–62
65. Richter BE, Jones BA, Ezzell JL, Porter NL, Avdalovic N, Pohl C (1996) Accelerated solvent extraction: a technique for sample preparation. Anal Chem 68(6):1033–1039. https://doi.org/10.1021/ac9508199
66. Munt O, Arias M, Hernandez M, Ritter E, Chulze Gronover, C, Prüfer D (2011) Fertilizer and planting strategies to increase biomass and improve root morphology in the natural rubber producer Taraxacum brevicorniculatum. Ind Crops Prod 36(2012):289–293